AF754766

Methods in Molecular Biology

Series Editor
John M. Walker
School of Life and Medical Sciences
University of Hertfordshire
Hatfield, Hertfordshire, UK

For further volumes:
http://www.springer.com/series/7651

For over 35 years, biological scientists have come to rely on the research protocols and methodologies in the critically acclaimed *Methods in Molecular Biology* series. The series was the first to introduce the step-by-step protocols approach that has become the standard in all biomedical protocol publishing. Each protocol is provided in readily-reproducible step-by-step fashion, opening with an introductory overview, a list of the materials and reagents needed to complete the experiment, and followed by a detailed procedure that is supported with a helpful notes section offering tips and tricks of the trade as well as troubleshooting advice. These hallmark features were introduced by series editor Dr. John Walker and constitute the key ingredient in each and every volume of the *Methods in Molecular Biology* series. Tested and trusted, comprehensive and reliable, all protocols from the series are indexed in PubMed.

Bacterial Multidrug Exporters

Methods and Protocols

Second Edition

Edited by

Kunihiko Nishino

The Institute of Scientific and Industrial Research, Osaka University, Ibaraki, Osaka, Japan

Mikio Tanabe

Institute of Materials Structure Science, High Energy Accelerator Research Organization (KEK), Tsukuba, Ibaraki, Japan

Humana Press

Editors
Kunihiko Nishino
The Institute of Scientific and Industrial Research
Osaka University
Ibaraki, Osaka, Japan

Mikio Tanabe
Institute of Materials Structure Science
High Energy Accelerator Research Organization (KEK)
Tsukuba, Ibaraki, Japan

ISSN 1064-3745 ISSN 1940-6029 (electronic)
Methods in Molecular Biology
ISBN 978-1-0716-5315-9 ISBN 978-1-0716-5316-6 (eBook)
https://doi.org/10.1007/978-1-0716-5316-6

© The Editor(s) (if applicable) and The Author(s), under exclusive license to Springer Science+Business Media, LLC, part of Springer Nature 2026

This work is subject to copyright. All rights are solely and exclusively licensed by the Publisher, whether the whole or part of the material is concerned, specifically the rights of translation, reprinting, reuse of illustrations, recitation, broadcasting, reproduction on microfilms or in any other physical way, and transmission or information storage and retrieval, electronic adaptation, computer software, or by similar or dissimilar methodology now known or hereafter developed.
The use of general descriptive names, registered names, trademarks, service marks, etc. in this publication does not imply, even in the absence of a specific statement, that such names are exempt from the relevant protective laws and regulations and therefore free for general use.
The publisher, the authors and the editors are safe to assume that the advice and information in this book are believed to be true and accurate at the date of publication. Neither the publisher nor the authors or the editors give a warranty, expressed or implied, with respect to the material contained herein or for any errors or omissions that may have been made. The publisher remains neutral with regard to jurisdictional claims in published maps and institutional affiliations.

This Humana imprint is published by the registered company Springer Science+Business Media, LLC, part of Springer Nature.
The registered company address is: 1 New York Plaza, New York, NY 10004, U.S.A.

If disposing of this product, please recycle the paper.

Preface

Multidrug exporters represent a critical component of bacterial physiology, serving as membrane-embedded molecular machines that protect cells from toxic compounds. Beyond their well-established role in antibiotic resistance, these transporters participate in diverse cellular processes including bile salt tolerance, biofilm formation, and virulence factor secretion. The clinical significance of multidrug exporters has intensified in recent years as antimicrobial resistance continues to spread globally, threatening the efficacy of current therapeutic strategies. Despite significant progress in understanding their mechanisms, the development of clinically applicable efflux pump inhibitors remains an unmet challenge in combating drug-resistant infections.

Bacterial efflux systems are classified into two major categories based on their energy sources. Primary transporters, represented by the ATP-binding cassette (ABC) superfamily, utilize ATP hydrolysis to drive substrate extrusion. While these transporters play predominant roles in eukaryotic multidrug resistance, they also serve important functions in bacterial resistance. Secondary transporters harness electrochemical gradients of protons or sodium ions and include the resistance-nodulation-division (RND) superfamily, the major facilitator superfamily (MFS), the multidrug and toxic compound extrusion (MATE) family, the small multidrug resistance (SMR) family, the proteobacterial antimicrobial compound efflux (PACE) family, and the *p*-aminobenzoyl-glutamate transporter (AbgT) family. In Gram-negative bacteria, the dual membrane architecture has driven the evolution of sophisticated molecular machines. Centered on RND-type efflux pumps and including certain MFS-type and ABC-type transporters, these systems form functional complexes with periplasmic adaptor proteins and outer membrane channels to expel substrates across both membranes. Furthermore, inner membrane transporters can operate cooperatively with RND systems. These findings highlight the need to understand multidrug resistance as an integrated system rather than as an isolated transporter function.

A defining characteristic of multidrug exporters is their remarkable promiscuity toward structurally diverse substrates. The archetypal AcrAB-TolC system in *Escherichia coli* exemplifies this versatility, recognizing compounds ranging from hydrophobic aromatics to charged amphiphiles, encompassing cationic dyes, anionic β-lactams, zwitterionic fluoroquinolones, and neutral macrolides. Yet this apparent promiscuity masks an underlying selectivity—subtle variations in molecular structure, size, or chemical properties can dramatically influence whether a compound serves as a substrate or remains unrecognized by the transporter. Elucidating the molecular basis of this polyspecific yet selective recognition, and understanding how conformational dynamics drive vectorial substrate translocation across membranes, represents a fundamental question at the intersection of structural biology, biophysics, and microbiology.

Since the first edition of this book, the field of multidrug exporter research has experienced remarkable advances. The development of cryo-electron microscopy (cryo-EM) and cryo-electron tomography (cryo-ET) has revolutionized our ability to visualize these complex molecular machines in near-native states. Advanced computational methods, including molecular dynamics simulations, now enable us to observe transporter dynamics at atomic resolution. Genetic tools such as CRISPR-Cas-based genome editing have expanded our capacity to dissect exporter function in clinically relevant bacterial strains. Furthermore, novel biochemical reconstitution approaches and single-molecule techniques have provided unprecedented insights into the mechanistic details of drug export. In this second edition, we have reorganized the content into four major sections that reflect the current state of the field and provide comprehensive methodological coverage.

Part I: Structural Analysis Methods presents cutting-edge approaches for determining high-resolution structures of multidrug exporters. This section covers the full spectrum of structural biology techniques, from solution NMR spectroscopy for studying PfMATE (Chap. 1) to crystallographic studies of RND- and ABC-type tripartite transporters (Chap. 2), single-particle cryo-EM analysis of TolC (Chap. 3), and in situ cryo-ET imaging of AcrAB-TolC complexes (Chap. 4). These chapters provide detailed protocols that have enabled breakthrough discoveries in understanding the structural basis of multidrug recognition and transport.

Part II: Biochemical Analysis and Reconstitution focuses on functional characterization of multidrug exporters. Methods for site-specific fluorescent labeling of AcrB (Chap. 5), reconstitution of membrane proteins using LMNG-mediated auto-insertion (Chap. 6), transport analysis of purified transporters (Chap. 7), and analytical ultracentrifugation for studying oligomeric states are presented (Chap. 8). These biochemical approaches are essential for validating structural findings and understanding the mechanistic principles underlying drug export.

Part III: Computational and Simulation Approaches showcases the power of in silico methods in multidrug exporter research. Molecular dynamics simulations of conformational changes in MdfA (Chap. 9) and comprehensive computer simulations of RND superfamily efflux pumps (Chap. 10) demonstrate how computational analyses complement experimental approaches to provide dynamic views of transporter function.

Part IV: Genetic Tools and Functional Studies presents a diverse array of methods for investigating multidrug exporters in their biological context. This section includes CRISPR-Cas-mediated genome editing for alanine-scanning analysis of MexB (Chap. 11), comprehensive genetic approaches for studying bacterial efflux pumps using knockout cells (Chap. 12), high-throughput gene mapping using TraDISort (Chap. 13), evaluation of efflux pump inhibitors through growth curve analysis (Chap. 14), cloning and characterization protocols (Chap. 15), and innovative silkworm infection models for assessing the role of efflux pumps in bacterial virulence (Chap. 16).

The field of multidrug exporter research continues to evolve rapidly, driven by technological innovations and the urgent clinical need to combat antimicrobial resistance. This collection of protocols represents state-of-the-art methodologies that span from atomic-resolution structural analysis to whole-organism functional studies. We hope this volume will serve as a comprehensive resource for researchers entering this field as well as for established investigators seeking to expand their methodological repertoire.

We would like to express our sincere gratitude to all contributing authors for sharing their expertise and valuable protocols with the research community. Their dedication to advancing our understanding of multidrug exporters is evident in the detailed and practical nature of their contributions. We also thank the series editor, Dr. John Walker, for his guidance and support throughout the preparation of this volume.

Ibaraki, Osaka, Japan ***Kunihiko Nishino***
Tsukuba, Ibaraki, Japan ***Mikio Tanabe***

Contents

Part IV Genetic Tools and Functional Studies of Multidrug Efflux Pumps

Contributors

MOHD ATHAR • *Physics Department, University of Cagliari, Monserrato, Italy*
DIRK BALD • *Amsterdam Institute for Life and Environment (A-LIFE), AIMMS, Vrije Universiteit Amsterdam, Amsterdam, The Netherlands*
AMY K. CAIN • *ARC Centre of Excellence in Synthetic Biology, Macquarie University, Sydney, NSW, Australia; School of Natural Sciences, Macquarie University, Sydney, NSW, Australia*
MEGAN R. CASWELL • *Department of Molecular and Cellular Biology, University of Guelph, Guelph, ON, Canada*
GEORGINA COX • *Department of Molecular and Cellular Biology, University of Guelph, Guelph, ON, Canada*
RENWICK C. J. DOBSON • *Biomolecular Interaction Centre, School of Biological Sciences, University of Canterbury, Christchurch, New Zealand; Bio21 Molecular Science and Biotechnology Institute, Department of Biochemistry and Pharmacology, University of Melbourne, Parkville, VIC, Australia*
KAZUYUKI FURUTA • *Graduate School of Medicine, Dentistry, and Pharmaceutical Sciences, Okayama University, Okayama, Japan*
CHRISTOPH GERLE • *Life Science Research Infrastructure Group RIKEN SPring-8 Center, Sayo, Japan*
SILVIA GERVASONI • *Physics Department, University of Cagliari, Monserrato, Italy*
EVAN GIBBS • *ARC Centre of Excellence in Synthetic Biology, Macquarie University, Sydney, NSW, Australia; School of Science, University of Newcastle, Callaghan, NSW, Australia*
KARL A. HASSAN • *ARC Centre of Excellence in Synthetic Biology, Macquarie University, Sydney, NSW, Australia; School of Natural Sciences, Macquarie University, Sydney, NSW, Australia; School of Science, University of Newcastle, Callaghan, NSW, Australia*
MIKI HIASA • *Department of Membrane Biochemistry, Graduate School of Medicine, Dentistry and Pharmaceutical Sciences, Okayama University, Okayama, Japan*
KINUKA HONGU • *Graduate School of Medicine, Dentistry, and Pharmaceutical Sciences, Okayama University, Okayama, Japan*
KAZUYA ISHIKAWA • *Graduate School of Medicine, Dentistry, and Pharmaceutical Sciences, Okayama University, Okayama, Japan*
CHIKARA KAITO • *Graduate School of Medicine, Dentistry, and Pharmaceutical Sciences, Okayama University, Okayama, Japan*
NAOKI KOBAYASHI • *Department of Food Science and Human Nutrition, Faculty of Agriculture, Setsunan University, Hirakata, Osaka, Japan*
AYANO KURIU • *Graduate School of Medicine, Dentistry, and Pharmaceutical Sciences, Okayama University, Okayama, Japan*
LIPING LI • *ARC Centre of Excellence in Synthetic Biology, Macquarie University, Sydney, NSW, Australia; School of Natural Sciences, Macquarie University, Sydney, NSW, Australia*
CLAIRE MAHER • *ARC Centre of Excellence in Synthetic Biology, Macquarie University, Sydney, NSW, Australia; School of Science, University of Newcastle, Callaghan, NSW, Australia*

GIULIANO MALLOCI • *Physics Department, University of Cagliari, Monserrato, Italy*
DUNCAN G. G. MCMILLAN • *School of Biological Sciences, Whiteknights, University of Reading, Reading, UK*
TOSHIO MORIYA • *Structural Biology Research Center, Institute of Materials Structure Science, High Energy Accelerator Research Organization (KEK), Tsukuba, Ibaraki, Japan*
SATOSHI MURAKAMI • *Department of Life Science and Technology, Institute of Science Tokyo, Yokohama, Japan*
JIAJUN NING • *School of Biological Sciences, The University of Hong Kong, Hong Kong, China*
KUNIHIKO NISHINO • *SANKEN (The Institute of Scientific and Industrial Research), The University of Osaka, IbarakiOsaka, Japan; Graduate School of Pharmaceutical Sciences, The University of Osaka, Osaka, Japan; Center for Infectious Disease Education and Research, The University of Osaka, Osaka, Japan; Bioinformatics Center, Research Institute for Microbial Diseases, The University of Osaka, Osaka, Japan*
RACHEL A. NORTH • *School of Medical Sciences, Faculty of Medicine and Health, University of Sydney, Sydney, NSW, Australia*
UI OKADA • *Department of Life Science and Technology, Institute of Science Tokyo, Yokohama, Japan*
HIROSHI OMOTE • *Department of Membrane Biochemistry, Graduate School of Medicine, Dentistry and Pharmaceutical Sciences, Okayama University, Okayama, Japan*
IAN T. PAULSEN • *ARC Centre of Excellence in Synthetic Biology, Macquarie University, Sydney, NSW, Australia; School of Natural Sciences, Macquarie University, Sydney, NSW, Australia*
PAOLO RUGGERONE • *Physics Department, University of Cagliari, Monserrato, Italy*
XIAODONG SHI • *Jiangsu Province Key Laboratory of Anesthesiology and Jiangsu Province Key Laboratory of Anesthesia and Analgesia Application, Xuzhou Medical University, Xuzhou, Jiangsu, China*
FRANCESCA SHORT • *School of Natural Sciences, Macquarie University, Sydney, NSW, Australia; Infection Program and Department of Microbiology, Biomedicine Discovery Institute, Monash University, Clayton, VIC, Australia*
KOH TAKEUCHI • *Graduate School of Pharmaceutical Sciences, The University of Tokyo, Hongo, Bunkyo, Tokyo, Japan; Molecular Profiling Research Center for Drug Discovery and Cellular Molecular Biotechnology Research Institute, National Institute of Advanced Science and Technology, Aomi, Koto, Tokyo, Japan*
MIKIO TANABE • *Institute of Materials Structure Science, High Energy Accelerator Research Organization (KEK), Tsukuba, Ibaraki, Japan*
TOHRU TERADA • *Department of Biotechnology, Graduate School of Agricultural and Life Sciences, The University of Tokyo, Tokyo, Japan*
ZACHARY D. TILLETT • *Biomolecular Interaction Centre, School of Biological Sciences, University of Canterbury, Christchurch, New Zealand*
ATTILIO V. VARGIU • *Physics Department, University of Cagliari, Monserrato, Italy*
WENRUI WAN • *School of Biological Sciences, The University of Hong Kong, Hong Kong, China*
ZHAO WANG • *Department of Biochemistry and Molecular Pharmacology, Baylor College of Medicine, Houston, TX, USA*
MALLORY WRIGHT • *Department of Molecular and Cellular Biology, University of Guelph, Guelph, ON, Canada*

SEIJI YAMASAKI • *SANKEN (The Institute of Scientific and Industrial Research), The University of Osaka, Ibaraki, Osaka, Japan; Graduate School of Pharmaceutical Sciences, The University of Osaka, Suita, Osaka, Japan; Center for Infectious Disease Education and Research, The University of Osaka, Suita, Osaka, Japan*
AIXIN YAN • *School of Biological Sciences, The University of Hong Kong, Hong Kong, China*
ZHILI YU • *Department of Biochemistry and Molecular Pharmacology, Baylor College of Medicine, Houston, TX, USA*
XUETING ZHOU • *Department of Biochemistry and Molecular Pharmacology, Baylor College of Medicine, Houston, TX, USA*
MARTIJN ZWAMA • *SANKEN (The Institute of Scientific and Industrial Research), The University of Osaka, IbarakiOsaka, Japan*

Part I

Structural Biology Approaches: NMR, X-RAY Crystallography, and Cryo-EM

Chapter 1

Expression and Purification of a Multidrug Transporter PfMATE for Structural Analyses Using Solution NMR

Koh Takeuchi

Abstract

Multidrug and toxin extrusion (MATE) family transporters excrete various toxic compounds coupled to Na^+/H^+ influx. In a recent breakthrough, a structural analysis of *Pyrococcus furiosus* MATE (PfMATE) using solution NMR has unveiled how substrate export is coupled to ion influx. This significant advancement was made possible by expressing a stable-isotope-labeled PfMATE in an *E. coli* in vivo expressing system and an in vitro expression system using a commercially available *E. coli*-based in vitro expression system. In this chapter, we will delve into the details of both expression systems and provide a comprehensive summary of how stable-isotope-labeled PfMATE was expressed, purified, and subjected to the NMR analyses.

Key words Expression, In vitro translation, Purification, Membrane proteins, Multidrug transporter, Solution NMR, Stable isotope labeling

1 Introduction

The multidrug and toxin extrusion (MATE) family is a group of monomeric 12-transmembrane (TM) multidrug-efflux transporters [1–4]. Compound efflux in the MATE transporters is linked to Na^+/H^+ influx, categorizing them as ion-coupled antiporters. Like other transporters, the MATE transporters are believed to function through an alternating access mechanism [5], in which substrate export is coupled to a whole-molecule conformational rearrangement between the outward-facing (OF) and inward-facing (IF) conformations. For *Pyrococcus furiosus* MATE (PfMATE), atomic resolution structures of both the OF and IF conformations are available [6, 7]. A study employing electron paramagnetic resonance spectroscopy (EPR) spectroscopy [8] measured critical distances in the PfMATE structure supporting the protonation-dependent alternating access mechanism and identified Glu163 as a crucial intracellular protonation site facilitating

Kunihiko Nishino and Mikio Tanabe (eds.), *Bacterial Multidrug Exporters: Methods and Protocols*, Methods in Molecular Biology, vol. 3044, https://doi.org/10.1007/978-1-0716-5316-6_1,
© The Author(s), under exclusive license to Springer Science+Business Media, LLC, part of Springer Nature 2026

rearrangements between the OF and IF conformations. However, the mechanism by which substrate excretion is coupled to proton release has not been solved until recently.

Nuclear magnetic resonance (NMR) analyses of PfMATE solved this issue by structurally analyzing conformational equilibrium and substrate interaction associated with the transport activity in functional-relevant states [9]. The authors used the stable-isotope-labeled PfMATE solubilized in n-dodecyl-β-D-maltoside (DDM) micelle and conducted NMR spectroscopy under various pH. It has been shown that PfMATE can exist in an equilibrium between the OF and IF conformations in a Glu163-dependent manner and predominantly adopts the OF conformation in a deprotonated state (pH 6.8). In contrast, it exhibits the IF conformation in a protonated state (pH 3.7). In addition, the NMR interaction analysis unveiled that the IF conformation exhibits an affinity sufficient to recognize the substrates. In contrast, the OF conformation shows insufficient affinity to retain the bound state. Thus, PfMATE operates as an affinity-directed substrate/H^+-antiporter, wherein substrates exclusively bind to the protonated IF conformation in the equilibrium, and subsequent proton release, coupled with the IF to OF conformational rearrangement, mechanistically ensures H^+-mediated substrate excretion due to the reduced substrate affinity to the OF conformation.

Here, we show the strategy for the NMR analyses of the structure of PfMATE by expressing and purifying the stable-isotope-labeled PfMATE solubilized in DDM micelles. While the paper mentioned above exclusively uses PfMATE expressed in the *E. coli* expression system, we recently found that a commercially available in vitro expression system using *E. coli* lysate can express PfMATE more efficiently and in a larger amount. Thus, we will introduce both expression systems in this chapter. The subsequent protocol for purification and the NMR measurement will also be discussed.

2 Materials

2.1 Expression of PfMATE in E. coli-Based In Vivo Expression System

1. Expression vector. The gene encoding an N-terminal hexa-histidine-tagged PfMATE embedded in a pET11b vector established in Nureki's group [7].
2. Expression cells. The commercially available chemically competent C41 (DE3) strain (Lucigen) [10].
3. LB–Amp broth (Miller). 1% tryptone, 0.5% yeast extract, 1% sodium chloride, 100 μg/mL ampicillin.
4. LB–Amp agar: LB–Amp broth with 1.5% agar.
5. D_2O LB–Amp broth: LB–Amp broth in 99.6% D_2O.

6. SOC medium: 2% tryptone, 0.5% yeast extract, 0.05% sodium chloride, 0.24% magnesium sulfate (anhydrous), 0.00186% potassium chloride, 0.4% glucose.
7. Incubator (20–37 °C) with shaker.
8. A heat block 42 °C.
9. Centrifuge. Beckman Avanti HP-20I for collecting cells.
10. Rotors. Beckman JA-12 for 50 mL centrifuge tubes and Beckman JS-5.3 for 500 mL centrifuge tubes.
11. Baffled flasks. 200–300 mL and 2 L.
12. D_2O M9–Amp Celton medium: 8.5 g/L anhydrous Na_2HPO_4, 3.0 g/L KH_2PO_4, 0.5 g/L NaCl, 1.0 g/L NH_4Cl, 3.0 g/L [$^2H_7$]-D-glucose, 0.24 g/L $MgSO_4$, 11 mg/L $CaCl_2$, 80 mg/L ampicillin, 1 mg/L biotin, 1 mg/L thiamine, 0.5 mL/L Trace element solution, 1.0 g/L ^2H-Celton base powder, and 99.6% D_2O (*see* **Note 1**).
13. Stable isotopically labeled amino acid precursors. [methyl-^{13}C, 3,3-$^2H_2$]-α-ketobutyric acid, 100 mg/L of [3-methyl-^{13}C, 3,4,4,4-$^2H_4$]-α-ketoisovaleric acid, and 50 mg/L of [ε-^{13}C] methionine.
14. Inducer. 0.5 mM isopropyl β-D-thiogalactopyranoside (IPTG).
15. Deep freezer (−80 °C).

2.2 Purification of PfMATE from E. coli-Based In Vivo Expression System

1. Lysis buffer: 20 mM Tris–HCl, pH 8.0, 300 mM NaCl.
2. Protease inhibitor. cOmplete™ ULTRA, EDTA-free protease inhibitor (*see* **Note 2**).
3. Sonicator. Probe sonicator.
4. Centrifuge. Beckman Avanti HP-20I for precipitating debris after sonication.
5. Rotor. Beckman JA-12 for 50 mL centrifuge tubes.
6. Ultracentrifuge. Beckman Optima X for precipitating membrane fraction.
7. Ultracentrifuge rotor. Rotor Type 45 Ti (Beckman).
8. Metal-chelating affinity purification resin.
9. His-tag digestion enzyme. Trypsin.
10. Solubilization buffer. 20 mM Tris–HCl, pH 8.0, 300 mM NaCl, 1.5% *n*-dodecyl-β-D-maltoside, and 20 mM imidazole.
11. Wash buffer. 20 mM Tris–HCl, pH 8.0, 300 mM NaCl, 0.2% *n*-dodecyl-β-D-maltoside, and 30 mM imidazole.
12. Elution buffer. 20 mM Tris–HCl, pH 8.0, 300 mM NaCl, 0.2% *n*-dodecyl-β-D-maltoside, and 300 mM imidazole.
13. Size-exclusion chromatography column. HiLoad 16/60 Superdex 200 prep grade.

14. Semi-automated purification system. AKTA explorer.
15. Size-exclusion chromatography buffer. 20 mM Tris–HCl, pH 8.0, 300 mM NaCl, and 0.2% *n*-dodecyl-β-D-maltoside.
16. Ultrafiltration devices. Molecular weight cut-off 30,000 Da (e.g., Amicon Ultra-14).
17. Protein molecular weight marker (e.g., SeeBlue prestained marker).
18. SDS–PAGE gel (e.g., NuPAGE 4–12% Bis–Tris gel).
19. NMR buffer. 10 mM sodium phosphate, pH 6.8, 100 mM NaCl, and 99.6% D_2O.

2.3 Expression of PfMATE in E. coli-Based In Vitro Expression System

1. PfMATE in vitro expression vector. pEX-K4J2 vector (Eurofins Genomics) embattled with a gene encoding an N-terminal NHH-tagged PfMATE (*see* **Note 3**).
2. In vitro expression kit. Musaibo Kun SI SS (A89-0127, Taiyo Nippon Sanso) (*see* **Note 4**).
3. Dialysis device. Provided in the in vitro expression kit Musaibo Kun SI SS (A89-0127, Taiyo Nippon Sanso).
4. Outer dialysis solution for in vitro expression (total 10,000 μL). 7250 μL Musaibo Kun kit outer solution (A89-0127, Taiyo Nippon Sanso), 750 μL labeled amino acid mix (SAT0401-2, Taiyo Nippon Sanso), 100 μL 1 M dithiothreitol, and 1900 μL Milli-Q water (*see* **Note 5**).
5. Inner dialysis solution for in vitro expression (total 1000 μL). 685 μL Musaibo Kun kit inner solution (A89-0127, Taiyo Nippon Sanso), 75 μL labeled amino acid mix (SAT0401-2, Taiyo Nippon Sanso), 10 μL 1 M dithiothreitol, 100 μL membrane protein expression additive Set-A M1 solution (A226-0290, Taiyo Nippon Sanso), 30 μL Milli-Q water, and 100 μL template plasmid DNA (50 ng/μL).

2.4 Purification of PfMATE from In Vitro Expression System

1. Centrifuge. Micro refrigerated centrifuge for 1.5 mL tubes.
2. Metal-chelating affinity purification resin.
3. HRV3C digestion enzyme.
4. Solubilization buffer. 20 mM Tris–HCl (pH 8.0), 300 mM NaCl, 1.5% n-dodecyl-β-D-maltoside, and 20 mM imidazole.
5. Wash buffer. 20 mM Tris–HCl (pH 8.0), 300 mM NaCl, 0.2% n-dodecyl-β-D-maltoside, and 30 mM imidazole.
6. Elution buffer: 20 mM Tris–HCl (pH 8.0), 300 mM NaCl, 0.2% n-dodecyl-β-D-maltoside, and 300 mM imidazole.
7. Size-exclusion chromatography column. HiLoad 16/60 Superdex 200 prep grade.
8. Semi-automated purification system. AKTA explorer.

9. Size-exclusion chromatography buffer. 20 mM Tris–HCl, pH 8.0, 300 mM NaCl, and 0.2% n-dodecyl-β-D-maltoside.
10. Ultrafiltration devices, molecular weight cut-off 30,000 Da (e.g., Amicon Ultra-14).
11. Protein molecular weight marker (e.g., SeeBlue prestained marker).
12. SDS–PAGE gel (e.g., NuPAGE 4–12% Bis–Tris gel).
13. NMR buffer. 10 mM sodium phosphate, pH 6.8, 100 mM NaCl, and 99.6% D2O.

2.5 NMR Experiments

1. NMR microtube. Shigemi 5 mm microtube.
2. NMR spectrometer (e.g., AVANCE III 600 MHz (Bruker Biospin)).
3. Cryogenic NMR probe (e.g., 600 MHz triple-resonance carbon inverse (TCI) cryoprobe (Bruker Biospin)).
4. NMR control software (e.g., Topspin ver. 3.5 (Bruker Biospin)).

3 Methods

3.1 Expression of PfMATE in E. coli Expression System

1. Preheat a heat block at 42 °C, 500 μL SOC medium at 37 °C, and an LB–Amp agar plate at 37 °C.
2. Thaw 10 μL of chemically competent C41 (DE3) cells on ice (about 5–10 min).
3. Add 0.5 μL of vector (~ 100 ng/μL) to 10 μL of chemically competent C41 (DE3) cells.
4. Mix by carefully flicking the tube with your finger.
5. Place on ice for 10 min.
6. Place the cell in the 42 °C heat block for 45 s.
7. Place on ice for 2 min.
8. Add 500 μL of prewarmed super optimal broth with catabolite repression (SOC) medium and shake at 37 °C for 1 h.
9. Plate 100 μL of cells on a prewarmed LB–Amp agar plate.
10. Place the plate upside down in the incubator and let the cells grow overnight at 37 °C (*see* **Note 6**).
11. The following day, pick ~10 colonies and inoculate them into 5 mL of D_2O LB–Amp broth.
12. Incubate the cell for 10 h at 37 °C, 200 rpm in a shaker incubator.
13. Spin down the cell at 3000 g for 10 min with a JA-12 rotor.

14. Discard the supernatant and resuspend with 50 mL of D_2O M9–Amp Celton medium.
15. Transfer the culture to 200–300 mL baffled flask.
16. Incubate the cell for overnight at 37 °C, 120 rpm in a shaker incubator.
17. Spin down the cell at 3000 *g* for 10 min with a JA-12 rotor.
18. Discard the supernatant and resuspend the cells in 1 L of D_2O M9–Amp Celton medium.
19. Check the optical density at 600 nm (OD_{600}), which would typically be 0.3.
20. Grow the cells in the shaker incubator at 37 °C, 120 rpm, until they reach OD_{600} 0.5.
21. Add 100 mg/L of [methyl-^{13}C, 3,3-$^2H_2$]-α-ketobutyric acid, 100 mg/L of [3-methyl-^{13}C, 3,4,4,4-$^2H_4$]-α-ketoisovaleric acid, and 50 mg/L of [ε-^{13}C] methionine (*see* **Note 7**).
22. Lower the culture temperature to 20 °C.
23. One hour after the lowering of the temperature, add 0.5 mM IPTG to initiate the expression of PfMATE.
24. Incubate the induced culture at 20 °C for 16 h.
25. Harvest the cells by centrifugation using 500 mL centrifuge tubes at 6000 g for 20 min.
26. Discard the supernatant and freeze the cell pellet at −80 °C until purification.

3.2 Purification of PfMATE Expressed in *E. coli*

1. Resuspend the cells with 80 mL lysis buffer and add the cOmplete™ ULTRA, EDTA-free protease inhibitor.
2. Disrupt the cell by sonication on ice for 3 min × 3 times.
3. Centrifuge at 15,000 g for 20 min at 8 °C and collect the supernatant.
4. Ultracentrifuge the collected supernatant at 125,000 g for 60 min at 6 °C to precipitate the membrane fraction containing PfMATE.
5. Discard the supernatant.
6. Resuspend the precipitated membrane fraction in 20 mL of the solubilization buffer by vortexing and sonication.
7. Dissolved solution was stirred for 1 h at 4 °C to completely solubilize the protein.
8. Ultracentrifuge the solution at 125,000 g for 30 min at 6 °C to remove the insolubilized debris.
9. Mix the supernatant with 1 mL of preequilibrated Ni-NTA resin and allow it to interact for 1 h at 4 °C under rotation.

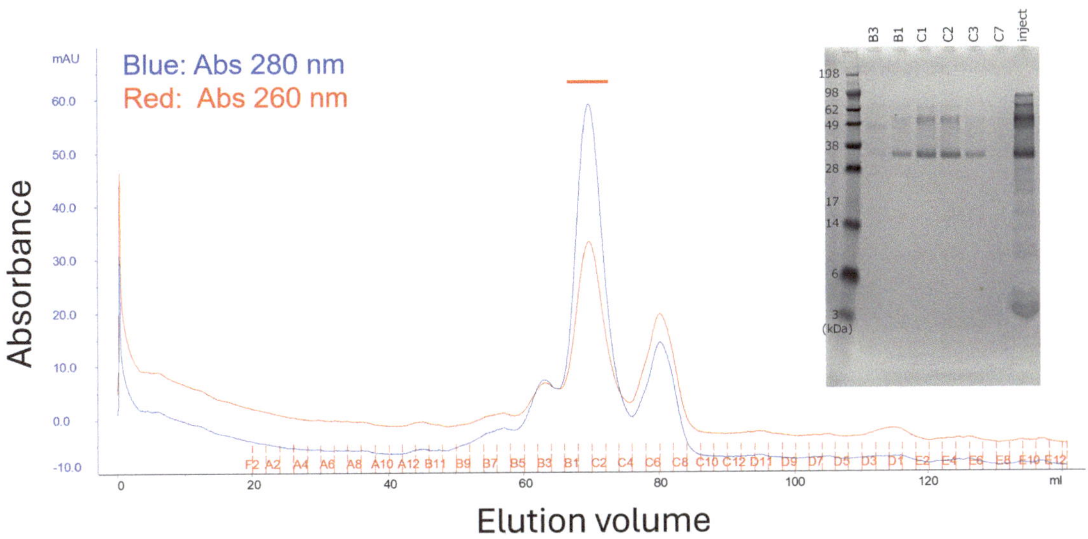

Fig. 1 Size-exclusion chromatography (SEC) profile of purified PfMATE obtained from in vivo expression. The fractions that contain monomeric PfMATE are indicated. The inset shows the SDS–PAGE of the purified protein. The sample was loaded without heating, and PfMATE typically yields a double band in the SDS–PAGE

10. Transfer the mix to the empty column and obtain the flow-through.
11. Wash the column twice with 20 mL of 20 mM Tris–HCl, pH 8.0, 300 mM NaCl, and 0.2% DDM solution.
12. Elute the PfMATE protein with 20 mL of 20 mM Tris–HCl, pH 8.0, 300 mM NaCl, 0.2% DDM, and 300 mM imidazole solution.
13. Concentrate the eluted fraction by ultrafiltration using a 30 kDa molecular weight cut-off (MWCO) membrane to 2 mL.
14. Exchange the buffer to 20 mM Tris–HCl, pH 8.0, 300 mM NaCl.
15. Add 1 μg/mL trypsin to the eluted fraction and incubate for 16 h at 4 °C to remove the hexahistidine tag.
16. Check the digest by SDS–PAGE to confirm that the digestion is finished.
17. Filter the sample through a 0.22 μm syringe filter.
18. Subject the digested sample to Size-exclusion chromatography (SEC) using a HiLoad 16/60 Superdex 200 prep grade (Cytiva) equilibrated in 20 mM Tris–HCl, pH 8.0, 300 mM NaCl, 0.2% DDM.
19. Collect the fractions corresponding to the monomeric PfMATE protein (Fig. 1).
20. Check the purified product by SDS–PAGE.

21. Exchange buffer to NMR buffer and concentrate to 20–50 μM. Typical yield of PfMATE from 1 L culture is 0.5 mg (*see* **Notes 8** and **9**).

3.3 Expression of PfMATE in E. coli-Based In Vitro Expression System

1. Prepare 5 μg PfMATE in vitro expression vector embattled by miniprep, followed by an extensive phenol–chloroform extraction and ethanol precipitation.
2. Dilute the vector with autoclaved ultrapure water at a concentration of 50 ng/μL.
3. Prepare the outer dialysis solution and put it in the outer chamber of the dialysis device.
4. Incubated the outer dialysis solution at 30 °C (*see* **Note 10**).
5. Prepare the inner dialysis solution right before the expression starts and put it into the inner chamber of the dialysis device.
6. Allow the protein expression for 16 h at 30 °C under gentle stirring of the outer chamber solution.
7. Recover the inner dialysis solution containing PfMATE from the inner dialysis chamber and put it on ice to stop the reaction.
8. Wash the inner dialysis chamber with 1 mL of 20 mM Tris–HCl, pH 8.0, 300 mM NaCl, and 0.2% DDM solution.
9. Centrifuge at 15,000 *g* for 1 min at 8 °C to obtain the supernatant.
10. Mix the supernatant with 0.5 mL of preequilibrated Ni-NTA resin.
11. Allow it to interact for 1 h at 4 °C under rotation.
12. Transfer the mix to the empty column and obtain the flow-through.
13. Wash the column twice with 10 mL of 20 mM Tris–HCl, pH 8.0, 300 mM NaCl, and 0.2% DDM solution.
14. Elute the PfMATE protein with 10 mL of 20 mM Tris–HCl, pH 8.0, 300 mM NaCl, 0.2% DDM, and 300 mM imidazole solution.
15. Concentrate the eluted fraction by ultrafiltration using a 30 kDa molecular weight cut-off (MWCO) membrane to 2 mL.
16. Exchange buffer to 20 mM Tris–HCl, pH 8.0, 300 mM NaCl, 0.2% DDM. The typical yield of the protein would be 2 mg from 1 mL reaction.
17. Add the 2 μg HRV3C protease to 100 μg of PfMATE and incubate for 20 h at 4 °C to remove the NHH tag.
18. Mix the digest with 0.5 mL of preequilibrated Ni-NTA resin and allow it to interact for 1 h at 4 °C under rotation.

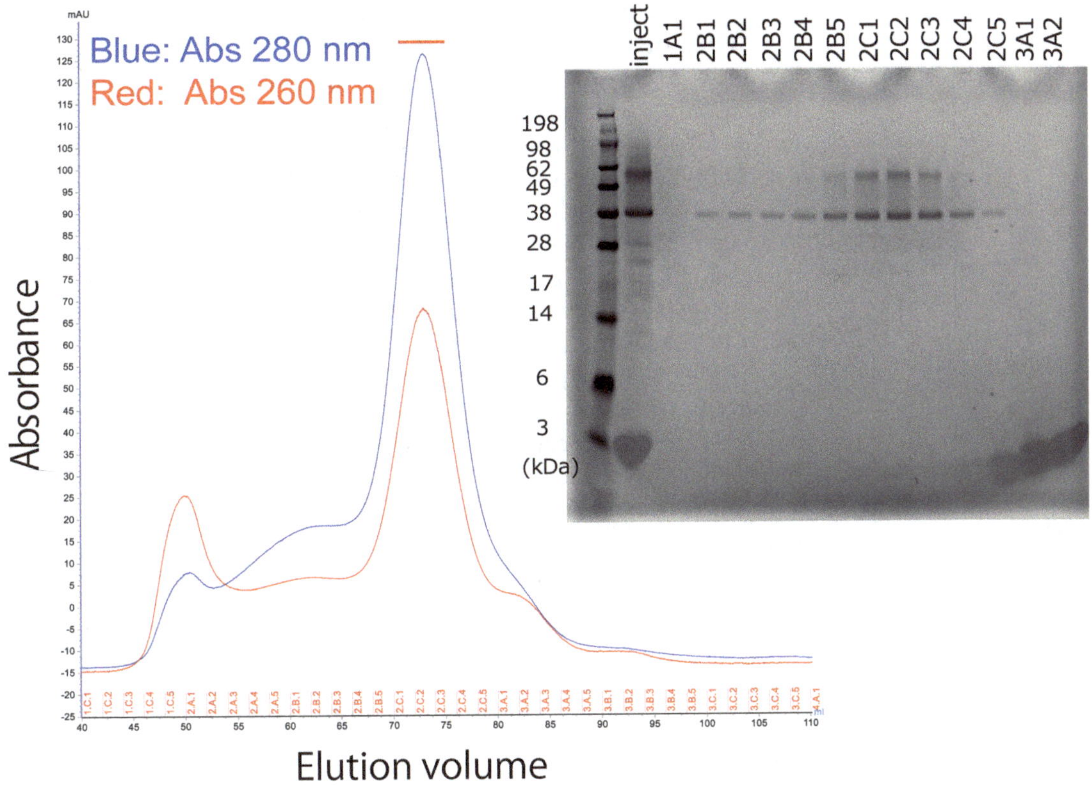

Fig. 2 SEC profile of purified PfMATE obtained from in vitro expression. The fractions that contain monomeric PfMATE are indicated. The inset shows the SDS–PAGE of the purified protein. The sample was loaded without heating, and PfMATE typically yields a double band in the SDS–PAGE

19. Transfer the mix to the empty column and obtain the flow-through, which contains PfMATE.
20. Wash the column twice with 7 mL of 20 mM Tris–HCl, pH 8.0, 300 mM NaCl, and 0.2% DDM solution to recover the protein fully (*see* **Note 11**).
21. Concentrate the flow-through and wash fractions by ultrafiltration using a 30 kDa molecular weight cut-off (MWCO) membrane to 2 mL.
22. Subject the digested sample to SEC using a HiLoad 16/60 Superdex 200 prep grade column (Cytiva) in 20 mM Tris–HCl, pH 8.0, 300 mM NaCl, and 0.2% DDM.
23. Correct the fractions corresponding to the monomeric PfMATE protein (Fig. 2).
24. Check the purified product by SDS–PAGE.
25. Exchange buffer to NMR buffer and concentrate to 20–50 μM by ultrafiltration using a 30 kDa molecular weight cut-off (MWCO) membrane. Typical yield of PfMATE from 1 mL in vitro reaction is 1.5 mg (*see* **Note 12**).

3.4 NMR Experiment of PfMATE

NMR experiments would be performed at Bruker Avance-600 MHz or Bruker Avance III-800 MHz spectrometers (Bruker Biospin) equipped with cryogenic triple resonance probes. The typical concentration of PfMATE for an NMR experiment would be 20–50 μM. To investigate the structure of PfMATE using NMR, the methyl resonances of Met ε1 and Ile δ1 can be recorded by the ^{1}H–^{13}C heteronuclear multiple quantum coherence (HMQC) [11] experiment. Since PfMATE is a protein from thermophilic bacteria, all NMR spectra were acquired at 50 °C, which is more physiologically relevant than previously performed structural studies of PfMATE (i.e., 18–20 °C for X-ray crystallography and 25 °C for EPR) [6–8, 12]. The elevated temperature also enhances the sensitivity of NMR detection, and almost all methionine and isoleucine resonances can be observed in ^{1}H-^{13}C HMQC spectra at relatively low concentrations (20–50 μM) with a reasonable amount of time (3–5 h) (Fig. 3) [9]. The leucine and valine resonances are less sensitive; thus, we did not use the resources for further analyses.

1. Transfer 250 μL of purified PfMATE to a Shigemi microtube.
2. Set the NMR temperature at 50 °C.
3. Put the PfMATE NMR sample in the NMR magnet.
4. Wait for 10 min for temperature stability.
5. Tune and match the cryogenic probe for ^{1}H and ^{13}C channels.
6. Optimize the shimm by TopShim GUI.
7. Determine 90° pulses for ^{1}H.
8. Read typical ^{1}H–^{13}C HMQC experimental parameters.

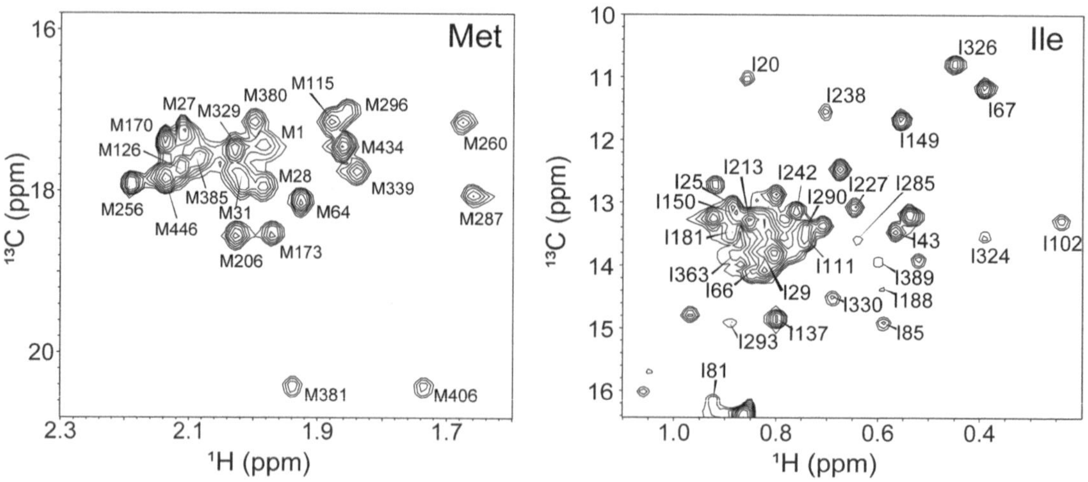

Fig. 3 HMQC spectra of PfMATE at neutral pH. The methionine ε1 and isoleucine δ1 regions of the ^{1}H^{13}C HMQC spectra of PfMATE are shown on the left and right, respectively. The residue numbers of the assigned resonances are indicated. (The spectra were adopted from [9])

9. Set the determined 90° pulse values.
10. Check the receiver gain.
11. Run ^1H–^{13}C HMQC experiment.
12. Process the data in TopSpin by Fourier transforming the data to yield NMR spectra.

4 Notes

1. Filter the D_2O M9 Celton media through the 0.22 μm filtration device after dissolving the materials mentioned for removal of undissolved materials and sterilization. The in vivo expression of PfMATE needs Celton base powder (Cambridge Isotope Laboratory) or similar supplements, such as an Algal amino acid mixture from Isotec. Expression of PfMATE without Celton base powder is extremely low.
2. Phenylmethylsulfonyl fluoride (PMSF) or other protease inhibitors can also be used.
3. Optimize the gene for the *E.coli* codon usage and place the HRV3C site between the NHH tag and PfMATE. The regular hexahistidine tag substantially reduces the expression of PfMATE in an in vitro protein synthesis. It is better to use the NHH tag instead.
4. For in vitro expression of PfMATE, use a commercially available *E. coli*-based cell-free protein synthesis kit, "Musaibo Kun" (Taiyo Nippon Sanso), which enables rapid and easy protein synthesis with high isotope labeling efficiency.
5. Use the stereo-array isotope labeling (SAIL) amino acid mixture [13, 14] that is specifically labeled at the desired position, for expression of a perdeuterated, Ile (δ1), Leu, Val, and Met methyl-$^1H^{13}C$ labeled PfMATE. The Stereo-Array Isotope Labeling (SAIL) amino acid mixtures are commercially available from Taiyo Nippon Sanso.
6. Do not use glycerol stock for the in vivo expression of PfMATE. Use freshly prepared C41 (DE3) colonies for stable expression.
7. Add [ε-^{13}C] methionine, [methyl-^{13}C, 3,3-$^2H_2$]–αketobutyric acid, and [3-methyl-^{13}C, 3,4,4,4-$^2H_4$]-α-ketoisovaleric acid directly to the medium as powder.
8. The typical time required for expressing and purifying PfMATE in an *E. coli*-based in vivo expression system is 5 days.
9. Additional DDM is not necessary when exchanging buffers, as ultrafiltration also accumulates DDM.

10. Prepare the outer dialysis solution before the inner dialysis solution and incubate at 30 °C to maximize the expression.
11. After washing the column, elute the NHH tag and HRV3C protease with 10 mL of 20 mM Tris–HCl (pH 8.0), 300 mM NaCl, 0.2% DDM, and 300 mM imidazole solution.
12. The typical time required for expressing and purifying PfMATE in an in vitro expression system is 3 days, and the yield was 1.5 mg/mL reaction. The total cost for expression material is less for in vitro expression. Therefore, expressing the PfMATE using an in vitro expression system is better regarding the time and resources required. The protein structure and stability are identical between in vivo and in vitro expression.
13. The SOFAST HMQC spectra [15] can also be used to record the NMR spectra of PfMATE. However, additional optimization of pulse length, etc., is needed to avoid unwanted excitation of water resonances.

Acknowledgments

We thank Dr. Tsukasa Kusakizako, Dr. Ryuichiro Ishitani, and Prof. Osamu Nureki for giving us the PfMATE expression plasmid. This work was supported by grants from Japan Society for the Promotion of Science (JSPS) KAKENHI (grant numbers JP22K18374 and JP20H03378 to K.T.) and the Japan Agency for Medical Research and Development (AMED) (grant number JP24gm161000 to K.T.).

Conflicts of Interest The authors have no conflicts of interest to declare that are relevant to the content of this chapter.

References

1. Du D, van Veen HW, Murakami S, Pos KM, Luisi BF (2015) Structure, mechanism and cooperation of bacterial multidrug transporters. Curr Opin Struct Biol 33:76–91. https://doi.org/10.1016/j.sbi.2015.07.015
2. Brown MH, Paulsen IT, Skurray RA (1999) The multidrug efflux protein NorM is a prototype of a new family of transporters. Mol Microbiol 31:394–395. https://doi.org/10.1046/j.1365-2958.1999.01162.x
3. Kusakizako T, Miyauchi H, Ishitani R, Nureki O (2020) Structural biology of the multidrug and toxic compound extrusion superfamily transporters. Biochim Biophys Acta (BBA) – Biomembr 1862:183154. https://doi.org/10.1016/j.bbamem.2019.183154
4. Kuroda T, Tsuchiya T (2009) Multidrug efflux transporters in the MATE family. Biochim Biophys Acta (BBA) – Proteins and Proteomics 1794:763–768. https://doi.org/10.1016/j.bbapap.2008.11.012
5. Drew D, Boudker O (2016) Shared molecular mechanisms of membrane transporters. Annu Rev Biochem 85:543–572. https://doi.org/10.1146/annurev-biochem-060815-014520
6. Zakrzewska S, Mehdipour AR, Malviya VN, Nonaka T, Koepke J, Muenke C, Hausner W, Hummer G, Safarian S, Michel H (2019)

Inward-facing conformation of a multidrug resistance MATE family transporter. Proc Natl Acad Sci 116:12275. https://doi.org/10.1073/pnas.1904210116

7. Tanaka Y, Hipolito CJ, Maturana AD, Ito K, Kuroda T, Higuchi T, Katoh T, Kato HE, Hattori M, Kumazaki K et al (2013) Structural basis for the drug extrusion mechanism by a MATE multidrug transporter. Nature 496:247–251. https://doi.org/10.1038/nature12014
8. Jagessar KL, Claxton DP, Stein RA, McHaourab HS (2020) Sequence and structural determinants of ligand-dependent alternating access of a MATE transporter. Proc Natl Acad Sci 117:4732. https://doi.org/10.1073/pnas.1917139117
9. Takeuchi K, Ueda T, Imai M, Fujisaki M, Shimura M, Tokunaga Y, Kofuku Y, Shimada I (2024) Affinity-directed substrate/H(+)-antiport by a MATE transporter. Structure. https://doi.org/10.1016/j.str.2024.05.004
10. Miroux B, Walker JE (1996) Over-production of proteins inEscherichia coli: mutant hosts that allow synthesis of some membrane proteins and globular proteins at high levels. J Mol Biol 260:289–298. https://doi.org/10.1006/jmbi.1996.0399
11. Tugarinov V, Hwang PM, Ollerenshaw JE, Kay LE (2003) Cross-correlated relaxation enhanced $^{1}H-^{13}C$ NMR spectroscopy of methyl groups in very high molecular weight proteins and protein complexes. J Am Chem Soc 125:10420–10428. https://doi.org/10.1021/ja030153x
12. Claxton DP, Jagessar KL, Steed PR, Stein RA, McHaourab HS (2018) Sodium and proton coupling in the conformational cycle of a MATE antiporter from Vibrio cholerae. Proc Natl Acad Sci 115:E6182. https://doi.org/10.1073/pnas.1802417115
13. Kainosho M, Guntert P (2009) SAIL–stereo-array isotope labeling. Q Rev Biophys 42:247–300. https://doi.org/10.1017/s0033583510000016
14. Kainosho M, Torizawa T, Iwashita Y, Terauchi T, Mei Ono A, Güntert P (2006) Optimal isotope labelling for NMR protein structure determinations. Nature 440:52–57. nature04525 [pii] 10.1038/nature04525
15. Schanda P, Kupce E, Brutscher B (2005) SOFAST-HMQC experiments for recording two-dimensional heteronuclear correlation spectra of proteins within a few seconds. J Biomol NMR 33:199–211. https://doi.org/10.1007/s10858-005-4425-x

Chapter 2

Crystallographic Studies of RND- and ABC-Type Tripartite Multidrug Transporters

Ui Okada and Satoshi Murakami

Abstract

Three-dimensional structural information is crucial for the precise and detailed investigation of protein functions and their molecular mechanisms. X-ray crystallography is an effective method for determining protein structures. Here, we describe our standardized protocols, including overexpression, purification, crystallization, and handling of crystals for the structural analysis of multidrug transporters. We have determined several crystal structures of resistance–nodulation–division (RND) and ATP-binding cassette (ABC)-type tripartite multidrug transporters using these methods. These results facilitate the understanding of the multidrug resistance molecular mechanisms, which are mediated by tripartite multidrug efflux complexes.

Key words Multidrug resistance, RND transporters, ABC transporters, Tripartite multidrug efflux complex, X-ray crystallography, Membrane protein

1 Introduction

The efflux of drugs by multidrug transporters is a primary cause of drug resistance in pathogenic bacteria [1–3]. In Gram-negative bacteria that have two membranes, an inner membrane (IM) and an outer membrane (OM), tripartite efflux complexes that are able to span these two membranes are used to export drugs out of cells [4]. The tripartite efflux complex is composed of three proteins, including the OM channel, the membrane fusion protein (MFP), and the IM transporter. The OM channel forms a chimney-like structure that penetrates part of the periplasmic space and OM and provides a pathway for substrate translocation [5]. The MFP forms an inverted funnel-like structure and connects the OM channel and IM transporter [6]. The IM transporter is located on the IM and provides substrate selectivity and the driving force for substrate translocation. The overall structure of the tripartite complex was determined by cryo-electron microscopy [7–9].

Kunihiko Nishino and Mikio Tanabe (eds.), *Bacterial Multidrug Exporters: Methods and Protocols*, Methods in Molecular Biology, vol. 3044, https://doi.org/10.1007/978-1-0716-5316-6_2,

© The Author(s), under exclusive license to Springer Science+Business Media, LLC, part of Springer Nature 2026

Here, we focus on the IM transporters that form part of the tripartite complexes. Transporters from the RND superfamily, some transporters from the ABC superfamily, and the major facilitator superfamily (MFS) form a tripartite efflux complex as an IM transporter. To date, we have determined several IM transporter crystal structures, mainly from pathogenic bacteria. Specifically, AcrB from *Escherichia coli* [10, 11], OqxB from *Klebsiella pneumoniae* [12], BpeB and BpeF from *Burkholderia pseudomallei* [13], and MacB from *Acinetobacter baumannii* [14] were subjected to structural analysis. Among these, only MacB is classified as an ABC-type IM transporter, whereas AcrB, OqxB, BpeB, and BpeF belong to the RND superfamily. From these experiences with crystal structural analysis, we established standardized protocols for protein preparation and crystallization for the structural analysis thereof. Finally, we describe the practical protocol and include notes herein, but we believe that the most important challenge to overcome the structural analysis of membrane proteins is to carefully continue experimenting without overlooking small changes for improvement.

2 Materials

All solutions were prepared with ultrapure water (Milli-Q grade) and molecular biology-grade reagents.

2.1 Overexpression

1. *Escherichia coli* BL21 (DE3) (Novagen) for protein expression.
2. Expression plasmid pET22b (Novagen) harboring the target gene fused with an N- or C-terminal hexa-histidine tag.
3. 25 mg/mL ampicillin solution, sterilized.
4. 2× YT medium: Weigh 1 g of yeast extract, 1.6 g of tryptone, and 0.5 g of NaCl in 100 mL of water, sterilized.
5. Davis minimal medium [15]: Weigh 7 g of K_2HPO_4, 3 g of KH_2PO_4, 0.5 g of sodium citrate, 0.1 g of $MgSO_4$, and 1 g of $(NH_4)_2SO_4$ in 1 L of water, sterilized.
6. Glucose solution (20%, w/v), sterilized.
7. Casamino acid solution (10%, w/v), sterilized.
8. Isopropyl-β-D-thiogalactopyranoside (250 mM) (IPTG).
9. Buffer A: 20 mM Tris–HCl pH 7.0, 1 mM $MgCl_2$, 0.5 mM EDTA, and 10% (v/v) glycerol.
10. Incubator shaker, Innova 44 (New Brunswick Scientific).
11. High-speed refrigerated centrifuge CR22E with continuous rotor R18C and fixed-angle rotor R9A (Hitachi Koki).
12. Excel autohomogenizer ED-7 (Nihonseiki Kaisha Ltd.).

2.2 Preparation of Membrane Fraction

1. Buffer B: 50 mM Tris–HCl pH 7.0, 1 mM $MgCl_2$, and 0.5 mM EDTA.
2. Buffer C: 5 mM Tris–HCl pH 7.0 and 0.5 mM EDTA.
3. Buffer D: 50 mM Tris–HCl pH 7.0 and 10% (v/v) glycerol.
4. Pierce BCA protein assay kit (Thermo Fisher Scientific).
5. Excel autohomogenizer ED-7 (Nihonseiki Kaisha Ltd.).
6. High-speed refrigerated centrifuge CR22E with the fixed-angle rotors R9A and R20A2 (Hitachi Koki).
7. Ultracentrifuge CP56GII with a P45AT rotor (Hitachi Koki).
8. Microfluidizer M-110EH (Microfluidics Corp.).
9. Dounce tissue grinder (WHEATON).

2.3 Purification

1. Buffer D: 50 mM Tris–HCl pH 7.0, 10% (v/v) glycerol.
2. *n*-Dodecyl-β-D-maltopyranoside (DDM) solution, 20% (w/v).
3. Chelating Sepharose Fast Flow (Cytiva).
4. 0.1 M $NiSO_4$ solution.
5. Buffer E: 20 mM Tris–HCl pH 7.5, 300 mM NaCl, 10% (v/v) glycerol, and 0.05% (w/v) DDM.
6. Buffer F: 20 mM Tris–HCl pH 7.5, 200 mM NaCl, 10% (v/v) glycerol, and 0.05% (w/v) DDM.
7. Falcon tube (50 mL).
8. Econo-Column chromatography column 1.0 × 10 cm (Bio-Rad).
9. Flow adaptor (Bio-Rad).
10. Ultracentrifuge CP56GII with a P50AT2 rotor (Hitachi Koki).
11. Peristaltic pump SJ-1211 (Atto).
12. Fraction collector FC-203B (Gilson).
13. Amicon Stirred Cell (Merck Millipore) with a 100 kDa molecular weight cutoff Omega ultrafiltration membrane disc filter (Pall Corp.).
14. Ultrafree-MC GV centrifugal filter (Merck Millipore).
15. Superdex 200 Increase 10/300 GL (Cytiva).
16. AKTA explore 10S (Cytiva).

2.4 Crystallization

1. Commercially available crystallization screening kits.
2. PS MRC 96-well 2-drop crystallization plate (Molecular Dimensions).
3. Titer Stick HC film (Kajixx, Kanagawa, Japan).
4. Sealing film applicator (Hampton Research).

5. Micro-reservoir strips (8-well) (Sptlabtech).
6. Nanoliter liquid handling system Mosquito LCP (Sptlabtech).
7. Incubator.
8. Stereomicroscope.

2.5 Handling of Crystals

1. Simulated mother liquor. Details will be provided later.
2. Cryoprotectant solution, such as glycerol or polyethylene glycol (PEG) 400.
3. Liquid nitrogen.
4. Cryo-tongs, 18 mm (Hampton Research) (Fig. 1a).
5. CrystalWand, magnetic (Hampton Research) (Fig. 1b).
6. Vial clamp, curved (Hampton Research) (Fig. 1c).
7. CrystalClear Strips with Frame (Douglas Instruments) (Fig. 1d, e).
8. CrystalCap with vial (Hampton Research) (Fig. 1f, g).
9. Mounted CryoLoop, 20 μm (Hampton Research).
10. CryoCane, 6-vial holder (Hampton Research) (Fig. 1h).
11. Storage dewar.
12. Cryostream system (Rigaku).

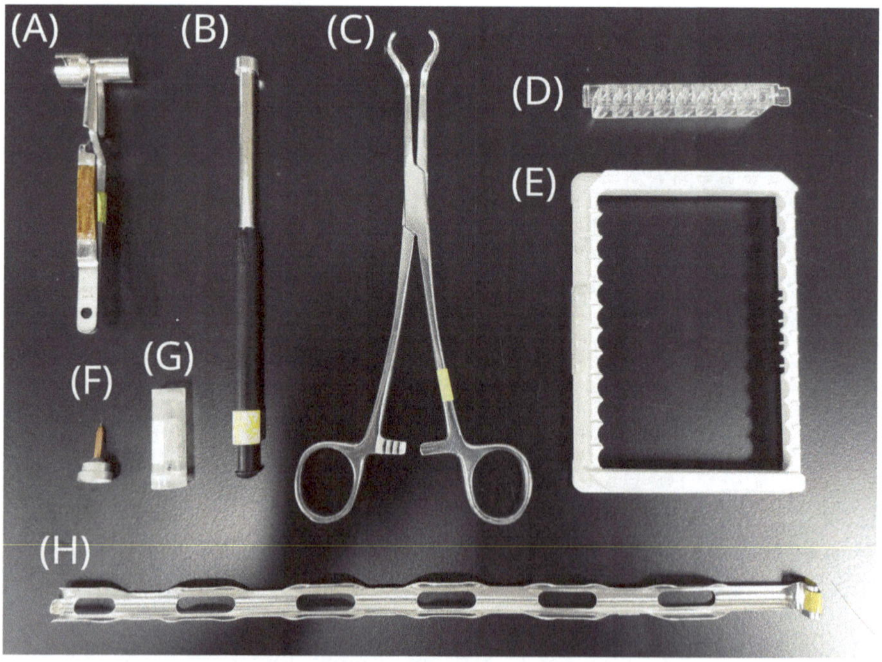

Fig. 1 Tools for the handling of crystals. (**a**) Cryo-tongs, (**b**) CrystalWand, (**c**) vial clamp, (**d**) CrystalClear Strips, (**e**) frame, (**f**) CrystalCap with CryoLoop, (**g**) cryo vial, and (**h**) CryoCane

2.6 Packing of Crystals for Shipping

1. Liquid nitrogen.
2. CrystalWand–magnetic (Hampton Research) (Fig. 1b).
3. Vial clamp–curved (Hampton Research) (Fig. 1c).
4. UniPuck (Molecular Dimensions) (Fig. 2a, b).
5. Angled cryo-tongs (Molecular Dimensions) (Fig. 2c).
6. Hooked handle (Molecular Dimensions) (Fig. 2d).
7. Puck wand (Molecular Dimensions) (Fig. 2e).
8. Locking rod (Molecular Dimensions) (Fig. 2f).
9. Shelved shipping cane (Molecular Dimensions) (Fig. 2g).
10. Puck-loading dewar (Molecular Dimensions).
11. Dry shipper.
12. Shipping case.

3 Methods

3.1 Overexpression

1. Transform *E. coli* BL21(DE3) with the pET22b-based expression vector harboring the target gene.
2. Normally, the main culture is performed on a 30 L scale. The following amounts of medium should be adjusted accordingly. After autoclaving, all media are allowed to cool and are stored at the cultivation temperature.

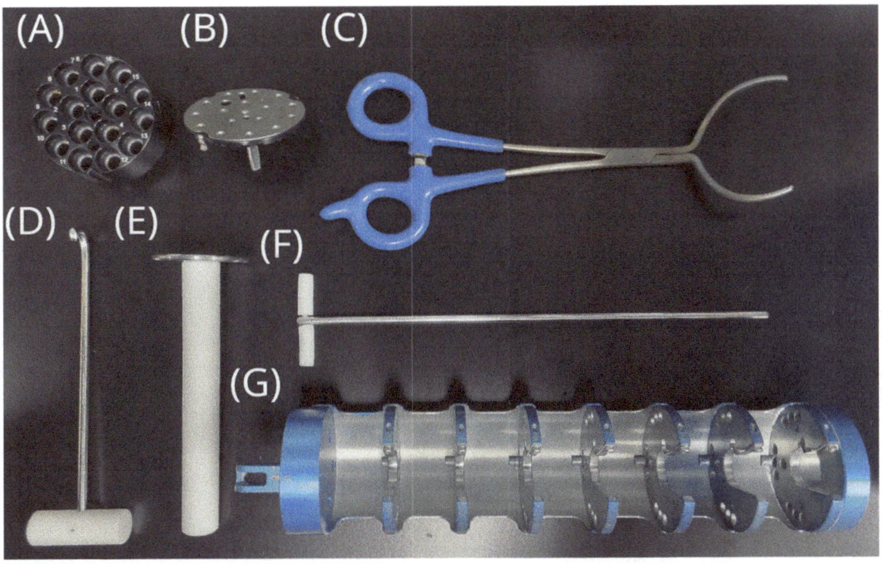

Fig. 2 Tools to pack the crystals for shipping. (**a**) UniPuck sample enclosure, (**b**) UniPuck sample base, (**c**) angled cryo-tongs, (**d**) hooked handle, (**e**) puck wand, (**f**) locking rod, and (**g**) shelved shipping cane

3. For the pre-pre-culture, inoculate a single colony of the transformant into 10 mL of 2× YT medium containing 0.1 mg/mL of ampicillin and cultivate at 37 °C with shaking until the OD_{600} reaches 0.6.
4. For the pre-culture, inoculate the pre-pre-culture into 630 mL of 2× YT medium in an Erlenmeyer flask and cultivate at 37 °C with rotation at 120 rpm until the OD_{600} reaches 0.6.
5. Inoculate 50 mL of the pre-culture into 12 Erlenmeyer flasks containing 2.5 L of Davis minimal medium supplemented with 0.2% (w/v) glucose and 0.1% (w/v) casamino acids (*see* **Note 1**). Cultivate these flasks at 37 °C with rotation at 120 rpm until the OD_{530} reaches 0.6.
6. For protein expression, add the IPTG solution to a final concentration of 0.1 mM and incubate at 25 °C with rotation at 120 rpm for 8–16 h (*see* **Note 2**).
7. Centrifuge the culture at 24,000× *g* to harvest the transformed cells using a continuous rotor R18C.
8. Resuspend and wash the cell pellets in ice-cold Buffer A using an autohomogenizer at 3000 rpm for 3 min with ice water cooling.
9. Centrifuge the suspended cells at 12,000× *g* for 20 min and remove the supernatant.
10. Weigh the collected cells.

3.2 Preparation of Membrane Fraction

1. Resuspend the collected cells in ice-cold Buffer B at a concentration of 0.2 g wet cells/mL using an autohomogenizer as described above.
2. Centrifuge the suspensions at 1700× *g* for 2 min to remove dust to prevent the microfluidizer from clogging.
3. Load the supernatants into the microfluidizer and disrupt the cells three times at 15,000 psi with ice water cooling.
4. Remove the cellular debris by centrifugation at 27,000× *g* for 10 min. Collect the supernatant and centrifuge it again under the same conditions to completely remove cellular debris.
5. Ultracentrifuge the supernatants at 214,000× *g* for 1 h. Collect the precipitates as the membrane fraction. Ultracentrifuge the supernatants again at 214,000× *g* for 1.5 h to increase the yield and merge the precipitates.
6. For washing, suspend the membrane fractions in ice-cold Buffer C using a Dounce tissue grinder. Then, ultracentrifuge the suspended samples at 214,000 × *g* for 1.5 h.
7. Resuspend the precipitates in ice-cold Buffer D using a Dounce tissue grinder.

8. Determine the concentration of the total protein in the prepared membrane fractions using a bicinchoninic acid (BCA) assay. Divide into appropriate amounts, flash freeze in liquid nitrogen, and store at −80 °C until use (*see* **Note 3**).

3.3 Purification

1. Perform all procedures at 4 °C unless otherwise specified.
2. Thaw the frozen membrane fraction containing approximately 200 mg of total protein on ice.
3. Dilute the membrane fraction to 10 mg/mL of total protein with Buffer D.
4. Add 20% (w/v) DDM solution to the diluted membrane fraction to a final concentration of 2% (w/v) and incubate on ice for 20 min (*see* **Note 4**).
5. Ultracentrifuge the solubilized membrane fraction at 172,000× *g* for 1 h.
6. Add the supernatant to 5 mL of Chelating Sepharose Fast Flow charged with Ni^{2+} and equilibrated with Buffer E. Stir the mixture gently for 1 h in a 50-mL Falcon tube.
7. Pack the protein–resin complex with an Econo-Column chromatography column and a flow adapter. Connect the packed column to a peristaltic pump and fraction collector.
8. Wash the column with 10 column volumes of Buffer E supplemented with 25 mM imidazole at a flow rate of 1 mL/min. Thereafter, wash the column with 10 column volumes of Buffer E supplemented with 100 mM imidazole (*see* **Note 5**).
9. Elute the His-tagged protein with 10 column volumes of Buffer E supplemented with 300 mM imidazole.
10. Pool the fractions containing the target protein and concentrate in an Amicon Stirred Cell with an ultrafiltration membrane disc filter (100 kDa molecular weight cutoff) rinsed with Buffer F under N_2 gas at a pressure of 0.1–0.2 MPa. Concentrate to an amount that can be injected into a size-exclusion chromatography (SEC) column.
11. Filter the concentrate using a 0.22-μm centrifugal filter rinsed with Buffer F.
12. Connect a Superdex 200 Increase 10/300 GL column to an AKTA explore system and equilibrate with 1.3 column volumes of Buffer F at a flow rate of 0.3 mL/min.
13. Inject the filtered protein solution into the SEC column and elute with 1.1 column volumes of Buffer F at a flow rate of 0.3 mL/min, and collect 0.5 mL fractions (*see* **Note 6**).
14. Pool the fractions containing the target protein and concentrate them as described above.

15. Filter the concentrate with a 0.22-μm centrifugal filter.
16. Determine the concentration of purified protein using a BCA assay and analyze the purity by SDS-PAGE (*see* **Note 7**).

3.4 Crystallization

1. Incubate the freshly prepared protein solution at the crystallization temperature.
2. Dispense 50 μL of a commercially available crystallization screening kit (*see* **Note 8**) or a homemade grid screen into each well of a 96-well crystallization plate as reservoirs.
3. Mix 20–30 mg/mL of the protein solution with the reservoir at a 1:1 ratio and dispense into the crystallization plate using a Mosquito nanoliter liquid handling system (*see* **Note 9**).
4. Seal the crystallization plate with a sealing film.
5. Incubate the crystallization plate (*see* **Note 10**).
6. Observe the crystallization droplets under a stereomicroscope. If crystals are observed, the reservoir conditions should be optimized further.

3.5 Handling of Crystals

1. Open the sealed crystallization plate at a droplet where the crystals are observed.
2. Add 5–10 μL of simulated mother liquor into the droplet. If the crystal melts or cracks, the composition of the simulated mother liquor needs to be optimized further (*see* **Note 11**).
3. Transfer the crystal, together with the simulated mother liquor, to a new well of the CrystalClear strips by using a micropipette. If necessary, cut off the pipette tip to avoid damaging the crystal. From this step to scooping the crystal in **step 5**, perform the operations under a stereomicroscope.
4. Remove the simulated mother liquor, ensuring that the crystal does not dry out; immediately add new simulated mother liquor supplemented with a cryoprotectant, and mix well, being careful not to damage the crystal. Repeat this procedure to increase the cryoprotectant concentration until it reaches a high enough concentration (*see* **Note 12**).
5. Scoop up the crystal with a CryoLoop mounted on a CrystalCap, and immediately expose it to flash cooling under a cryostream.
6. Transfer the CrystalCap holding the crystal into liquid nitrogen by using a cryo-tongs cooled with liquid nitrogen.
7. Connect a CrystalWand to the CrystalCap and keep the CrystalWand in a position where the crystal does not protrude from the liquid nitrogen until the end of **step 9**.
8. Open and remove the cryo-tongs.

9. Place a cryo vial over the CrystalCap using a vial clamp, being careful not to touch the CryoLoop holding the crystal. The cryo vial and the vial clamp should be cooled with liquid nitrogen before the procedure.
10. Remove the CrystalWand by releasing the magnetic force by pushing the bottom button. The vial clamp is currently attached to the cryo vial and CrystalCap.
11. Take out the CrystalCap covered by the cryo vial from the liquid nitrogen, remove the vial clamp, and immediately set it in a CryoCane, which is then submerged in a storage dewar filled with liquid nitrogen.
12. Store the crystals in liquid nitrogen until data collection. When it is time for analysis using a synchrotron beam, prepare to send the crystals according to the procedure detailed below.

3.6 Packing of Crystals for Shipping

1. Cool a UniPuck sample enclosure with liquid nitrogen in a puck-loading dewar. Align the protrusion of the sample enclosure with the dent of the dewar to prevent the enclosure from shifting during the subsequent procedures (*see* **Note 13**).
2. Remove the vial-covered CrystalCap from the cane in the storage dewar and place it into the liquid nitrogen in the puck-loading dewar. All procedures from this step to **step 8** should be performed while keeping the crystal in liquid nitrogen.
3. Clasp the cryo vial with the vial clamp and face the base of the CrystalCap upward.
4. Press the bottom button of the CrystalWand to release the magnetic force, bring the tip of the wand close to the base of the CrystalCap, touch it, then slowly push the button back to activate the magnetic force and attach the CrystalCap to the wand.
5. Remove the cryo vial with the vial clamp while holding the CrystalCap with the wand.
6. Insert the CrystalCap into the appropriate numbered hole of the sample enclosure using the wand. Insert it straight to prevent the crystal from touching the walls.
7. Remove the CrystalWand from the CrystalCap by releasing the magnetic force.
8. Repeat **steps 2–7** to fill the holes of the sample enclosure with the crystals to be shipped.
9. Cool a UniPuck sample base mounted on a puck wand with liquid nitrogen.
10. Align the guide rod of the sample base, place it over the sample closure, and push it rapidly until it is fitted in.

11. If multiple UniPucks need to be shipped, repeat **steps 1–10**.
12. Cool a shelved shipping cane and a locking rod with liquid nitrogen.
13. Insert the UniPuck into the shelved shipping cane using an angled cryo-tongs and align the notch on the side of the UniPuck with the guide on the cane. Repeat this procedure to insert all the UniPucks into the cane.
14. Lift the cane slightly using a hooked handle and insert the locking rod into the cane.
15. Using the hooked handle, place the cane into a dry shipper that has been cooled with liquid nitrogen.
16. Place the dry shipper in a shipping case and send it to the synchrotron radiation facility. Perform X-ray diffraction experiments and determine the protein structure.

4 Notes

1. In some instances, a ZY medium [16] can be used in the main culture instead of Davis minimal medium. Using a ZY medium may improve the target protein expression level or the yield when solubilized with a detergent
2. The cultivation temperature of the main culture also affects the target protein yield. Although incubating at 37 °C may be acceptable, we found that lower temperatures (20–25 °C) provide better results for membrane proteins. In the case of *E. coli* that grow too rapidly, it is possible that the quality control systems of the host cells may not be able to adequately manage the overexpression of the membrane proteins
3. Normally, we divide the prepared membrane fractions into approximately 200 mg of total protein. This is the amount that is used for one purification batch, which yields a few mg of the final purified products
4. The first choice of solubilizing detergent for membrane proteins is DDM. However, if the solubilization efficiency is poor or the target protein is unstable during purification, it is necessary to use other detergents, such as sucrose monododecanoate, n-decyl-β-D-maltopyranoside, or Cymal 6
5. The concentration of imidazole in the washing buffer should be modified according to the affinity between the target protein and the resin
6. SEC is both a purification procedure and an evaluation of the prepared protein sample. If there is a high peak near the void volume of the column, it is an indication that the target protein tends to cause non-specific aggregation. It is necessary to optimize the buffer conditions, such as the pH, salt concentration,

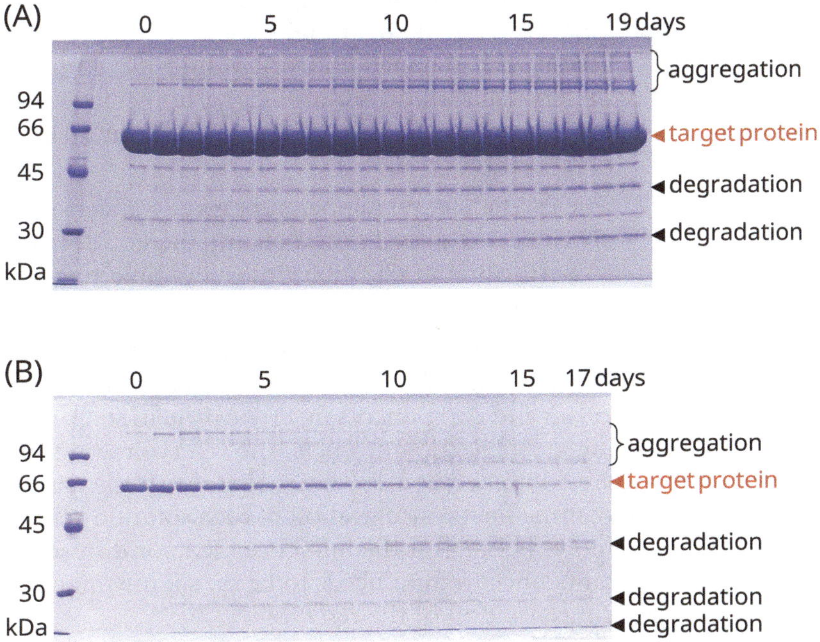

Fig. 3 Example of a prepared protein stability check. Details are described in **Note 7**. (**a**) A good example. Although some aggregation and degradation can be observed, the prepared protein is generally stable. A large amount of sample is loaded to observe small changes. (**b**) A bad example. The target protein is unstable and degrades within a few days. Crystal formation and growth cannot be expected under these conditions; therefore, it is necessary to adjust the purification conditions, such as the buffer pH, salt concentration, and/or the detergent used

and detergent used. Furthermore, it can be decided whether the target protein maintains a functional oligomer based on the molecular weight estimated from the elution position of the main peak. However, it should be noted that the estimated molecular weight is a complex of the protein and detergent

7. When preparing the protein for the first time, store the concentrated protein solution, which will be used for the crystallization trials, at the crystallization temperature to assess its stability. Every day, sample a small amount of the protein solution, combine it with a sampling buffer for SDS-PAGE, and store it at −30 °C. On the final day, electrophorese the samples to determine the presence of any aggregation or degradation (Fig. 3)

8. Crystallization screening kits are available from multiple manufacturers, such as Molecular Dimensions, Hampton Research, and Jena Bioscience; however, it is important to carefully check the conditions of each kit. There are instances where the composition of each condition is the same but in a different order; therefore, trying both such differently named but essentially identical kits would be a waste of samples

9. For initial screening using crystallization screening kits, combine 50–100 nL of protein solution and reservoir. If initial crystals are obtained and you would like to further optimize the crystallization conditions in a homemade reservoir, combine the protein solution and reservoir in a larger volume to obtain larger crystals
10. The crystallization temperature greatly affects whether crystals are obtained as well as the quality thereof. It is recommended to perform crystallization under multiple temperature conditions, such as 4 °C and 25 °C
11. Simulated mother liquor reproduces the composition of the crystallization droplet, including the precipitant from the reservoir and components from the protein solution. When forming crystallization droplets, the reservoir and protein solution are combined at a 1:1 ratio; therefore, it is important to note that the final concentration of each solution is halved. Furthermore, because of the influence of vapor diffusion, the precipitant concentration needs to be set slightly higher than half the concentration of the reservoir
12. If the concentration of the cryoprotectant is not sufficient, ice crystals will grow, making it challenging to perform diffraction experiments on the protein crystals (Fig. 4). The amount of cryoprotectant that should be added varies depending on the composition of the simulated mother liquor. Thus, it is necessary to evaluate whether it freezes with only the simulated mother liquor supplemented with cryoprotectant without picking up the protein crystal
13. The puck-loading dewar must be filled with sufficient liquid nitrogen for the crystal packing procedure. Fill it to about 2 cm above the top of the submerged sample enclosure. Add more liquid nitrogen if the level gets low during the procedure

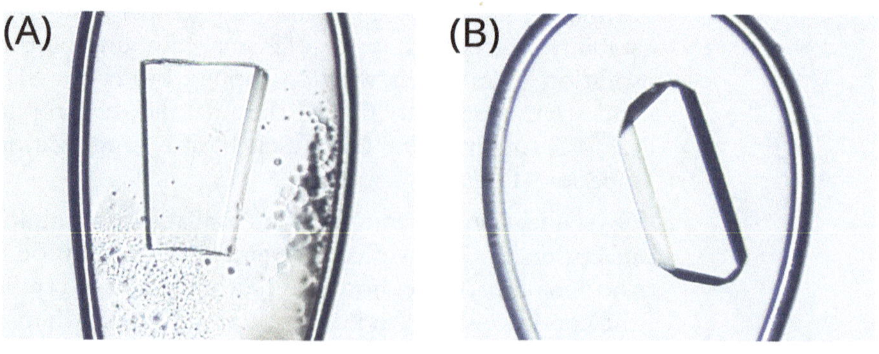

Fig. 4 Example of flash-cooled crystals. (**a**) A bad example. Ice crystals form because the cryoprotectant concentration is not sufficient. (**b**) A good example. There is no ice formation

Acknowledgments

This work was supported by the Japan Society for the Promotion of Science (JSPS) KAKENHI, the Platform Project for Supporting Drug Discovery and Life Science Research, the Basis for Supporting Innovative Drug Discovery and Life Science Research (BINDS) from the Japan Agency for Medical Research and Development (AMED), and the Joint Research Committee of the Institute for Protein Research, Osaka University.

Conflicts of Interest The authors have no conflicts of interest to declare that are relevant to the content of this chapter.

References

1. Li XZ, Nikaido H (2004) Efflux-mediated drug resistance in bacteria. Drugs 64:159–204
2. Munita JM, Arias CA (2016) Mechanisms of antibiotic resistance. Microbiol Spectr 4. https://doi.org/10.1128/microbiolspec.VMBF-0016-2015
3. Zheng M, Lupoli TJ (2023) Counteracting antibiotic resistance enzymes and efflux pumps. Curr Opin Microbiol 75:102334
4. Neuberger A, Du D, Luisi BF (2018) Structure and mechanism of bacterial tripartite efflux pumps. Res Microbiol 169:401–413
5. Koronakis V, Sharff A, Koronakis E, Luisi B, Hughes C (2000) Crystal structure of the bacterial membrane protein TolC central to multidrug efflux and protein export. Nature 405: 914–919
6. Mikolosko J, Bobyk K, Zgurskaya HI, Ghosh P (2006) Conformational flexibility in the multidrug efflux system protein AcrA. Structure 14: 577–587
7. Du D, Wang Z, James NR, Voss JE, Klimont E, Ohene-Agyei T, Venter H, Chiu W, Luisi BF (2014) Structure of the AcrAB-TolC multidrug efflux pump. Nature 509:512–515
8. Wang Z, Fan G, Hryc CF, Blaza JN, Serysheva II, Schmid MF, Chiu W, Luisi BF, Du D (2017) An allosteric transport mechanism for the AcrAB-TolC multidrug efflux pump. Elife 6:e24905
9. Fitzpatrick AW, Llabrés S, Neuberger A, Blaza JN, Bai XC, Okada U, Murakami S, van Veen HW, Zachariae U, Scheres SH, Luisi BF, Du D (2017) Structure of the MacAB-TolC ABC-type tripartite multidrug efflux pump. Nat Microbiol 2:17070
10. Murakami S, Nakashima R, Yamashita E, Yamaguchi A (2002) Crystal structure of bacterial multidrug efflux transporter AcrB. Nature 419:587–593
11. Murakami S, Nakashima R, Yamashita E, Matsumoto T, Yamaguchi A (2006) Crystal structures of a multidrug transporter reveal a functionally rotating mechanism. Nature 443: 173–179
12. Bharatham N, Bhowmik P, Aoki M, Okada U, Sharma S, Yamashita E, Shanbhag AP, Rajagopal S, Thomas T, Sarma M, Narjari R, Nagaraj S, Ramachandran V, Katagihallimath N, Datta S, Murakami S (2021) Structure and function relationship of OqxB efflux pump from Klebsiella pneumoniae. Nat Commun 12:5400
13. Kato T, Okada U, Hung LW, Yamashita E, Kim HB, Kim CY, Terwilliger TC, Schweizer HP, Murakami S (2023) Crystal structures of multidrug efflux transporters from. Proc Natl Acad Sci USA 120:e2215072120
14. Okada U, Yamashita E, Neuberger A, Morimoto M, van Veen HW, Murakami S (2017) Crystal structure of tripartite-type ABC transporter MacB from Acinetobacter baumannii. Nat Commun 8:1336
15. Davis BD (1949) The isolation of biochemically deficient mutants of bacteria by means of penicillin. Proc Natl Acad Sci USA 35:1–10
16. Studier FW (2005) Protein production by auto-induction in high density shaking cultures. Protein Expr Purif 41:207–234

Chapter 3

Purification and Sonication-Assisted Cryo-EM Grid Preparation of TolC

Mikio Tanabe and Toshio Moriya

Abstract

TolC is a multifunctional outer membrane protein essential for bacterial physiology. As the outer membrane component of tripartite efflux complexes, TolC mediates antibiotic resistance while also serving critical roles in protein secretion and metabolite transport. Understanding these diverse functions through structural and functional studies is critical for developing new antimicrobial therapies and overcoming drug resistance. Structural analysis of TolC has been predominantly conducted using X-ray crystallography. However, since TolC functions through complex formation with numerous periplasmic and secreted proteins in cells, cryogenic electron microscopy (cryo-EM) is more suitable for analyzing these transient complexes. This chapter describes purification strategies for TolC and its protein complexes, as well as cryo-EM sample preparation protocols, ***including sonication treatment to improve particle distribution.*** We also address challenges specific to membrane protein structural biology and discuss structural analysis and validation methods with detailed methodological updates.

Key words Multidrug efflux system, Outer membrane protein, Purification, Cryo-EM single-particle analysis, Sonication, Refolding

1 Introduction

Bacteria employ multidrug efflux systems to eliminate toxic substances, including antibiotics, from the cell. These efflux systems represent a major mechanism of antibiotic resistance, enabling bacteria to survive exposure to multiple antimicrobial compounds [1]. Some efflux systems in Gram-negative bacteria are composed of multiple subunits, forming the "tripartite efflux complexes," where an integral inner-membrane protein associates with periplasmic and outer membrane proteins to expel drugs across the two membranes using the energy of ATP hydrolysis or electrochemical gradient [2, 3]. TolC serves as an essential component of multidrug tripartite efflux complexes that play a central role in the acquisition of drug resistance in Gram-negative bacteria [4]. Beyond this primary function in drug efflux, TolC demonstrates remarkable

Kunihiko Nishino and Mikio Tanabe (eds.), *Bacterial Multidrug Exporters: Methods and Protocols*, Methods in Molecular Biology, vol. 3044, https://doi.org/10.1007/978-1-0716-5316-6_3,

© The Author(s), under exclusive license to Springer Science+Business Media, LLC, part of Springer Nature 2026

functional versatility across diverse cellular processes [5]. The protein facilitates the secretion of various virulence factors, including hemolysin [6, 7], colicin V [8], STB enterotoxin [9], and the antimicrobial peptide microcin J25 [10]. TolC also mediates both influx and efflux pathways, functioning as a receptor for colicin E1 toxin uptake [11], also exporting endogenous metabolites such as the siderophore [12], enterobactin, indole [13], and porphyrins [14]. These diverse functions underscore TolC's integration into multiple cellular pathways beyond drug resistance [5]. The protein is intimately connected to outer membrane structure and function, cell envelope integrity, acid resistance, and persister cell formation. Mutant variants of TolC exhibit pleiotropic phenotypes, reflecting its critical role in maintaining bacterial fitness under diverse environmental conditions. This multifaceted functionality positions TolC as a central hub in bacterial physiology, making it an attractive target for therapeutic intervention.

Understanding TolC's multiple functions at the atomic level is critical for developing novel antimicrobial agents and overcoming drug resistance. This requires high-resolution three-dimensional structures obtained through advanced structural biology techniques. While X-ray crystallography has provided fundamental structural insights into TolC [15], this method has limitations in capturing the dynamic conformational states and transient complexes that are essential for understanding TolC's diverse functions.

Recent advances in cryogenic electron microscopy single-particle analysis (cryo-EM SPA) have enabled near-atomic resolution analysis, including the atomic model building of membrane proteins (higher than 3.5 Å). TolC's homotrimeric structure (monomer—approximately 52 kDa) with an elongated architecture extending over 100 Å into the periplasmic space [15] makes it particularly suitable for cryo-EM analysis. Nevertheless, sample quality is still the primary determinant of success in protein structure determination. Several methods of purification of TolC has been established [15, 16]. Membrane proteins like TolC require detergents for isolation and purification, which can introduce contaminants or promote aggregation. Such impurities significantly degrade signal-to-noise ratios and particle homogeneity in cryo-EM images, compromising high-resolution structure acquisition. In this chapter, we present a detailed protocol for the structural determination of TolC using cryo-EM SPA. Our findings demonstrate that seemingly minor technical variations: (1) purification from inclusion body and (2) sonication-assisted grid preparation. First, we have developed an optimized purification strategy using inclusion body refolding, which offers advantages in yield and purity compared to membrane extraction methods. Second, we have found that sonication treatment effectively dissociates non-physiological aggregates without damaging the protein, significantly improving particle distribution on cryo-EM grids. These two small modifications can profoundly influence the quality of the resulting three-dimensional reconstructions.

2 Materials

2.1 Equipment

1. Shaker incubator (e.g., New Brunswick Innova44).
2. FPLC (e.g., ÄKTA system—Cytiva).
3. Superdex 200 10/300 GL (Cytiva) for size-exclusion chromatography (SEC).
4. Ultrafiltration devices, molecular weight cut-off 100,000 Da (e.g., Amicon Ultra).
5. 2.5 L Tunair flask.
6. Ultrasonicator (chip type).
7. Ultrasonicator (water bath type).
8. Refrigerated centrifuge capable of $12{,}000\times g$.
9. Gravity chromatography columns (e.g., Bio-Rad, Econo-Column, $\varphi = 2.5$ cm).
10. FPLC system (such as ÄKTA prime plus or equivalent).
11. Protein concentrators (Amicon, 100 kDa molecular weight cut-off).
12. Sterile filters (0.22 μm cut-off).
13. Rotating mixer.

2.2 Expression and Purification

1. Expression vector TolC-pET24a.
2. *E. coli* BL21(*DE3*) for TolC expression.
3. LB medium (1% tryptone, 0.5% yeast extract, 0.5% NaCl).
4. Isopropyl-β-D-thiogalactopyranoside (IPTG).
5. Kanamycin monosulfate.
6. Tris–HCl (pH 7.0).
7. 8 M urea.
8. *n*-Dodecyl-β-D-maltoside (DDM).
9. Sodium chloride.
10. Imidazole.
11. Phenylmethylsulfonyl fluoride (PMSF).
12. EDTA-free protease inhibitor.
13. Ni^{2+}-NTA superflow resin.
14. $Co(NH_3)_6Cl_3$.
15. SDS–PAGE gel (e.g., NuPAGE 4–12% Bis–Tris gel).
16. Protein molecular weight marker.

17. SDS–PAGE Sample Buffer, non-reducing (e.g., 4× LDS sample buffer).
18. NuPAGE 4–12% Bis–Tris Gel (Invitrogen #NP0323).
19. 1x MES–SDS Running Buffer: provided as a 20× stock (Invitrogen #NP0002).
20. Inclusion body (IB) prep buffer: 20 mM Tris–HCl pH 7.5, 200 mM NaCl, 5 mM EDTA, 2% Triton X-100.
21. Refolding buffer: 20 mM Tris–HCl, pH 7.5, 1% (w/v) DDM.
22. Purification buffer: 20 mM Tris–HCl, pH 7.5, 150 mM NaCl, 0.02% (w/v) DDM.
23. Size-exclusion buffer: 20 mM Tris–HCl, pH 7.5, 150 mM NaCl, 0.02% (w/v) DDM.

2.3 Cryo-EM Grid Preparation

1. Vitrobot Mark IV System (Thermo Fisher Scientific).
2. 100 holey carbon film R 1.2/1.3 grids (Gold, 300 mesh) (Quantifoil).
3. Glow discharge system PIB-10 (vacuum device).
4. Paper filter for blotting, φ = 55/20 mm (Ted Pella, Inc.).
5. C-clip ring, C-clip insertion tool.
6. The loading station.
7. Cryo-EM grid box, AutoGrid box.
8. Liquid nitrogen storage tank (e.g., Taylor Wharton 35L).
9. Ultrasonicator (bath type).

2.4 Cryo-EM Data Collection and Analysis

1. Titan Krios G4 with Selectris and Selectris-X imaging filters pairing with a Falcon 4i detector (Thermo Fisher Scientific).
2. EPU (version 3.1.0.4506REL) (Thermo Fisher Scientific).
3. RELION 4.0.1 [17].
4. UCSF ChimeraX-1.4 [18].

3 Methods

3.1 Protein Expression and Inclusion Body Preparation

1. Inoculate a single colony of *E. coli* BL21(DE3) containing a His-tagged TolC (UniProt P02930, residues 23–493) in the pET24a vector into LB medium supplemented with 50 μg/mL kanamycin. Incubate at 37 °C for 10–12 h.
2. Dilute the pre-culture 1:200 in 800 mL LB medium supplemented with 50 μg/mL kanamycin. Grow the cells at 30 °C until OD_{600} = 0.4–0.5, then increase the temperature to 37 °C to induce expression of the TolC by the addition of 320 mL of 1 M IPTG.

3. Harvest cells 4 h after induction by centrifugation at 4000 × *g* for 10 min at 4 °C. Store isolated cell pellets at −80 °C.
4. Resuspend the cell pellets in IB prep buffer (10 g wet cell/ 100 mL) supplemented with 1 mM PMSF.
5. Disrupt the cells by high-pressure homogenization or sonication (chip type). Collect pellets by centrifugation at 10,000× *g* for 15 min.
6. Repeat the sonication and centrifugation process (**steps 4** and **5**) for three additional cycles.
7. Divide the final inclusion body pellet into equal aliquots (1–1.5 g) (*see* **Note 1**).
8. Store unused aliquots at −80 °C until needed.

3.2 Purification of TolC

1. Dissolve inclusion body pellet (~1.0–1.5 g) in 4 mL denaturation buffer (8 M urea).
2. Vortex vigorously for 30 s and then sonicate in a bath-type ultrasonicator for 5 min to ensure complete solubilization.
3. Prepare 200 mL of refolding buffer in a beaker on ice.
4. Slowly add the dissolved TolC-urea solution to 200 mL of the refolding buffer while stirring (1:50 dilution ratio) (*see* **Note 2**).
5. Continue stirring for 1 h at 4 °C.
6. Centrifuge the supernatant at 12,000× *g* for 20 min at 4 °C.
7. Remove any visible pellet by gentle decanting, and collect the clear supernatant (~200 mL) for further purification.
8. Incubate the solubilized fraction with 25 mL Ni^{2+}–NTA resin pre-equilibrated in purification buffer for 2 h to allow batch binding.
9. Load the resin into a gravity column and wash with 1 column volume (CV) of purification buffer containing 5 mM imidazole, followed by 15 CV of purification buffer containing 60 mM imidazole.
10. Elute TolC with 60 mL of elution buffer containing 250 mM imidazole.
11. Immediately concentrate eluted protein fractions using Amicon concentrators (100 kDa cut-off). Filter concentrated protein through a 0.22 μm filter.
12. Submit a concentrated sample containing TolC on a Superdex 200 10/300 GL column equilibrated with SEC buffer (Fig. 1a).
13. Analyze fractions by SDS–PAGE (Fig. 1b).

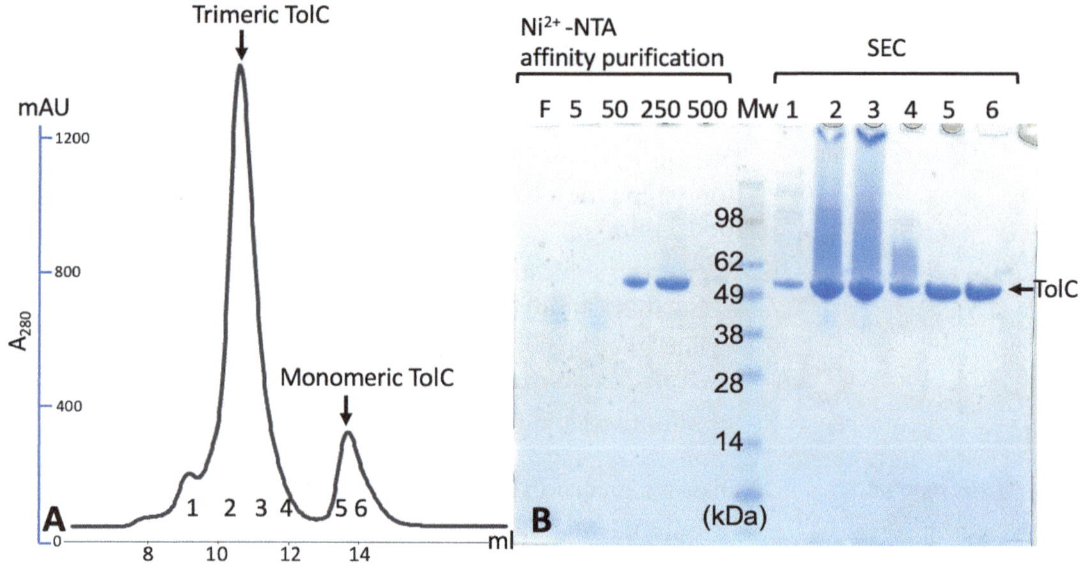

Fig. 1 Purification of TolC. (**a**) Size-exclusion chromatography of TolC with absorbance at 280 nm (A_{280}). Trimeric and monomeric TolC fractions (arrow) are identified. mAU: milli-absorbance unit. (**b**) SDS–PAGE analysis of TolC using a 4–12% NuPAGE Bis–Tris gel in the absence of reducing agents. Left: Ni^{2+}–NTA purification showing flow-through (unbound fraction) and eluted fractions with 5 mM, 50 mM, 250 mM, and 500 mM imidazole. Right: Samples from the fraction numbers indicated in the SEC profile (**a**). Mw: molecular weight marker (SeeBlue™ Plus2 Pre-stained Protein Standard)

3.3 Grid Preparation of TolC

1. Hydrophilize the grid with PIB-10 (Hard mode, 1.5 min).
2. Turn the grid over and hydrophilize the other side (Hard mode, 1.5 min).
3. The Vitrobot is preset to 18 °C (or the desired temperature) with 100% humidity.
4. Pinch the grid with the Vitrobot tweezers, secure it with the black clasp, and attach the tweezers to the Vitrobot.
5. Immediately before grid preparation, sonicate the purified protein sample using a water bath-type ultrasonicator with alternating cycles of 30 s sonication at 20 °C and 30 s rest on ice for a total sonication time of 2.5 min (five cycles). Apply the TolC sample onto the grid (*see* **Note 3**).
6. Apply 3 μL of the protein sample (6 mg/mL TolC) onto the grid (mesh side), blot the grid (5 s, force = 10), and plunge it into the liquid ethane (*see* **Note 4**).
7. Move the grid quickly from the ethane cup to the liquid nitrogen.
8. Store the grid in the cryo-EM grid box and prepare AutoGrid under liquid nitrogen.

9. Insert the C-clip into the C-clip insertion tool, place the C-clip rings into the loading station, and let them cool.
   (All procedures below are performed in the loading station. The tools should be pre-cooled before each step.)
10. Put a C-clip ring in the center of the AutoGrid alignment tool.
11. Take the grid out of the grid box and place it in the center of the C-clip ring.
12. Snap the C-clip onto the C-clip ring (AutoGrid is now ready).
13. Store experiment-ready AutoGrids in the AutoGrid container under liquid nitrogen until the data acquisition using a Titan Krios G4 microscope.
14. Load the AutoGrid into the Titan Krios G4 microscope equipped with Selectris-X, and a Falcon 4i detector, followed by screening of the prepared sample grids.
15. Screen the prepared grids and collect data on suitable grids by following a standard cryo-EM protocol using the EPU software (*see* **Note 5**).

3.4 Sample and Grid Preparation for TolC–Inhibitor Complex

1. Dissolve hexaamminecobalt(III) chloride $[Co(NH_3)_6]Cl_3$ in SEC buffer and prepare a 5 mM solution.
2. Add 5 mM $[Co(NH_3)_6]Cl_3$ to purified TolC (6 mg/mL) at a 1: 4 (v/v) ratio and mix gently (*see* **Note 6**).
3. Incubate for 30 min at 4 °C.
4. Sonicate the sample for 30 s × 2 times at 20 °C using a bath-type ultrasonicator. Keep the sample on ice between sonication.
5. Apply the cryo-EM sample to the grid. For subsequent steps, follow **steps 5–16** of Subheading 3.3.

4 Notes

1. Through repeated sonication cycles, the detergent (Triton) solubilizes *E. coli* membranes and associated membrane proteins, progressively increasing the purity of the inclusion bodies. This purification process is visually evident as the pellet transitions from a brownish color, indicative of membrane protein contamination, to a nearly white appearance. For practical handling in this protocol, we recommend preparing inclusion bodies from approximately 800 mL of LB medium culture. The resulting cell pellet can be divided into two aliquots (each equivalent to 400 mL culture), with each aliquot yielding approximately 1–1.5 g of inclusion bodies suitable for a single purification procedure

2. In this refolding protocol, the quick dilution method is used, where the dilution volume represents a critical parameter. The principle of rapid dilution is to rapidly pass through the intermediate denaturant concentration range, which corresponds to the region of maximum protein aggregation propensity. Although an empirical starting dilution of approximately 100-fold is recommended for first-time refolding experiments, we use 50-fold dilution for TolC with good results. The key consideration is ensuring that the final protein concentration is sufficiently low to prevent aggregation during the refolding process, allowing the protein to refold into its native conformation
3. The duration of sonication should be adjusted for each sample. Given that bath-type ultrasonicators vary in intensity, optimization is essential. Sonication should be performed at an intensity that prevents liquid splashing inside the tube. TolC can withstand a total of 2.5 min of sonication due to its relatively stable beta-barrel structure, but we are currently evaluating conditions for other samples
4. The blotting force of a Vitrobot can vary significantly depending on the specific machine used. Therefore, it is important to establish suitable values for both blotting time and force for the machine used in grid preparation
5. Achieving consistent grid preparation can be a daunting task. Thus, it is recommended to prepare a minimum of two, preferably four, grids under the same conditions. In this experiment, images were taken at an acceleration voltage of 300 kV and a nominal magnification of ×165,000, which corresponds to a calibrated pixel size of 0.723 Å/pixel at the specimen level. We used a 10-eV energy filter slit for zero-loss imaging. The total electron dose was set to $50\ e^{-}/\text{Å}^{2}$ at the specimen level. The data was collected with the help of the aberration-free image shift (AFIS) feature of the EPU software. The data subsets were mainly analyzed using RELION 4.0.1 (Figs. 2 and 3), according to the procedure detailed in our previous report [19]
6. Since TolC has diverse functions, it is essential to confirm that the protein retains relevant functional activity according to the experimental objectives. In addition to activity assays, verification of structural integrity is equally important. The hexaamminecobalt(III) chloride complex serves two purposes in this protocol: (1) it validates the structural integrity of purified TolC by confirming that the periplasmic entrance can accommodate the inhibitor at its known binding site [20], and (2) it provides a reference structure for understanding substrate transport inhibition. The 1:4 (v/v) ratio ensures saturating concentrations of the inhibitor (approximately 1 mM final

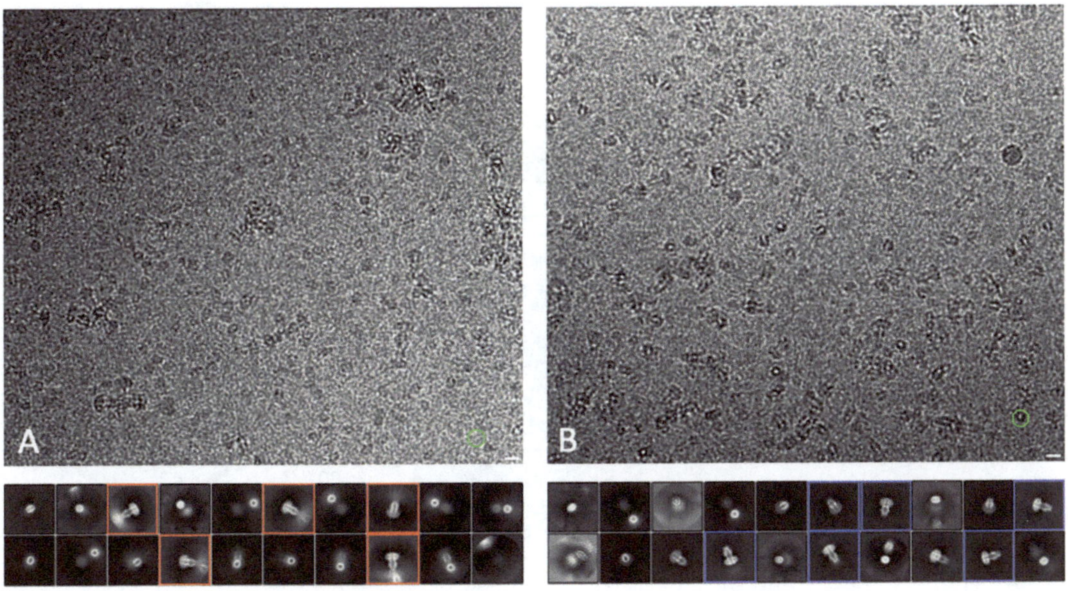

Fig. 2 2D class averages of TolC. TolC samples without (**a**) and with (**b**) sonication treatment. Representative motion-corrected micrographs (top) and 2D class averages from the first 2D classification using the EM algorithm in RELION (bottom) are shown. Sonication treatment improved particle separation; in almost all side views in A (red outline), a blurred shadow of a nearby particle can be seen on the periplasmic side, whereas it is not observed in B (blue outline). The scale bar and green circle represent 100 Å

concentration). For other TolC functional studies, different ligands or binding partners can be used with appropriate optimization of incubation conditions

5 Summary

TolC is a multifunctional protein that plays diverse roles in bacterial physiology, and determining its structure in interaction with various partners in the native cell environment is critically important. Therefore, in situ imaging techniques such as cryogenic electron tomography (cryo-ET) and super-resolution microscopy represent the future for visualizing TolC complexes in bacterial cells. Nevertheless, detailed reference structures obtained through in vitro cryo-EM SPA are still essential for interpreting and understanding these cellular imaging data. In this chapter, we have presented an optimized protocol for TolC structural determination, incorporating inclusion body-based purification and sonication-assisted cryo-EM grid preparation. Our results demonstrate that these methodological improvements significantly enhance particle distribution and sample quality, leading to high-resolution structure determination. While the general applicability of sonication to all proteins requires further evaluation, our findings establish it as a

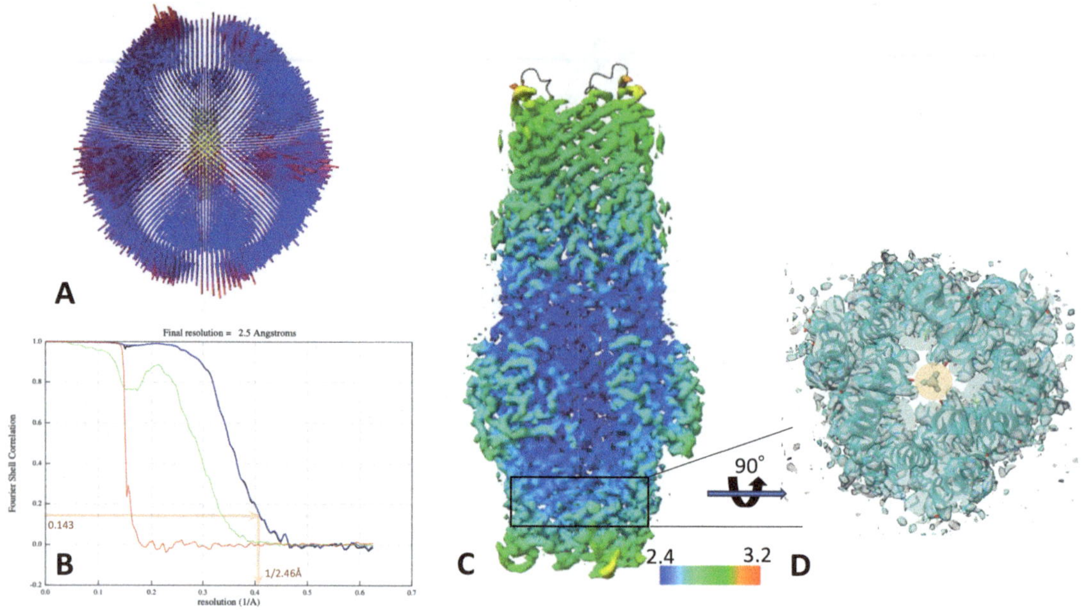

Fig. 3 Final 3D map of TolC in complex with $[Co(NH_3)_6]Cl_3$. (**a**) 3D angular distribution of particles in the final cryo-EM map. (**b**) Gold-standard Fourier shell correlation (FSC) curves of the final 3D reconstruction (resolution determined at FSC = 0.143). The black, green, blue, and red lines indicate FSC corrected, unmasked, masked, and phase-randomized masked maps, respectively. (**c**) The refined 3D map colored by the estimated local resolution calculated in RELION 4.0. (**d**) Zoomed-in view of the cryo-EM density map showing the Cobalt ion binding region (orange circle) from the periplasmic side

practical and readily accessible method to dissociate non-physiological aggregates and improve cryo-EM SPA outcomes. These protocols provide a foundation for future structural studies of TolC in complex with diverse binding partners, ultimately contributing to the development of novel antimicrobial strategies.

Acknowledgments

We thank Toshiya Senda (KEK) for general discussion and support for the research. We also thank Tamie Aoki and Kazuyo Takada for supporting the sample preparation of TolC. This work was supported by JSPS KAKENHI Grant Numbers 24 K01688 (to MT) and the Shionogi Infectious Disease Research Promotion Foundation (to MT). This work was also partially covered under the Cooperative Research Programme of the Network Joint Research Center for Materials and Devices (MEXT) (to MT). The cryo-EM data were collected at KEK (Tsukuba, Japan) and supported by the Research Support Project for Life Science and Drug Discovery (Basis for Supporting Innovative Drug Discovery and Life Science

Research (BINDS)) from AMED under Grant Number 24ama121001 (TS). The authors declare no conflict of interest.

Conflicts of Interest The authors have no conflicts of interest to declare that are relevant to the content of this chapter.

References

1. Darby EM, Trampari E, Siasat P, Gaya MS, Alav I, Webber MA, Blair JMA (2023) Molecular mechanisms of antibiotic resistance revisited. Nat Rev Microbiol 21:280–295
2. Du D, Wang-Kan X, Neuberger A, van Veen HW, Pos KM, Piddock LJV, Luisi BF (2018) Multidrug efflux pumps: structure, function and regulation. Nat Rev Microbiol 16:523–539
3. Henderson PJF, Maher C, Elbourne LDH, Eijkelkamp BA, Paulsen IT, Hassan KA (2021) Physiological functions of bacterial "multidrug" efflux pumps. Chem Rev 121: 5417–5478
4. Hinchliffe P, Symmons MF, Hughes C, Koronakis V (2013) Structure and operation of bacterial tripartite pumps. Ann Rev Microbiol 67: 221–242
5. Wright M, Kaur M, Thompson LK, Cox G (2025) A historical perspective on the multifunctional outer membrane channel protein TolC in Escherichia coli. npj Antimicrob Resist 3:6
6. Wandersman C, Delepelaire P (1990) TolC, an Escherichia coli outer membrane protein required for hemolysin secretion. Proc Natl Acad Sci 87:4776–4780
7. Zhao H, Lee J, Chen J (2022) The hemolysin A secretion system is a multi-engine pump containing three ABC transporters. Cell 185: 3329–3340.e13
8. Gilson L, Mahanty HK, Kolter R (1990) Genetic analysis of an MDR-like export system: the secretion of colicin V. EMBO J 9:3875–3884
9. Foreman DT, Martinez Y, Coombs G, Torres A, Kupersztoch YM (1995) TolC and DsbA are needed for the secretion of STB, a heat-stable enterotoxin of Escherichia coli. Mol Microbiol 18:237–245
10. Delgado MA, Solbiati JO, Chiuchiolo MJ, Farías RN, Salomón RA (1999) Escherichia coli outer membrane protein TolC is involved in production of the peptide antibiotic Microcin J25. J Bacteriol 181:1968–1970
11. Budiardjo SJ, Stevens JJ, Calkins AL, Ikujuni AP, Wimalasena VK, Firlar E, Case DA, Biteen JS, Kaelber JT, Slusky JS (2022) Colicin E1 opens its hinge to plug TolC. elife 11:e73297
12. Bleuel C, Große C, Taudte N, Scherer J, Wesenberg D, Krauß GJ, Nies DH, Grass G (2005) TolC is involved in Enterobactin efflux across the outer membrane of Escherichia coli. J Bacteriol 187:6701–6707
13. Kawamura-Sato K, Shibayama K, Horii T, Iimuma Y, Arakawa Y, Ohta M (1999) Role of multiple efflux pumps in Escherichia coli in indole expulsion. FEMS Microbiol Lett 179: 345–352
14. Tatsumi R, Wachi M (2008) TolC-dependent exclusion of porphyrins in Escherichia coli. J Bacteriol 190:6228–6233
15. Koronakis V, Sharff A, Koronakis E, Luisi B, Hughes C (2000) Crystal structure of the bacterial membrane protein TolC central to multidrug efflux and protein export. Nature 405: 914–919
16. Budiardjo SJ, Ikujuni AP, Firlar E, Cordova A, Kaelber JT, Slusky JSG (2021) High-yield preparation of outer membrane protein efflux pumps by in vitro refolding is concentration dependent. J Membr Biol 254:41–50
17. Kimanius D, Dong L, Sharov G, Nakane T, Scheres SHW (2021) New tools for automated cryo-EM single-particle analysis in RELION-4.0. Biochem J 478:4169–4185
18. Goddard TD, Huang CC, Meng EC, Pettersen EF, Couch GS, Morris JH, Ferrin TE (2018) UCSF ChimeraX: meeting modern challenges in visualization and analysis. Protein Sci 27:14–25
19. Inaba-Inoue S, Moriya T, Nomura N, Tanabe M (2025) Stabilization of multidrug efflux transporter MdfA and antibody fragments complex by cross-linking for Cryo-EM single-particle analysis. Methods Enzymol 722:397
20. Higgins MK, Eswaran J, Edwards P, Schertler GFX, Hughes C, Koronakis V (2004) Structure of the ligand-blocked periplasmic entrance of the bacterial multidrug efflux protein TolC. J Mol Biol 342:697–702

Chapter 4

Overexpression of the Multidrug Efflux Pump AcrAB-TolC in *Escherichia coli* for In Situ Cryo-ET Study

Xueting Zhou, Zhili Yu, Xiaodong Shi, and Zhao Wang

Abstract

The emergence of antibiotic-resistant bacteria has become a crucial clinical challenge. Bacteria acquire resistance by employing multidrug efflux pumps to expel toxic substances and survive under antibiotic pressure. Among the efflux pumps, AcrAB-TolC represents a major type of resistance-nodulation-cell division efflux pump existing in Gram-negative bacteria, which requires an intact cellular environment and intricate conformational changes to effectively expel antibiotics. To date, cryo-electron tomography and subtomogram averaging stand as a unique structural biology method capable of achieving nanometer resolution 3D structural details in native cells. These advantages make it the preferred approach for investigating complex in situ assemblies and conformations within physiological cellular contexts. Here, we outline a detailed protocol for studying the in situ structure of the AcrAB-TolC efflux pump in *Escherichia coli*. The protocol applies to other tripartite systems in various bacterial strains, providing a comprehensive pipeline for in situ structural studies and enhancing the development of potential therapeutics.

Key words Multidrug efflux pump, AcrAB-TolC, cryo-ET, In situ structure, Drug resistance

1 Introduction

Efflux pumps are bacterial transport proteins that can export toxic substances, which makes them clinically important targets for opposing multidrug resistance [1–3]. In Gram-negative bacteria, efflux pumps are more complicated due to the multilayered cell envelope, which contributes to forming tripartite channels [4]. The tripartite resistance-nodulation-cell division (RND) pumps are only found in Gram-negative bacteria [5] and play an important role in antibiotic resistance. AcrAB-TolC is a representative RND-type efflux pump that is widely conserved across Gram-negative bacteria and has many homologs [6–9]. It is characterized by spanning the inner and outer membranes in the cell envelope [1–3] and transporting a broad range of compounds [10], thereby conferring resistance to various antibiotics [11–14]. Additionally, natural

Kunihiko Nishino and Mikio Tanabe (eds.), *Bacterial Multidrug Exporters: Methods and Protocols*, Methods in Molecular Biology, vol. 3044, https://doi.org/10.1007/978-1-0716-5316-6_4,
© The Author(s), under exclusive license to Springer Science+Business Media, LLC, part of Springer Nature 2026

overexpression of this pump has been observed to enhance the antibiotic resistance capability of *Escherichia coli* (*E. coli*) strains [3, 9, 15, 16].

AcrAB-TolC consists of the periplasmic membrane fusion protein AcrA, the inner membrane transporter AcrB, and the outer membrane channel TolC [17]. As the efflux pump is formed by 12 protomers (3 TolC, 6 AcrA, and 3 AcrB), proper assembly of these three components is essential to initiate bacterial resistance to antibiotics [18]. The membranes and cellular environment are critical to the membrane complex structure and function, but these important interactions are lost in studies on purified pumps [17, 19]. Furthermore, investigating the efflux pump structures in the context of intact bacterial cells reflects real physiological conditions, permitting structural studies for the design of potential therapeutics. Although the in vitro structures of AcrAB-TolC have been elucidated in both apo and transporting states [17, 19], it is not clear how the native bacterial environment influences conformational changes in AcrAB-TolC during the transport of substances to the exterior. Thus, cryo-electron tomography (cryo-ET) is a powerful imaging approach capable of capturing conformational variability in intact bacterial cells with the potential to reach near-atomic resolution [20, 21]. Recently, we have achieved sub-nanometer resolution for the in situ structure of AcrAB-TolC in *E. coli* using cryo-ET [21]. *A significant benefit of cryo-ET is precisely resolving structures of interest within native cellular environments.* This finding has enabled us to propose a model elucidating the allosteric mechanism in AcrAB-TolC in the presence of antibiotics.

This protocol outlines the experimental details for overexpressing efflux pump components, testing expression, confirming the successful assembly of the pump by functioning assay, and preparing cryo-grids, which allow in situ structural resolution of efflux pumps in Gram-negative bacteria using cryo-ET. We established this whole pipeline from bacteria sample preparation to cryo-ET data collection. This protocol uses AcrAB-TolC in *E. coli* as an example, but this pipeline can apply to other tripartite pump systems in *E. coli* or other bacterial strains.

2 Materials

2.1 Reagents

1. Plasmid pAcBH, which carries the *acrAB* locus and co-expresses AcrA and His-tagged AcrB, was a gift from Dr. Akihito Yamaguchi (Osaka University, Suita, Japan).
2. Plasmid pRSFDuet-*tolC*, which expresses Flag-tagged TolC, was a gift from Drs. Ben Luisi and Dijun Du (University of Cambridge, UK).

3. *E. coli* BL21 (DE3) cells.
4. LB broth powder: 10 g/L NaCl, 10 g/L tryptone, 5 g/L yeast extract.
5. LB agar powder: 10 g/L peptone 140, 5 g/L yeast extract, 5 g/L NaCl, and 12 g/L agar.
6. 1 M ampicillin sodium salt stock solution (store at −20 °C).
7. 1 M kanamycin sulfate stock solution (store at −20 °C).
8. 1 M isopropyl-β-D-galactopyranoside (store at −20 °C).
9. Dulbecco's phosphate-buffered saline (pH 7.4).
10. Puromycin dihydrochloride.
11. MBX3132.
12. Urea.
13. SDS–PAGE gel.
14. Gel loading dye.
15. Coomassie Plus protein assay reagent.
16. 6 nm BSA Gold tracers. Store at 4 °C.

2.2 Materials

1. Round-bottom tube.
2. Disposable cuvette.
3. 200-mesh R3.5/1 holey carbon grids with a 2 nm continuous carbon layer (QUANTIFOIL).

3 Method

3.1 Expression of the AcrAB-TolC Complex in E. coli

3.1.1 Transformation of the pAcBH and pRSFDuet-tolC Plasmids

1. Dilute plasmids pAcBH and pRSFDuet-*tolC* to 100–200 ng/μ L using filtered distilled water (*see* **Note 1**). Add 1 μL of each plasmid to 80 μL of competent *E. coli* BL21 (DE3) cells. Gently mix the solution by tapping the Eppendorf tube. Place the mixture on ice for 30 min.
2. Place the mixture on a 42 °C heat block for 1 min. After heat shock, immediately put the mixture on ice for 5 min. Add 920 μL LB at room temperature. Revive the *E. coli* in a 37 °C incubator, shaking at 220 rpm for 1 h.
3. Warm the LB plates with ampicillin (100 μg/mL) and kanamycin (50 μg/mL) to room temperature while reviving transformed competent *E. coli* BL21 (DE3) cells.
4. Centrifuge the revived cell mixture at 10,000 g for 1 min, remove 950 μL supernatant, and resuspend the pellet using the remaining volume. Spread the entire 50 μL mixture on the selection plates and incubate at 37 °C overnight.

3.1.2 Amplify Selected Colonies

1. Pick a single colony (*see* **Note 2**) of *E. coli* BL21 (DE3) transformed with pAcBH and pRSFDuet-*tolC* and inoculate into 3 mL LB with ampicillin (100 μg/mL) and kanamycin (50 μg/mL). Incubate in a 37 °C incubator and shake at 220 rpm overnight.

3.1.3 Induce Overexpression of AcrAB-TolC

1. Dilute the overnight growth with fresh LB with ampicillin (100 μg/mL) and kanamycin (50 μg/mL) 100–500 fold. Grow the bacteria in a 37 °C incubator and shake at 220 rpm. When OD_{600} reaches 0.8, induce protein expression with 0.1–0.5 mM isopropyl 1-thio-β-D-galactopyranoside (IPTG). Induce in a 20 °C incubator, shaking at 220 rpm for 16–24 h.
2. Measure and record the OD_{600} of the medium. Centrifuge 1 mL of bacteria growth at 4000 g for 8 min and resuspend using 1 mL filtered PBS. Centrifuge again at 4000 g for 8 min. Remove the supernatant and resuspend the bacteria in PBS to an $OD_{600} = 60–90$.
3. For a puromycin state, add 600 μg puromycin for every 1 mL of $OD_{600} = 1$ bacteria growth. For example, add 2400 μg for 2 mL of $OD_{600} = 2$ bacteria. Incubate at 37 °C and shake at 220 rpm for 30 min. Centrifuge at 10,000 g for 8 min, then directly resuspend the pellet without changing the supernatant.
4. For a MBX3132 state, add 1.4 μg MBX3132 for every 1 mL of $OD_{600} = 1$ bacteria growth. Incubate at 37 °C and shake at 220 rpm for 30 min. Centrifuge at 10,000 g for 8 min, then directly resuspend the pellet without changing the supernatant. The goal of this centrifugation step is to flatten the bacteria a little, so they are thinner on the grids.

3.1.4 Examine Protein Expression Using Coomassie Blue Staining and Western Blotting

1. Dilute the bacteria growth yielded from Subheading 3.1.2 100-fold and inoculate into 3 mL LB with 100 μg/mL ampicillin and 50 μg/mL kanamycin. Incubate in a 37 °C incubator and shake at 220 rpm until $OD_{600} = 0.8$ (bacteria need to be in the exponential phase to remain healthy).
2. Add 0.1–0.5 mM IPTG to the bacteria flask. Induce in a 20 °C incubator and shake at 220 rpm overnight.
3. Measure the OD_{600} of the bacterial growth. Take the bacteria growth that is equivalent to 200 μL of $OD_{600} = 2$. Centrifuge the mixture at 10,000 g for 1 min. Remove the supernatant and dispose.
4. Add 50 μL 8 M urea to the pellet. Put the mixture in a 100 °C heat block for 10 min to lyse the bacteria. Mix 30 μL lysed sample with 10 μL SDS–PAGE gel loading buffer and add 8 μL of the resulting mixture to the gel.

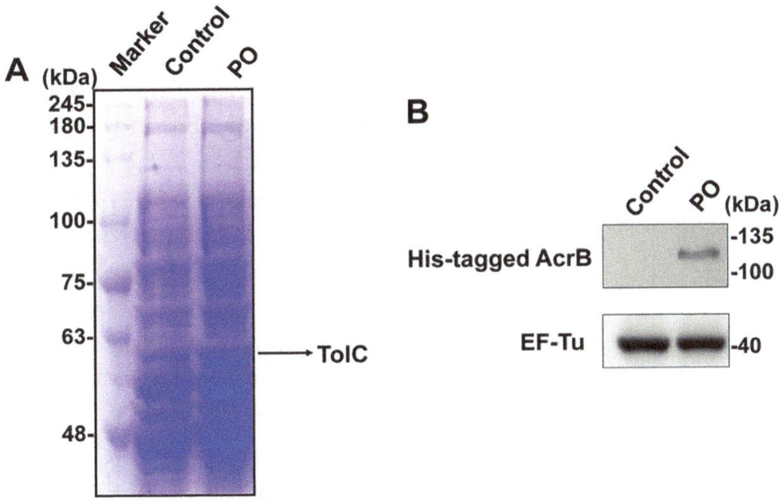

Fig. 1 Determination of the overexpression of AcrB and TolC in BL21 (DE3) cells [20]. pAcBH and pRSF-*tolC* were cotransformed into BL21 (DE3) cells. Protein expression was induced by 0.1 mM IPTG at 20 °C overnight. BL21 (DE3) cells without plasmid transformation were used as the control. (**a**) SDS–PAGE analysis result of TolC overexpression. (**b**) Immunoblotting results of AcrB overexpression using anti-His antibodies. The internal control EF-Tu was immunoblotted with anti-EF-Tu monoclonal antibodies. PO pump overexpression

Table 1
Puromycin susceptibility of *E. coli* BL21 (DE3) cells overexpressing AcrA, AcrB, and TolC [20]

Proteins expressed	MIC (μg/mL)
–	32
AcrA, AcrB, TolC	64–128

5. Run the SDS–PAGE gel using a normal protocol and analyze the gel using Coomassie blue staining and western blotting (Fig. 1). In our AcrAB-TolC project, the overexpression of TolC (Fig. 1a) and AcrB (Fig. 1b) can be clearly observed when compared to the control group without plasmid transformation [20].

3.2 Functional Assay for Bacteria

Minimal inhibitory concentration (MIC) is the minimum concentration of antibiotics that can completely inhibit the visible growth of the test strain [22]. It indicates the susceptibility of the bacteria to antibiotics. MIC in our experimental groups (*E. coli* with overexpression of AcrAB-TolC) is higher compared with the control group (*E. coli* without overexpression of AcrAB-TolC) (Table 1). This indicates that AcrAB-TolC overexpression is increasing the

E. coli's resistance to antibiotics. Below is the protocol for our functional assay.

1. Inoculate the strain from -80 °C stock to LB plates with 100 μg/mL ampicillin and 50 μg/mL kanamycin. Incubate overnight at 37 °C.
2. After 12–24 h, take out the plates and inoculate single colonies into liquid LB and shake at 37 °C, 220 rpm until OD_{600} reaches 0.6–1.0.
3. For the MIC experiment, in each round-bottom tube, add 1 mL LB, and add bacteria so that the final OD_{600} is about 10^{-4} to 10^{-5}. Add 100 μg/mL ampicillin, 50 μg/mL kanamycin, and 0.1 mM IPTG.
4. Add the drug intended for testing the MIC and design a concentration gradient. Add the drug to each round-bottom tube according to the gradient. Incubate all the tubes in a 37 °C incubator and shake at 220 rpm for 20 h.
5. After 20 h, visually observe each tube. Record if the growth is clear or turbid.

3.3 Prepare Samples for a Cryo-ET Study

Concentrated *E. coli* cells overexpressing the AcrAB-TolC complex are embedded in a thin layer of vitreous ice. Cryo-electron microscopy (cryo-EM) is used for initial screening and data collection.

3.3.1 Plunge Freeze

1. Mix resuspended *E. coli* with 6 nm BSA gold fiducials in a 3:1 (sample to gold) volume ratio.
2. Glow discharge R3.5/1 holey carbon grids with a 2 nm continuous carbon layer using PELCO easiGlow glow discharge cleaning system for 20–30 s.
3. Apply 2.5–3.5 μL of the *E. coli*–gold mixture to a glow-discharged grid. Blot with filter paper for 3–5 s to remove excess solution and quickly plunge into liquid ethane using the Thermo Fisher Vitrobot Mark IV system.
4. Store grids in liquid nitrogen until screening.

3.3.2 Screening Grids and Data Collection

1. Use cryo-EM for initial screening (*see* **Note 3**) in low magnification for good ice conditions (e.g., thin and vitreous) and intact cells. *E. coli* cells (*see* **Note 4**) with clear outer and inner membranes are good candidates for data collection, and broken cells are excluded (Fig. 2).
2. Collect the tilt series at high magnification with a pixel size of 1–3 Å. The tilt range is from −51 to +51° with a 3-degree increase, using a bidirectional scheme with a starting angle of −20 to −30°. It takes about 20 min for one tilt series data collection.

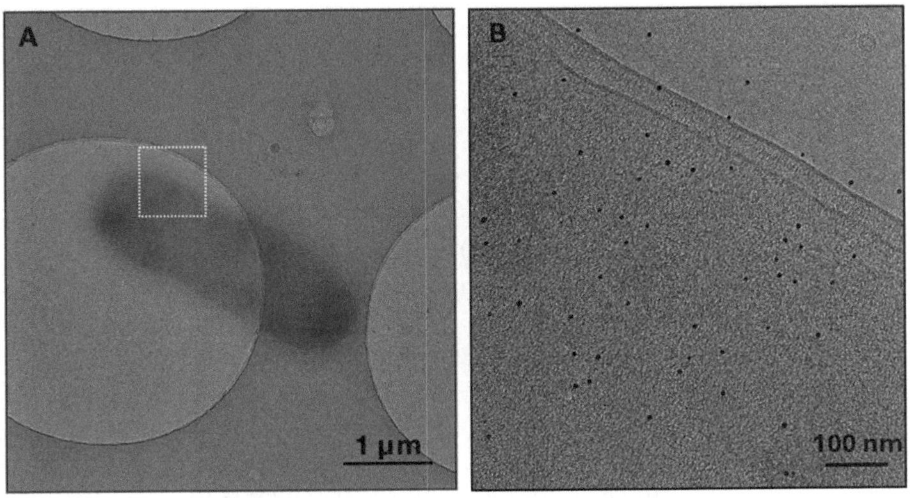

Fig. 2 A cryo-EM image at low magnification. (**a**) A representative example of a thin Gram-negative bacterial cell with clear outer and inner membranes. (**b**) A zoomed-in view of the thin bacteria with a clear view of the entire cell envelope

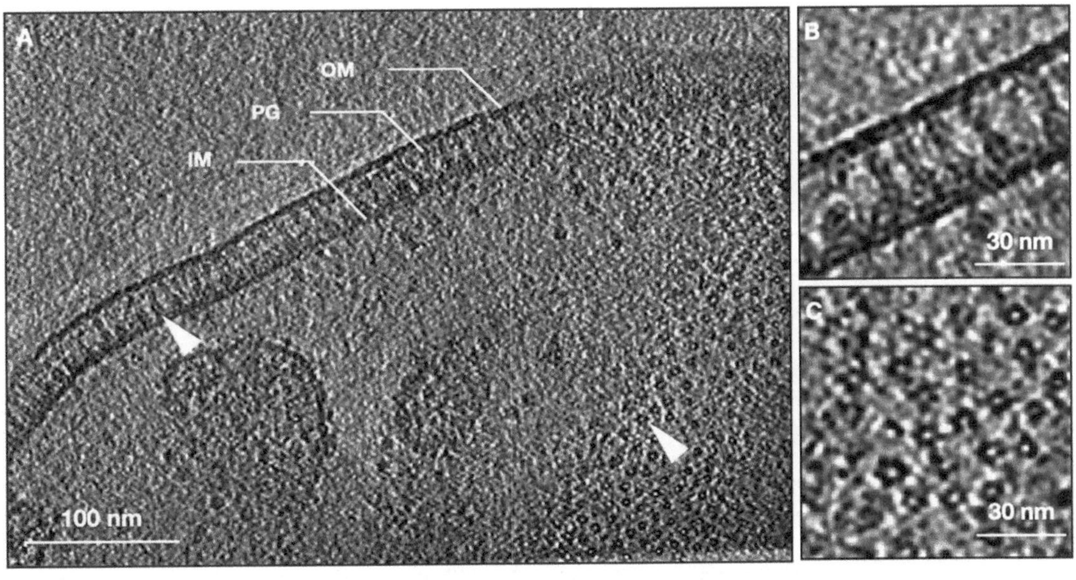

Fig. 3 A cryo-ET slice of an *E. coli* cell overexpressing the AcrAB-TolC multidrug efflux pump after reconstruction [21]. (**a**) A single slice from a tomogram of *E. coli* overexpressing AcrAB-TolC. (**b**) A zoomed-in image of the side view of AcrAB-TolC. (**c**) A zoomed-in image for the top view of AcrAB-TolC. OM outer membrane, IM inner membrane, PG peptidoglycan

3. In a good tomogram, clear top views (small circle) and side views (the density of two parallel lines spanning the cell envelope) are indicated in Fig. 3.

4 Notes

1. In this protocol, all instruments and buffers need to be sterile.
2. When inoculating single colonies, pick at least four replicates to avoid colony variance.
3. When we are screening for optimization, we will use a view mode with 4800× (Glacios). A good bacteria concentration for a cryo-EM sample is about 30–50 cells in a mesh (for a 200-mesh grid). We expect a good population of cells on the grids without overlapping.
4. Using our protocol, AcrAB-TolC found on the *E. coli* cell envelope should be observable in cryo-ET using the parameters provided. However, parameters can vary due to different bacterial conditions, and the parameters may require further optimization. Cryo-EM can be used to confirm bacterial conditions, and a large population of thin bacteria indicates a successful sample preparation.

Acknowledgments

This work was supported by NIH R01GM143380, R01HL162842, R01GM080139, R01GM072804, and P01GM121203; the Welch Foundation Q-2173-20230405, Q-1967-20180324, and AU-2014-2019330; BCM BMB department seed funds to Z.W.

Funding

This study was funded by NIAID R01AI179879 and Welch Foundation Q-2173-20230405.

References

1. Cao H, Xia T, Li Y, Xu Z, Bougouffa S, Lo YK et al (2019) Uncoupled quorum sensing modulates the interplay of virulence and resistance in a multidrug-resistant clinical Pseudomonas aeruginosa isolate belonging to the MLST550 clonal complex. Antimicrob Agents Chemother 63(4). https://doi.org/10.1128/AAC.01944-18
2. Ge Q, Yamada Y, Zgurskaya H (2009) The C-terminal domain of AcrA is essential for the assembly and function of the multidrug efflux pump AcrAB-TolC. J Bacteriol 191(13): 4365–4371. https://doi.org/10.1128/JB.00204-09
3. Swick MC, Morgan-Linnell SK, Carlson KM, Zechiedrich L (2011) Expression of multidrug efflux pump genes acrAB-tolC, mdfA, and norE in Escherichia coli clinical isolates as a function of fluoroquinolone and multidrug resistance. Antimicrob Agents Chemother 55(2):921–924. https://doi.org/10.1128/AAC.00996-10
4. Sharma A, Gupta VK, Pathania R (2019) Efflux pump inhibitors for bacterial pathogens: from bench to bedside. Indian J Med Res 149(2): 129–145. https://doi.org/10.4103/ijmr.IJMR_2079_17
5. Alvarez-Ortega C, Olivares J, Martinez JL (2013) RND multidrug efflux pumps: what

are they good for? Front Microbiol 4:7. https://doi.org/10.3389/fmicb.2013.00007

6. Sharff A, Fanutti C, Shi J, Calladine C, Luisi B (2001) The role of the TolC family in protein transport and multidrug efflux. From stereochemical certainty to mechanistic hypothesis. Eur J Biochem 268(19):5011–5026. https://doi.org/10.1046/j.0014-2956.2001.02442.x
7. Nikaido H (2009) Multidrug resistance in bacteria. Annu Rev Biochem 78:119–146. https://doi.org/10.1146/annurev.biochem.78.082907.145923
8. Schuldiner S (2018) The Escherichia coli effluxome. Res Microbiol 169(7–8):357–362. https://doi.org/10.1016/j.resmic.2018.02.006
9. Li XZ, Plesiat P, Nikaido H (2015) The challenge of efflux-mediated antibiotic resistance in Gram-negative bacteria. Clin Microbiol Rev 28(2):337–418. https://doi.org/10.1128/CMR.00117-14
10. Okusu H, Ma D, Nikaido H (1996) AcrAB efflux pump plays a major role in the antibiotic resistance phenotype of Escherichia coli multiple-antibiotic-resistance (Mar) mutants. J Bacteriol 178(1):306–308. https://doi.org/10.1128/jb.178.1.306-308.1996
11. Murakami S, Nakashima R, Yamashita E, Matsumoto T, Yamaguchi A (2006) Crystal structures of a multidrug transporter reveal a functionally rotating mechanism. Nature 443(7108):173–179. https://doi.org/10.1038/nature05076
12. Schindler BD, Kaatz GW (2016) Multidrug efflux pumps of Gram-positive bacteria. Drug Resist Updat 27:1–13. https://doi.org/10.1016/j.drup.2016.04.003
13. Nikaido H (1996) Multidrug efflux pumps of gram-negative bacteria. J Bacteriol 178(20):5853–5859. https://doi.org/10.1128/jb.178.20.5853-5859.1996
14. Masuda N, Sakagawa E, Ohya S, Gotoh N, Tsujimoto H, Nishino T (2000) Substrate specificities of MexAB-OprM, MexCD-OprJ, and MexXY-oprM efflux pumps in Pseudomonas aeruginosa. Antimicrob Agents Chemother 44(12):3322–3327. https://doi.org/10.1128/AAC.44.12.3322-3327.2000
15. Andersen JL, He GX, Kakarla P, K CR, Kumar S, Lakra WS et al (2015) Multidrug efflux pumps from Enterobacteriaceae, Vibrio cholerae and Staphylococcus aureus bacterial food pathogens. Int J Environ Res Public Health 12(2):1487–1547. https://doi.org/10.3390/ijerph120201487
16. Fitzpatrick AWP, Llabres S, Neuberger A, Blaza JN, Bai XC, Okada U et al (2017) Structure of the MacAB-TolC ABC-type tripartite multidrug efflux pump. Nat Microbiol 2:17070. https://doi.org/10.1038/nmicrobiol.2017.70
17. Du D, Wang Z, James NR, Voss JE, Klimont E, Ohene-Agyei T et al (2014) Structure of the AcrAB-TolC multidrug efflux pump. Nature 509(7501):512–515. https://doi.org/10.1038/nature13205
18. Du D, van Veen HW, Luisi BF (2015) Assembly and operation of bacterial tripartite multidrug efflux pumps. Trends Microbiol 23(5):311–319. https://doi.org/10.1016/j.tim.2015.01.010
19. Wang Z, Fan G, Hryc CF, Blaza JN, Serysheva II, Schmid MF et al (2017) An allosteric transport mechanism for the AcrAB-TolC multidrug efflux pump. elife 6. https://doi.org/10.7554/eLife.24905
20. Shi X, Chen M, Yu Z, Bell JM, Wang H, Forrester I et al (2019) In situ structure and assembly of the multidrug efflux pump AcrAB-TolC. Nat Commun 10(1):2635. https://doi.org/10.1038/s41467-019-10512-6
21. Chen M, Shi X, Yu Z, Fan G, Serysheva II, Baker ML et al (2022) In situ structure of the AcrAB-TolC efflux pump at subnanometer resolution. Structure 30(1):107–113 e103. https://doi.org/10.1016/j.str.2021.08.008
22. European Committee for Antimicrobial Susceptibility Testing of the European Society of Clinical, M., & Infectious, D (2000) EUCAST Definitive Document E.Def 1.2, May 2000: Terminology relating to methods for the determination of susceptibility of bacteria to antimicrobial agents. Clin Microbiol Infect 6(9):503–508. https://doi.org/10.1046/j.1469-0691.2000.00149.x

Part II

Biochemical Characterization and Reconstitution Methods

Chapter 5

Labeling of Specific Residues of the Multidrug Transporter AcrB with Its Fluorescent Substrate

Naoki Kobayashi

Abstract

Gram-negative bacteria possess resistance nodulation cell division (RND)-type drug efflux transporters in the inner membrane that confer a high degree of intrinsic resistance to various drugs. *Escherichia coli* expresses the most powerful RND-type drug efflux transporter, AcrB, which extrudes various chemical compounds, including antibiotics, by forming a complex with the periplasmic membrane fusion protein AcrA and the outer membrane channel TolC. To date, various reported structures of RND-type drug efflux transporters and AcrA–AcrB–TolC ternary complex have indicated substrate-binding sites and substrate translocation pathways, but their biochemical verification is insufficient.

In this chapter, I describe the labeling, purification, and detection of AcrB Cys mutants using tetramethylrhodamine-5-iodoacetamide, a thiol-reactive fluorescent reagent recognized by AcrB, as a substrate. The labeling of substrate-binding sites and substrate translocation pathways of the AcrA–AcrB–TolC system using this method will be useful for elucidating the multidrug transport mechanism of this ternary complex system.

Key words multidrug transporter, RND-type, Cysteine mutant, Labeling,, Tetramethylrhodamine-5-iodoacetamide, Thiol-reactive, Fluorescence

1 Introduction

The RND-type transporter AcrB is the most powerful proton-driven multidrug efflux pump in *Escherichia coli* and transports various structurally different chemical compounds, including antibiotics, dyes, and surfactants [1, 2]. *Pseudomonas aeruginosa*, a gram-negative bacterium that commonly causes hospital-acquired infections, also expresses the RND-type multidrug efflux transporters MexB and MexY; thus, treating patients with *P. aeruginosa* infections is difficult because of its high intrinsic resistance to various antibiotics [3, 4].

One characteristic of RND-type multidrug efflux transporters, which function in the inner membrane of gram-negative bacteria, is that they require two other components to express their activity. In

Kunihiko Nishino and Mikio Tanabe (eds.), *Bacterial Multidrug Exporters: Methods and Protocols*, Methods in Molecular Biology, vol. 3044, https://doi.org/10.1007/978-1-0716-5316-6_5,
© The Author(s), under exclusive license to Springer Science+Business Media, LLC, part of Springer Nature 2026

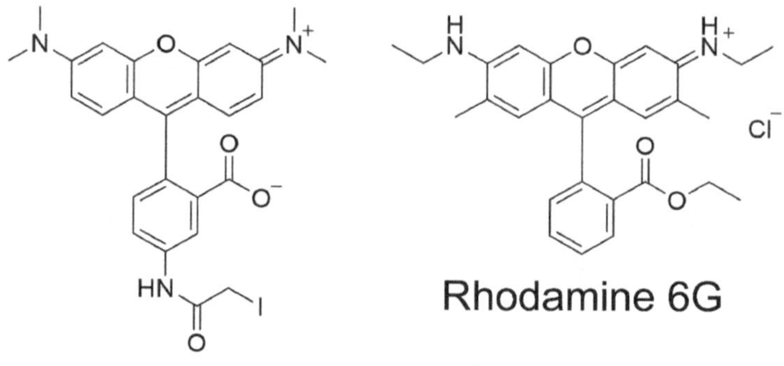

5-TMRIA

Rhodamine 6G

Fig. 1 Chemical structure of tetramethylrhodamine-5-iodoacetamide (5-TMRIA) and rhodamine 6G

E. coli AcrB, the drug is expelled by forming a ternary complex with the periplasmic membrane fusion protein AcrA and the outer membrane channel TolC. Cryoelectron microscopy studies of the AcrA–AcrB–TolC complex have revealed assembled forms of the ternary complex [5–10]. These reports show that the substrates carried by AcrB pass through channels formed by AcrA and TolC and reach the extracellular space.

AcrB is a trimeric multidrug efflux transporter, and its symmetric structure has been elucidated [11]. Subsequently, asymmetric structures of AcrB have been reported, mainly in the substrate-bound forms [12–15]. These structural analyses revealed that AcrB contains multiple substrate-binding sites, and each of these binding sites has multiple channels for substrate incorporation. These structural features provide AcrB with a wide range of substrate specificities. However, the biochemical evidence to explain how the formation of the AcrA–AcrB–TolC complex is linked to substrate translocation within the complex is insufficient.

Fluorescent dyes such as ethidium bromide, rhodamine 6G, and berberine, which are substrates of AcrB, have been used to measure the transport activity of AcrB [16]. Husain et al. have previously reported fluorescent labeling of AcrB substrate-binding sites and substrate translocation pathways using BODIPY FL maleimide [17–19]. However, whether BODIPY FL maleimide is a substrate of AcrB, or the labeling of BODIPY FL maleimide is inhibited by AcrB substrates, has not been shown. We have previously found that the thiol-reactive fluorescent probe tetramethylrhodamine-5-iodoacetamide (5-TMRIA) (Fig. 1) is a substrate for AcrB (*see* **Note 1**). In this chapter, we describe the methods for the specific labeling of AcrB cysteine mutants with 5-TMRIA and the purification and detection of labeled AcrB. This method will be a useful tool for elucidating how AcrA–AcrB–TolC complex formation and substrate movement within the complex are linked by labeling specific residues that constitute the substrate translocation pathway.

2 Materials

Prepare all buffers and solutions with analytical grade reagents and ultrapure water (resistivity of >18 MΩ.cm at 25 °C) at 25 ± 2 °C. All buffers and solutions were stored at 25 ± 2 °C, unless otherwise indicated.

2.1 Strains and Plasmids

1. *E. coli* W3104ΔacrB, an *acrB* gene-deletion derivative of *E. coli* W3104.
2. The plasmid pACRBHLR, a derivative of pACRBLR [20], carries *acrR* and His-tagged Cys-less *acrB* genes.
3. Plasmid pACBRN [21] that carries the sequence on the 5′ side of half of Cys-less *acrB* gene.
4. Plasmid pACBHLRC [21] that carries the sequence on the 3′ side of half of Cys-less *acrB* gene.

2.2 Site-Directed Mutagenesis

1. QuikChange Site-Directed Mutagenesis Kit (Agilent).

2.3 E. coli Strain Cultures

1. 2× YT broth: 1% yeast extract, 1.6% tryptone, and 0.5% NaCl. Sterile aqueous solution.
2. Potassium phosphate buffer: Mix 50 mM KH_2PO_4 solution and 50 mM KH_2PO_4 solution to adjust the pH to 7.5. Store at 4 °C.
3. 100 mg/mL ampicillin solution: Sterile aqueous solution. Store at −20 °C.
4. 50 mL conical tubes.
5. Shaker incubator (such as Taitec BR-23UM).
6. Spectrophotometer (such as Shimadzu UV-1280).
7. Centrifuge (such as Eppendorf 5425 R).

2.4 Labeling of AcrB Cysteine Mutants

1. 3 mM tetramethylrhodamine-5-iodoacetamide (5-TMRIA): solution in DMSO. Store at −20 °C.
2. 3 mM rhodamine 6G (Fig. 1) solution (optional): aqueous solution.
3. 40% glucose solution: sterile aqueous solution.
4. 1 M cysteine solution: aqueous solution.
5. Stop buffer: 0.5 mM cysteine in potassium phosphate buffer.

6. Centrifuge (such as Eppendorf 5425 R).
7. Deep freezer (−80 °C).

2.5 Purification of His-Tagged AcrB

1. Sonicator (such as Qsonica Q55) with a 1/8-in. diameter probe.
2. Ultracentrifuge (such as Optima MAX-TL).
3. Sodium phosphate buffer: mix 100 mM NaH_2PO_4 solution and 100 mM Na_2HPO_4 solution to adjust the pH to 7.4. Store at 4 °C.
4. 2× PBS: 40 mM sodium phosphate, pH 7.4, and 1 M NaCl. Add 120 mL sodium phosphate buffer to 100 mL water, dissolve 17.53 g NaCl, and adjust the volume to 300 mL using water.
5. 10% Triton X-100: aqueous solution.
6. 10% SDS: aqueous solution.
7. Ultrasonic bath (such as Branson Bransonic 1800).
8. Ni Sepharose resin.
9. 2 M imidazole solution: aqueous solution.
10. Tube rotator (such as SCINICS RVM-101).
11. 2× sample buffer: 0.125 M Tris–HCl, pH 6.8, 4% SDS, 20% glycerol, 0.04% bromophenol blue (reducing agent free). Aqueous solution.

2.6 SDS–PAGE

1. 10% polyacrylamide gel cast in gel plates (1 mm thick).
2. Running buffer: 25 mM Tris–192 mM glycine, pH 8.4, 0.1% SDS. Aqueous solution.
3. Protein molecular weight size marker.
4. Electrophoresis cell.
5. Power supply.

2.7 Imaging and Staining of Protein

1. Tupperware container.
2. Imaging instrument (such as Cytiva FLA-3000).
3. Staining solution: 0.1% (w/v) Coomassie Brilliant Blue R-250, 50% methanol, 10% acetic acid. Aqueous solution.
4. Shaker (such as BIO CRAFT BC-730).
5. Destaining solution: 7% methanol and 7% acetic acid. Aqueous solution.
6. Kimwipes.
7. Gel dryer (such as Atto AE-3711).

3 Methods

3.1 Site-Directed Mutagenesis of acrB *Gene and* E. coli *W3104Δ acrB Transformation*

1. Use the QuikChange Site-Directed Mutagenesis Kit to introduce cysteine mutations into the *acrB* gene using pACBRN or pACBHLRC as templates.
2. Verify the mutated plasmid sequences by DNA sequencing.
3. After cleavage of pACRBHLR, pACBRN, and pACBHLRC by restriction enzymes, replace the pACRBHLR fragment with the corresponding fragment of mutant pACBRN or pACBHLRC by a ligation reaction.
4. Prepare competent *E. coli* W3104ΔacrB cells for electroporation using the standard method.
5. Transform pACRBHLR carrying the mutant *acrB* gene into *E. coli* W3104ΔacrB competent cells by electroporation.

3.2 E. coli *W3104Δ acrB-Expressing AcrB Culture and Preparation*

1. Pick up *E. coli* strains from glycerol stocks to 2 mL 2× YT broth containing 2 μL 100 mg/mL ampicillin (f.c. 100 μg/mL) using toothpicks.
2. Incubate at 37 °C for 12 h.
3. Add 1 mL overnight culture to 20 mL 2× YT broth in a 50 mL conical tube (*see* **Note 2**). Grow the cells at 37 °C for about 3 h in a shaker incubator until the OD_{610} reaches 0.9. Periodically, sample 800 μL culture medium and measure the OD_{610} using a spectrophotometer.
4. Transfer 10 mL of cell culture to a new 50 mL conical tube on ice and collect the cells by centrifugation at 8000× *g* for 10 min at 4 °C.
5. Discard the supernatant and add 10 mL potassium phosphate buffer to the precipitate.
6. Resuspend the cell pellets by vortex mixing.
7. Centrifuge at 8000× *g* for 10 min at 4 °C.
8. Discard the supernatant and add 1 mL potassium phosphate buffer to the precipitate.
9. Resuspend the cell pellets by vortex mixing.
10. Transfer the cell suspension to a 1.5 mL tube.
11. After resuspending, sample 25 μL cell suspension to 975 μL potassium phosphate buffer in a new 1.5 mL tube, mix well, and measure OD_{530}.
12. Transfer a portion of the cell suspension prepared in step 10 to a new 1.5 mL tube according to the following formula (*see* **Note 3**):

$$\text{Volume of cell suspension to be transferred (mL)} = 6/(40 \times OD_{530} \text{ of the diluted suspension})$$

13. Centrifuge at 8000× *g* for 2 min at 4 °C.
14. Remove a portion of the supernatant, leaving 100 μL in the tube (*see* **Note 4**).
15. Resuspend the cells in the residual supernatant by gentle vortex mixing.

3.3 Labeling of Cysteine Mutants of AcrB with 5-TMRIA

1. Prewarm the 100 μL cell suspension at 30 °C for 10 min.
2. Prepare labeling buffer (20 μL per sample) (*see* **Note 5**). Mix 0.2 μL of 3 mM 5-TMRIA (*see* **Note 6**), 0.6 μL of 40% glucose (*see* **Note 6**), and 19.2 μL of potassium phosphate buffer, pH 7.5.
3. Add 20 μL of labeling buffer to 100 μL cell suspension and vortex mix gently.
4. Incubate the cells for the appropriate time (typically 1–30 min) at 30 °C in a water bath.
5. To terminate the reaction, add 900 μL of ice-cold stop buffer to the cell suspension and vortex mix immediately.
6. Centrifuge at 8000× *g* for 5 min at 4 °C.
7. Remove the supernatant completely.
8. Store at −80 °C overnight.

3.4 His-Tagged AcrB Purification

1. Prepare the lysis buffer and place it on ice. Mix 20 mL of 2× PBS, 15.6 mL of water, 4 mL of 10% Triton X-100, and 0.4 mL of 10% SDS (*see* **Note 7**).
2. Prepare binding buffer on ice by mixing 5 μL 1 M cysteine solution with 10 mL lysis buffer.
3. Prepare wash buffer on ice by mixing 1.25 mL 2 M imidazole solution with 48.8 mL lysis buffer.
4. Prepare the elution buffer and place it on ice. Mix 11.7 μL of 2× sample buffer, 7.5 μL of 2× PBS, 7.5 μL of 2 M imidazole solution, 3 μL of 10% Triton X-100, and 0.3 μL of 10% SDS (*see* **Note 7**).
5. Place the cell pellets on ice and add 1 mL ice-cold stop buffer.
6. Resuspend the cells by vortex mixing.
7. Disrupt the cells by sonication for 30 s with a sonication probe, followed by 15 s rest on ice twice.
8. Centrifuge at 17,000× *g* for 10 min at 4 °C.
9. Transfer the supernatant to a tube for ultracentrifugation.
10. Centrifuge at 213,000× *g* for 30 min at 4 °C (*see* **Note 8**).

11. Remove the supernatant completely and add 200 μL binding buffer to the membrane pellets.
12. Seal the tube top with Parafilm M® (*see* **Note 9**).
13. Add ice-cold tap water to the ultrasonic bath.
14. Sonicate sample tubes in the ice-cold ultrasonic bath to completely suspend the membrane pellets in the buffer (*see* **Note 10**).
15. Wipe water from the tube surface, and remove the Parafilm M® from the top of the tube.
16. Centrifuge at 210,000× *g* for 30 min at 4 °C to prepare a clear membrane lysate.
17. During the centrifugation, prepare 1.5 mL tubes containing 20 μL of Ni Sepharose slurry (bed volume is 10 μL) (*see* **Note 11**).
18. Transfer the membrane lysate immediately after centrifugation to the 1.5 mL tubes prepared in **step 17** (*see* **Note 12**).
19. Mix gently by shaking rotationally for 2 h at 4 °C using a tube rotator.
20. Centrifuge at 2000× *g* for 15 s at 4 °C.
21. Allow the tube to stand on ice for 1 min and remove the supernatant.
22. Add 800 μL ice-cold wash buffer and thoroughly suspend the Sepharose resin.
23. Centrifuge at 2000× *g* for 15 s at 4 °C and allow the tube to stand on ice for 1 min.
24. Remove the supernatant.
25. Wash the Sepharose resin by repeating **steps 22–24**.
26. Add 30 μL ice-cold elution buffer and vortex mix the suspension gently for 1 min.
27. Centrifuge at 2000× *g* for 15 s at 4 °C and allow the tube to stand on ice for 3 min.
28. Transfer the supernatant (eluate containing His-tagged AcrB) to new 1.5 mL tubes.

3.5 SDS-PAGE

1. Centrifuge the tubes at 8000× *g* for 1 min at 4 °C.
2. Apply 25 μL of the sample to a 10% polyacrylamide gel.
3. Perform electrophoresis according to conventional procedures.
4. Remove the gel from the gel plates and wash it in ultrapure water in a Tupperware container (*see* **Note 13**).

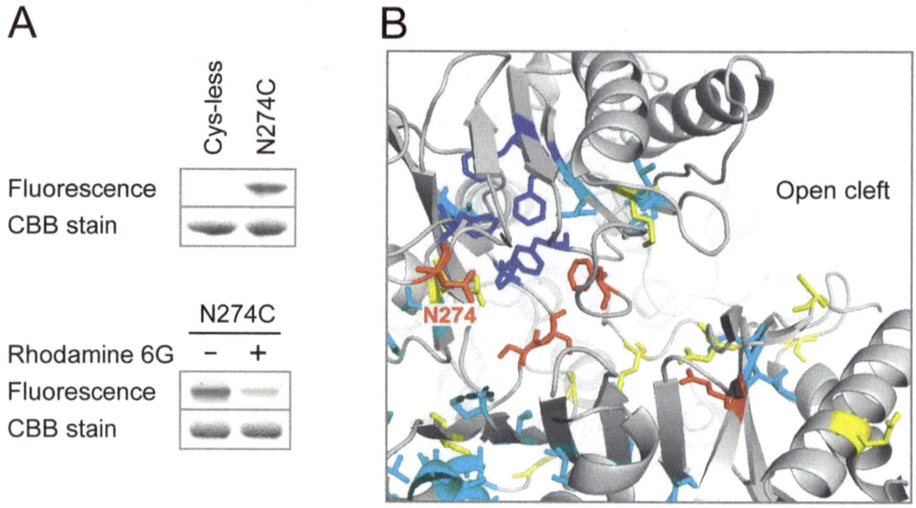

Fig. 2 Labeling of the Cys-less AcrB mutants by 5-TMRIA. (**A**) Labeling of the N274C mutant of Cys-less AcrB and its inhibition by rhodamine 6G. W3104ΔacrB-expressing Cys-less AcrB or its N274C mutant was incubated in labeling buffer containing 5 μM 5-TMRIA with or without 50 μM rhodamine 6G at 30 °C for 1 min. AcrB was purified from the bacterial cell membrane fraction. SDS–PAGE of purified AcrB was performed and the fluorescence derived from 5-TMRIA was detected (fluorescence). The gel was then stained with Coomassie Brilliant Blue to detect the purified AcrB (CBB stain). W3104ΔacrB carrying the N274C mutant of Cys-less AcrB retains rhodamine 6G resistance comparable to W3104ΔacrB carrying Cys-less AcrB (data not shown). (**B**) 5-TMRIA labeling of the cysteine-mutated residues in the substrate-binding pocket of AcrB. Each residue colored in red, yellow, or light blue indicates high, low, or no 5-TMRIA labeling, respectively, when each residue was substituted with cysteine. The phenylalanine cluster in the distal binding pocket is shown in blue

3.6 Imaging and Staining of Protein

1. Visualize 5-TMRIA-labeled AcrB cysteine mutants using an imaging instrument with excitation at 532 nm and a 580 nm longpass emission filter (Fig. 2).
2. Place the gel in a Tupperware container with the staining solution and incubate at 25 ± 2 °C for 20 min with gentle shaking on a shaker.
3. Transfer the gel to a Tupperware container containing destaining solution and incubate overnight at 25 ± 2 °C with gentle agitation on a shaker (*see* **Note 14**).
4. Discard destaining solution and wash the gel with ultrapure water for 10 min.
5. Dry the gel in a gel dryer and observe the purified AcrB band (Fig. 2).

4 Notes

1. To identify the thiol-reactive fluorescent reagents that can be transported by AcrB, we examined the growth differences between AcrB-expressing (WT) and AcrB-non-expressing (KO) strains in culture media containing various fluorescent reagents. Of the 12 reagents examined, only 5-TMRIA inhibited cell growth of the KO strain, but not that of the WT strain (Fig. 3), indicating that 5-TMRIA is a substrate of AcrB.
2. Loosen the tube cap and secure it with plastic tape (do not seal).
3. When testing for inhibition of 5-TMRIA labeling by AcrB substrates, double the amount of suspension to be collected to prepare two samples.
4. Determine in advance how much supernatant should be removed to prepare 100 μL cell suspension.
5. Prepare the labeling buffer immediately before use because 5-TMRIA is unstable in aqueous solutions. Subsequently, avoid exposing the sample to light as much as possible. To test for the inhibition of 5-TMRIA labeling by AcrB substrates, add AcrB substrates at any concentration to the labeling buffer.
6. The final concentrations of 5-TMRIA and glucose are 5 μM and 0.2%, respectively, after the addition of labeling buffer to the bacterial suspension.

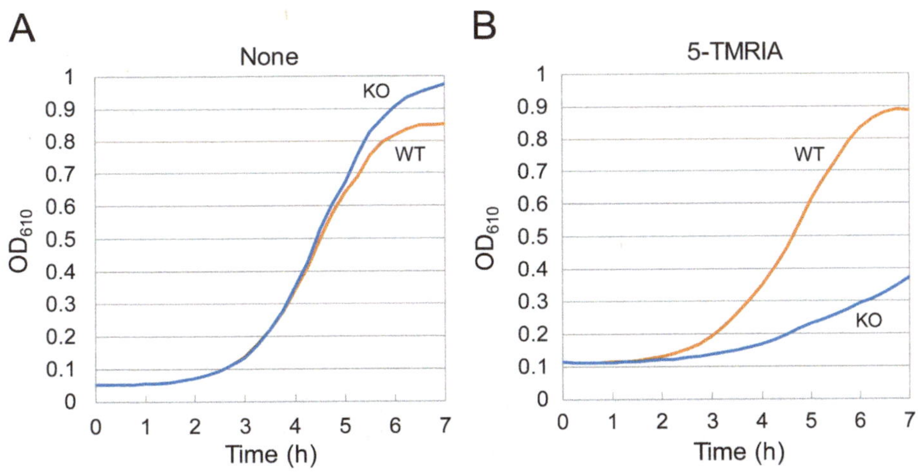

Fig. 3 Growth ability of AcrB-expressing (WT) and AcrB-non-expressing (KO) *Escherichia coli* strains. *E. coli* W3104ΔacrB (KO) and W3104ΔacrB expressing Cys-less AcrB (WT) were cultured at 37 °C in a medium with or without 100 μg/mL 5-TMRIA with shaking. The OD_{610} of the culture medium was measured continuously every 15 min. Blue and orange lines indicate KO and WT, respectively. The growth of W3104ΔacrB was similar to that of W3104ΔacrB expressing Cys-less AcrB in a medium without 5-TMRIA (**A**) but was more inhibited than that of W3104ΔacrB expressing Cys-less AcrB in a medium containing 5-TMRIA (**B**)

7. Add 10% SDS last to avoid precipitation.
8. Precipitate the membrane fraction of the cells in this step.
9. This step prevents water from accidentally getting inside the tube.
10. Do not place too much ice in the ultrasonic bath, as this weakens ultrasonic wave power. When the water in the ultrasonic bath was warm, replace it with fresh ice-cold water.
11. Wash Ni Sepharose resin twice with 50 volumes of binding buffer and prepare a 50% slurry with binding buffer prior to use. Resuspend the beads before use and dispense using a widebore tip.
12. Perform this step quickly to avoid carryover of the insoluble precipitate.
13. After removing one of the gel plates, rinse the gel well with deionized water using a wash bottle to prevent carryover of the Sepharose beads to the next step.
14. To accelerate destaining, put a Kimwipe inside the Tupperware container.

Acknowledgments

I thank Akihito Yamaguchi and Satoshi Murakami for their discussions and advice regarding this research. I am indebted to Tsuyoshi Nishi for the careful proofreading and thoughtful recommendations regarding this manuscript. I also thank Shoki Hoshikawa for preparing the chemical structures shown in Fig. 1. This work was supported by CREST from the Japan Science and Technology Agency and by grants-in-aid from the Ministry of Education, Culture, Sports, Science, and Technology of Japan.

Conflicts of Interest The authors have no conflicts of interest to declare that are relevant to the content of this chapter.

References

1. Kobylka J, Kuth MS, Muller RT, Geertsma ER, Pos KM (2020) AcrB: a mean, keen, drug efflux machine. Ann N Y Acad Sci 1459(1): 38–68
2. Zwama M, Yamaguchi A (2018) Molecular mechanisms of AcrB-mediated multidrug export. Res Microbiol 169(7–8):372–383
3. Zwama M, Nishino K (2021) Ever-adapting RND efflux pumps in gram-negative multidrug-resistant pathogens: a race against time. Antibiotics (Basel) 10(7):774
4. Avakh A, Grant GD, Cheesman MJ, Kalkundri T, Hall S (2023) The art of war with Pseudomonas aeruginosa: targeting Mex efflux pumps directly to strategically enhance antipseudomonal drug efficacy. Antibiotics (Basel) 12(8):1304
5. Du D, Wang Z, James NR, Voss JE, Klimont E, Ohene-Agyei T, Venter H, Chiu W, Luisi BF

(2014) Structure of the AcrAB-TolC multidrug efflux pump. Nature 509(7501):512–515

6. Shi X, Chen M, Yu Z, Bell JM, Wang H, Forrester I, Villarreal H, Jakana J, Du D, Luisi BF, Ludtke SJ, Wang Z (2019) In situ structure and assembly of the multidrug efflux pump AcrAB-TolC. Nat Commun 10(1):2635
7. Daury L, Orange F, Taveau JC, Verchere A, Monlezun L, Gounou C, Marreddy RK, Picard M, Broutin I, Pos KM, Lambert O (2016) Tripartite assembly of RND multidrug efflux pumps. Nat Commun 7:10731
8. Jeong H, Kim JS, Song S, Shigematsu H, Yokoyama T, Hyun J, Ha NC (2016) Pseudoatomic structure of the tripartite multidrug efflux pump AcrAB-TolC reveals the intermeshing cogwheel-like interaction between AcrA and TolC. Structure 24(2):272–276
9. Kim JS, Jeong H, Song S, Kim HY, Lee K, Hyun J, Ha NC (2015) Structure of the tripartite multidrug efflux pump AcrAB-TolC suggests an alternative assembly mode. Mol Cells 38(2):180–186
10. Wang Z, Fan G, Hryc CF, Blaza JN, Serysheva II, Schmid MF, Chiu W, Luisi BF, Du D (2017) An allosteric transport mechanism for the AcrAB-TolC multidrug efflux pump. elife 6:e24905
11. Murakami S, Nakashima R, Yamashita E, Yamaguchi A (2002) Crystal structure of bacterial multidrug efflux transporter AcrB. Nature 419(6907):587–593
12. Murakami S, Nakashima R, Yamashita E, Matsumoto T, Yamaguchi A (2006) Crystal structures of a multidrug transporter reveal a functionally rotating mechanism. Nature 443(7108):173–179
13. Nakashima R, Sakurai K, Yamasaki S, Nishino K, Yamaguchi A (2011) Structures of the multidrug exporter AcrB reveal a proximal multisite drug-binding pocket. Nature 480(7378):565–569
14. Eicher T, Cha HJ, Seeger MA, Brandstatter L, El-Delik J, Bohnert JA, Kern WV, Verrey F, Grutter MG, Diederichs K, Pos KM (2012) Transport of drugs by the multidrug transporter AcrB involves an access and a deep binding pocket that are separated by a switch-loop. Proc Natl Acad Sci USA 109(15):5687–5692
15. Sennhauser G, Amstutz P, Briand C, Storchenegger O, Grutter MG (2007) Drug export pathway of multidrug exporter AcrB revealed by DARPin inhibitors. PLoS Biol 5(1):e7
16. Zwama M, Yamasaki S, Nakashima R, Sakurai K, Nishino K, Yamaguchi A (2018) Multiple entry pathways within the efflux transporter AcrB contribute to multidrug recognition. Nat Commun 9(1):124
17. Husain F, Nikaido H (2010) Substrate path in the AcrB multidrug efflux pump of Escherichia coli. Mol Microbiol 78(2):320–330
18. Husain F, Bikhchandani M, Nikaido H (2011) Vestibules are part of the substrate path in the multidrug efflux transporter AcrB of Escherichia coli. J Bacteriol 193(20):5847–5849
19. Husain F, Nikaido H (2018) Determining ligand path through a major drug transporter, AcrB, in Escherichia coli. Methods Mol Biol 1700:167–175
20. Tamura N, Murakami S, Oyama Y, Ishiguro M, Yamaguchi A (2005) Direct interaction of multidrug efflux transporter AcrB and outer membrane channel TolC detected via site-directed disulfide cross-linking. Biochemistry 44(33): 11115–11121
21. Murakami S, Tamura N, Saito A, Hirata T, Yamaguchi A (2004) Extramembrane central pore of multidrug exporter AcrB in Escherichia coli plays an important role in drug transport. J Biol Chem 279(5):3743–3748

Chapter 6

Reconstitution of Membrane Proteins into Preformed Lipid Bilayers via LMNG-Mediated Auto-Insertion

Christoph Gerle, Dirk Bald, and Duncan G. G. McMillan

Abstract

A major challenge for the successful implementation of an in vitro transporter assay is the reconstitution of a given transporter into a lipid bilayer for measurements of vectorial transport. Here, we present the application of lauryl-maltose neopentyl glycol LMNG-mediated direct membrane insertion, termed LAiR (LMNG Auto-insertion Reintegration), for its use on preformed lipid bilayers such as liposomes or planar bilayers. Its ease of use, the stability and tightness of the resulting reconstituted lipid bilayer membrane protein system make this novel approach an attractive choice for the establishment of robust membrane transport assays.

Key words Membrane reconstitution, Proteoliposome, Proton pumping, Activity assay, Channel, F-ATPase, Mammalian membrane protein

1 Introduction

A common problem with the realization of assays that probe the functionality of integral membrane proteins (IMPs), such as multi-drug efflux transporters, is that reconstitution into a given in vitro membrane system often only yields unsatisfactory results. Common issues are low efficiency, instability, and leakiness of the reconstituted membrane or even instability of the target IMP due to the harshness of the detergent required for membrane reconstitution. Direct auto-insertion into preformed membranes using conventional detergents leads to membrane instability and, consequently, is typically not used. The double acyl chain tailed lipid-like detergent lauryl-maltose neopentyl glycol (LMNG) was originally developed as a milder alternative to the commonly used non-ionic detergent dodecyl-maltoside (DDM) for the stabilization of fragile GPCR G-protein complexes for structural studies by X-ray crystallography [1]. LMNG not only forms protein–detergent complexes (PDCs) of high enough stability that free detergent can be removed

Kunihiko Nishino and Mikio Tanabe (eds.), *Bacterial Multidrug Exporters: Methods and Protocols*, Methods in Molecular Biology, vol. 3044, https://doi.org/10.1007/978-1-0716-5316-6_6,
© The Author(s), under exclusive license to Springer Science+Business Media, LLC, part of Springer Nature 2026

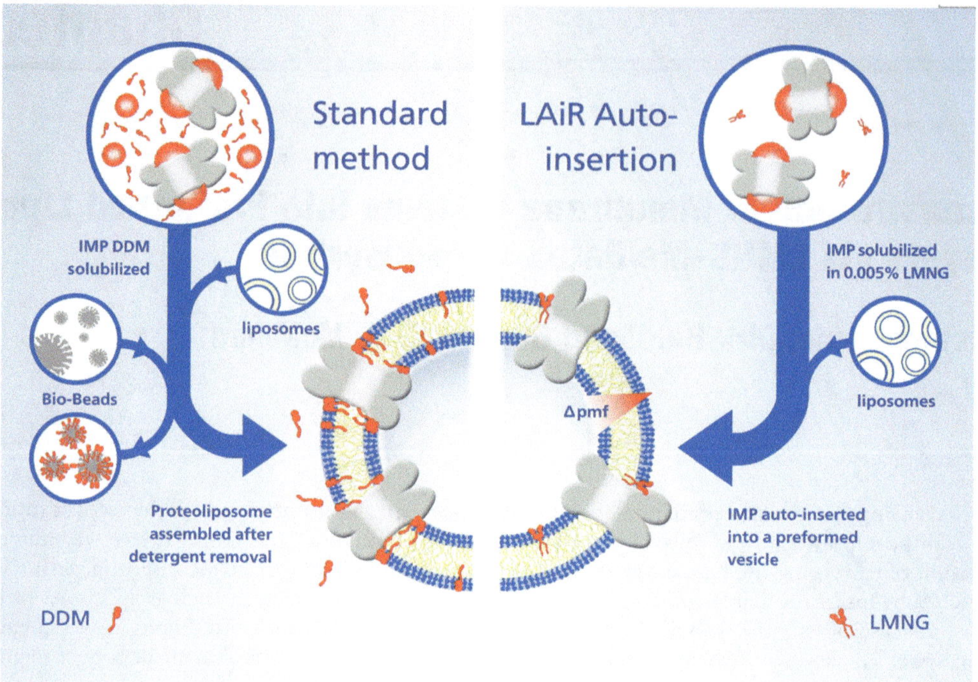

Fig. 1 LMNG-mediated auto-insertion (LAiR). Reconstitution of integral membrane proteins into lipid bilayers typically involves the presence of excess free detergent that needs to be removed by Bio-Beads adsorption or other means (left side panel). In the LAiR approach, the lipid-like high-affinity detergent LMNG allows for both the virtual absence of free detergent and promotes the proton-tight auto-insertion into preformed lipid bilayers

without causing membrane protein aggregation [2] but also promotes rapid and efficient membrane auto-insertion [3] as illustrated in the cartoon of Fig. 1. These favorable properties of LMNG for membrane auto-insertion have previously already been successfully exploited in functional studies of bovine F-ATP synthase using both liposomes [4], which are suitable for ACMA fluorescence quenching experiments such as described in Fig. 2, and also planar bilayers that can be used for measurements of electric currents elicited by the auto-inserted membrane protein [5], and the characterization of human TM16F scramblase by the ALBiC system [6, 7]. Recently, we have systematically investigated the properties of LMNG Auto-insertion Reintegration (LAiR) for bacterial as well as mammalian integral membrane proteins [3] and present here protocols for bovine F-ATP synthase [8] and the *Caldalkalibacillus thermarum* sodium channel [9] for direct auto-insertion into preformed liposomes. We expect that ease and efficiency of LMNG-driven membrane auto-insertion and the simplicity of the experimental protocol, as recently demonstrated for the human lysosomal cholesterol sensor LYCHOS [10], will make LAiR a valuable tool for establishing in vitro membrane transport assays for the investigation of bacterial multidrug exporters.

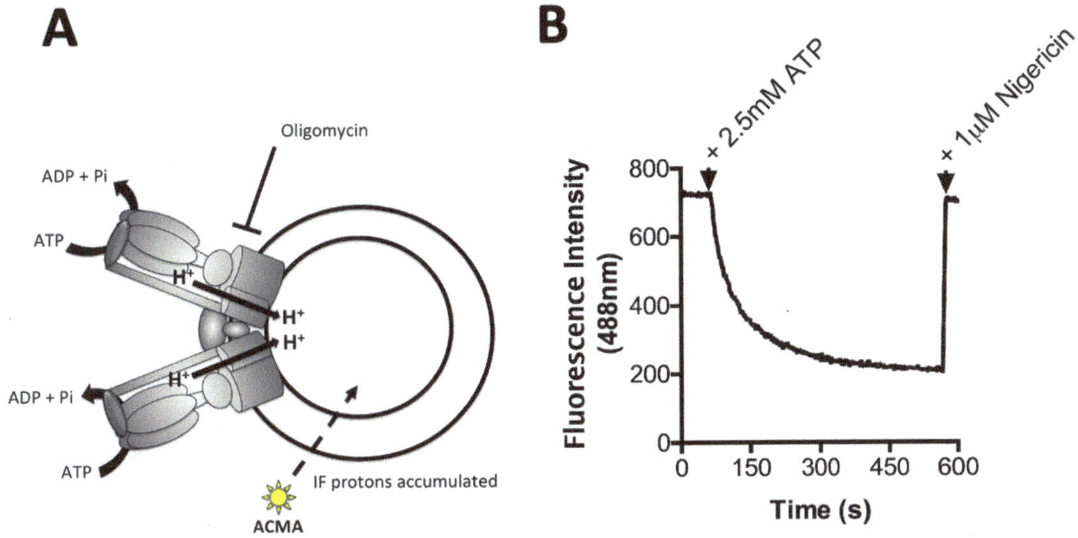

Fig. 2 ATP-dependent liposome acidification (ACMA fluorescence quenching). (**a**) Cartoon of the experimental design of bovine F-ATP synthase reconstituted into liposomes, in which acidification leads to ACMA accumulation and consequent quenching of fluorescence. (**b**) A typical proton pumping assay fluorescence curve, where the addition of ATP leads to protons being pumped into the liposome lumen, causing acidification that is released upon the addition of the ionophore nigericin

2 Materials

2.1 Materials for Purification of Sodium Channel TA2NavCt

1. *Escherichia coli* strain C43 carrying a pTrc99A plasmid expressing TA2NavCt [9].
2. Lauryl-maltose neopentyl glycol (LMNG; Anatrace).
3. Nickel-Sepharose High Performance column (GE Healthcare).
4. Ultracentrifugal filter device (molecular weight cut-off of 100,000 Da) (Amicon).

2.2 Materials for Purification of Bovine FoF1 ATP Synthase

1. Fresh *Bos taurus* hearts from the slaughterhouse.
2. Homogenizer (Nihon Seiki).
3. Decyl-maltoside (DM; Anatrace).
4. Poros-20HQ ion-exchange column (Thermo Fisher).
5. Lauryl-maltose neopentyl glycol (LMNG; Anatrace).

2.3 Materials for Liposome Formation and Membrane Protein Reintegration

1. Extruder (Avanti Polar Lipids, Inc.).
2. ECPL (Avanti Polar Lipids, Inc.).
3. BHPL (Avanti Polar Lipids, Inc.).
4. PD Minitrap™ G-25 column (GE Healthcare).

3 Methods

3.1 Purification of Caldalkalibacillus Thermarum Strain TA2.A1 Sodium Channel

This protocol for purification of a voltage-gated sodium channel from *Caldalkalibacillus thermarum* (TA2NavCt) is based on [9].

1. Grow *Escherichia coli* strain BL21 carrying a pTrc99A plasmid expressing TA2NavCt aerobically to mid-log phase at 37 °C in Luria–Bertani (LB) medium supplemented with 500 μM $CuSO_4$ and 100 μg mL^{-1} carbenicillin.
2. Then wash the pellet, and re-pellet twice with buffer A (20 mM Tris-(hydroxymethyl)aminomethane hydrochloride (Tris–HCl, Sigma), pH 8.0, 5 mM $MgCl_2$, 10% glycerol).
3. Resuspend cells in buffer A containing 0.1 mM phenylmethylsulfonyl fluoride (PMSF), 0.1 mg pancreatic DNase per mL, and lyse by two passages through a French press at 20,000 psi.
4. Remove remaining debris and unbroken cells by centrifugation at 10,000 × *g* for 20 min (Sorvall RC-5C with a F21S-8x50y rotor).
5. Ultracentrifuge the supernatant (180,000× *g*, 45 min, 4 °C in a Beckman Model-7 using a Ti 45 rotor) and resuspend the membrane pellet in buffer B (20 mM Tris–HCl, pH 8.0, 5 mM $MgCl_2$, and 25% w/w sucrose).
6. Apply the resulting suspension to the top of a 30% w/w to 55% w/w sucrose gradient and ultracentrifugate (130,000× *g*, 16 h, 4 °C in a Beckman Model-7 using a Ti 45 rotor) with no deceleration or breaking to separate inner membrane from outer membrane.
7. Remove the inner membrane fraction from the sucrose gradient and wash three times with buffer B by ultracentrifugation (200,000× *g*, 45 min, 4 °C in a Beckman Model-7 using a Ti50.2 rotor).
8. Resuspend inner membranes in buffer B and either use immediately for purification or store in aliquots at −80 °C until further use.
9. Extract TA2NavCt inner membrane vesicles by dilution to 5 mg of protein per mL in solubilization buffer (20 mM Tris–HCl, pH 8.0, 5 mM $MgCl_2$, 10% glycerol, 0.5% lauryl-maltose neopentyl glycol (LMNG; Anatrace), 300 mM NaCl, 10 mM imidazole) and incubate at 30 °C for 30 min with gentle inversion every 5 min.
10. Remove unsolubilized material by ultracentrifugation (180,000× *g*, 45 min, 4 °C in a Beckman Model-7 using a Ti50.2 rotor), and apply the supernatant to a Nickel-Sepharose High Performance (GE Healthcare) column that was previously washed with water and equilibrated with IMAC buffer

(20 mM Tris–HCl, pH 8.0, 5 mM $MgCl_2$, 10% glycerol, 0.005% LMNG, 300 mM NaCl, and 10 mM imidazole).

11. Remove contaminating proteins by washing the resin with IMAC buffer containing 30 mM imidazole and 150 mM NaCl, and TA2NaCh is eluted with IMAC buffer containing 250 mM imidazole, 150 mM NaCl, and 20% glycerol.
12. Pool TA2NavCt containing fractions and concentrate using an Amicon Ultra centrifugal filter device (molecular weight cut-off of 100,000 Da).

3.2 Reconstitution of TA2NavCt into Preformed Vesicles

1. Dry 10 mg of *E. coli* polar lipids (ECPLs) in chloroform under an N_2 stream and subsequently further dry under vacuum for 2 h.
2. Resuspend ECPL in 1 mL PL buffer (20 mM Tris–HCl, pH 8.0, 5 mM $MgCl_2$) at a final concentration of 10 mg/mL and extrude 13 times through a 400 nm polycarbonate membrane using an extruder (Avanti Polar Lipids, Inc., Alabaster, AL).
3. Mix the liposome solution with the purified TA2NaCh (solubilized in in 0.005% LMNG).
4. The final concentration of protein is 0.2 mg/mL (5 μL protein used) and 9.75 mg/mL lipids (200 μL used), resulting in a final concentration of 0.00015% LMNG.
5. Invert the mixture slowly for 30 min at 20 °C.
6. Pass the reconstituted sample through a PD Minitrap™ G-25 (GE Healthcare) column according to the manufacturer's instructions with PL buffer.

3.3 Purification of Bovine F_1F_o Complexes

This protocol for purification of F_oF_1-ATP synthase from bovine heart is based on [3].

1. Obtain fresh *Bos taurus* hearts immediately after slaughter from an authorized slaughterhouse.
2. Carefully remove fat and connective tissues, allowing the preparation of 1000 g of minced meat.
3. Suspend 500 g each in 3250 mL of 23 mM sodium phosphate buffer (pH 7.4 at 0 °C) and homogenize for 5 min at 13,000 rpm in a homogenizer (Nihon Seiki), followed by centrifugation for 20 min at 2800 rpm at 4 °C in a Kubota Model 9810 using a RS-6600 rotor.
4. Suspend the precipitate in 3375 mL of 22.2 mM sodium phosphate buffer (pH 7.4) and re-homogenize, followed by centrifugation for 20 min at 2800 rpm.

5. Combine supernatants and centrifuge for a further 30 min at 10,000 rpm with a refrigerated centrifuge (Beckman Model Avanti HP-30I) using a JLA-10.500 rotor.
6. Suspend the precipitate in ~40 mL 50 mM Tris–HCl buffer (pH 8.0) and pellet for 30 min at 30,000 rpm with an ultracentrifuge (Beckman Model-7 using a 45 Ti rotor).
7. Suspend the pellet in 50 mM Tris–HCl buffer (pH 8.0) containing 660 mM sucrose to a final protein concentration of ~23 mg/mL.
8. Keep the suspension in a 40 mM 4-(2-hydroxyethyl)-1-piperazine-ethane-sulfonic acid (HEPES) buffer (pH 7.8) containing 2 mM $MgCl_2$, 0.1 mM ethylenediaminetetraacetic acid (EDTA), and 0.1 mM DTT and solubilize on ice via slow addition of deoxycholate and decyl-maltoside (Anatrace) to final concentrations of 0.7% (wt/vol) and 0.4% (wt/vol), respectively.
9. Centrifuge the suspension at 176,000× *g* for 50 min and apply the supernatant to a sucrose step gradient (40 mM HEPES, pH 7.8, 0.1 mM EDTA, 0.1 mM DTT, 0.2% wt/vol decyl-maltoside and 2.0 M, 1.1 M, 1.0 M, or 0.9 M sucrose) and centrifuge at 176,000× *g* for 15.5 h.
10. Load fractions exhibiting ATP hydrolysis activity onto a Poros-20HQ ion-exchange column.
11. Exchange the detergent to LMNG (Anatrace) using a double gradient from 0.2% to 0% decyl-maltoside and 0–0.05% LMNG for 80 min at 1 mL/min.
12. Elute complexes by a linear concentration gradient of 0–240 mM KCl in 40 mM HEPES, pH 7.8, 150 mM sucrose, 2 mM $MgCl_2$, 0.1 mM EDTA, 0.1 mM dithiothreitol (DTT), and 0.005% (wt/vol) LMNG.

3.4 Reconstitution of Bovine F_oF_1 into Preformed Vesicles

1. Dry 10 mg of bovine heart polar lipids (BHPLs) in chloroform under an N_2 stream and subsequently dry under vacuum for 2 h.
2. Resuspend BHPL in reintegration buffer (20 mM 3-(*N*-morpholino)propanesulfonic acid (MOPS, pH 7.4, 30 mM Na_2SO_4, 100 mM KCl) and extrude 13 times through a 400 nm polycarbonate membrane using an extruder (Avanti Polar Lipids, Inc., Alabaster, AL).
3. Mix this liposome solution with the purified bovine F_1F_o ATP synthase (solubilized in in 0.005% LMNG).
4. The final concentration of protein is 0.2 mg/mL (5 μL protein used), and 9.75 mg/mL lipids (200 μL used), resulting in a final concentration of 0.00015% LMNG.

5. Slowly invert the mixture for 30 min at 20 °C and then pass through a PD Minitrap™ G-25 (GE Healthcare) column according to the manufacturer's instructions with 20 mM MOPS (pH 7.4, 30 mM Na_2SO_4, 100 mM KCl).
6. Immediately use the reconstituted sample then for assays.

3.5 ATP-Dependent Liposome Acidification (ACMA Fluorescence Quenching)

1. Determine ATP-dependent proton translocation at 38 °C based on the quenching of 9-amino-6-chloro-2-methoxyacridine (ACMA).
2. Prepare a 1.5-mL reaction mixture containing 50 mM MOPS (pH 7.4), 30 mM NaCl, 100 mM KCl, 3 mM phospho*enol*-pyruvate, 1.5 mM $MgCl_2$, 0.25 mM NADH, 0.57 U/mL pyruvate kinase, 3.2 U/mL lactate dehydrogenase, 1 μM ACMA, and 1 μM valinomycin.
3. Add 10 μg F_1F_o ATP synthase complexes reintegrated into bovine heart lipid liposomes to the reaction mixture.
4. After the fluorescence signal has stabilized, initiate the reaction by the addition of 2.5 mM neutralized ATP.
5. Measure fluorescence with an excitation wavelength of 410 nm and an emission wavelength of 480 nm (slit width, 10 nm) in a modified Cary Eclipse spectrophotometer (Agilent).
6. Release any established gradient by the addition of nigericin at the end of the assay for assay validation.
7. See Fig. 2 for the experimental design cartoon and an exemplary reaction ACMA quenching curve.

4 Notes

1. The usage of LMNG as a detergent for IMP solubilization prior to liposome reintegration is critical; other detergents with low critical micelle concentration that we tested, such as dodecyl-maltoside, glycol-diosgenin, and digitonin, yielded clearly inferior performance of the reintegrated IMP. For details on this topic, also refer to our LAiR paper [3].
2. In our hands, it is not critical to use the purified IMP directly after purification, and auto-insertion can be performed successfully with snap-frozen samples stored at −80 °C over a longer time span, according to our experience from eight different IMPs so far. Our recommendation is to shock-freeze small aliquots in the presence of 10–15% glycerol.
3. For the preparation of preformed liposomes, we prefer the use of an extruder over sonication due to the suboptimal liposome size resulting from sonication.

4. Lipids must not be stored dried, unless they are stored under a nitrogen or argon atmosphere, or they will oxidize and change the kinetics of IMP auto-insertion into preformed lipid bilayers.
5. A highly saturated lipid composition will likely decrease the auto-insertion efficiency of a given IMP but note that integration efficiency is also a feature influenced by the individual biophysical properties of the IMP used in the experiment.
6. For achieving high reproducibility of the auto-insertion process, it is advisable to not only consider the final protein concentration, but rather the detergent to lipid ratio and protein to lipid ratios used in the experiment.
7. Pure lipids behave very differently from lipid mixtures, and unexpected results should be taken into account when working with pure lipids.
8. It is critical to eliminate ALL the solvent from a given lipid mixture before using it. For example, if chloroform-solubilized lipids are dried down in a glass vial for subsequent buffer wetting and extrusion, a too short drying process by either vacuum or nitrogen stream can result in residual chloroform that is highly detrimental to the experimental outcome.

Acknowledgments

This work was funded by a BINDS grant from AMED (JP24ama121001) to Masaki Yamamoto and C.G.

Conflict of Interest Statement The authors have no conflicts of interest to declare that are relevant to the content of this chapter.

References

1. Chae PS, Rasmussen SG, Rana RR, Gotfryd K, Chandra R, Goren MA, Kruse AC, Nurva S, Loland CJ, Pierre Y, Drew D (2010) Maltose–neopentyl glycol (MNG) amphiphiles for solubilization, stabilization and crystallization of membrane proteins. Nat Methods 7(12): 1003–1008
2. Hauer F, Gerle C, Fischer N, Oshima A, Shinzawa-Itoh K, Shimada S, Yokoyama K, Fujiyoshi Y, Stark H (2015) GraDeR: membrane protein complex preparation for single-particle cryo-EM. Structure 23(9):1769–1775
3. Godoy-Hernandez A, Asseri AH, Purugganan AJ, Jiko C, de Ram C, Lill H, Pabst M, Mitsuoka K, Gerle C, Bald D, McMillan DGG (2023) Rapid and highly stable membrane reconstitution by LAiR enables the study of physiological integral membrane protein functions. ACS Central Science 9(3): 494–507
4. Quintana-Cabrera R, Quirin C, Glytsou C, Corrado M, Urbani A, Pellattiero A, Calvo E, Vázquez J, Enríquez JA, Gerle C, Soriano ME (2018) The cristae modulator optic atrophy 1 requires mitochondrial ATP synthase oligomers to safeguard mitochondrial function. Nat Commun 9(1):3399
5. Urbani A, Giorgio V, Carrer A, Franchin C, Arrigoni G, Jiko C, Abe K, Maeda S, Shinzawa-Itoh K, Bogers JF, McMillan DGG

(2019) Purified F-ATP synthase forms a Ca2+-dependent high-conductance channel matching the mitochondrial permeability transition pore. Nat Commun 10(1):4341

6. Watanabe R, Sakuragi T, Noji H, Nagata S (2018) Single-molecule analysis of phospholipid scrambling by TMEM16F. Proc Natl Acad Sci USA 115(12):3066–3071
7. Watanabe R, Soga N, Ohdate SY, Noji H (2018) Single-molecule analysis of membrane transporter activity by means of a microsystem. In: Bacterial multidrug exporters: methods and protocols, pp 321–330
8. Jiko C, Davies KM, Shinzawa-Itoh K, Tani K, Maeda S, Mills DJ, Tsukihara T, Fujiyoshi Y, Kühlbrandt W, Gerle C (2015) Bovine F1Fo ATP synthase monomers bend the lipid bilayer in 2D membrane crystals. elife 4:e06119
9. Tsai CJ, Tani K, Irie K, Hiroaki Y, Shimomura T, McMillan DGG, Cook GM, Schertler GF, Fujiyoshi Y, Li XD (2013) Two alternative conformations of a voltage-gated sodium channel. J Mol Biol 425(22):4074–4088
10. Bayly-Jones C, Lupton CJ, Keen AC, Dong S, Mastos C, Luo W, Qian C, Jones GD, Venugopa H, Chang YG, Clarke RJ, Halls ML, Ellisdon AM (2024) LYCHOS is a human hybrid of a plant-like PIN transporter and a GPCR. Nature 634:1238–1244

Chapter 7

Reconstitution and Transport Analysis of Multidrug Transporters

Hiroshi Omote and Miki Hiasa

Abstract

The purification and reconstitution technique is a powerful method to determine the functions of multidrug transporters. Elucidation of the transport properties of multidrug transporters is important to understand the pharmacokinetics of drugs; however, their detailed analysis is difficult due to various factors, such as drug hydrophobicity and direction of transport. In this chapter, we describe protocols based on spin label and spin column methods to assess drug transport using purified and reconstituted human multidrug and toxin extrusion protein 1 and P-glycoprotein.

Key words Transporter, Purification, Reconstitution, Proteoliposomes, Driving force, EPR (Electron Paramagnetic Resonance), P-glycoprotein, MATE (Multidrug and Toxin Extrusion), Spin column

1 Introduction

Transporters are involved in various biological activities by translocating substances across membranes. Therefore, elucidation of their transport functions is essential to understand various biological phenomena. Drug transporters are involved in the excretion of drugs from the body and exhibit different structures and molecular mechanisms. In addition to approximately 50 ATP-binding cassette (ABC)-type transporters, various other transporters, including organic cation transporter (OCT)-type, multidrug and toxin extrusion (MATE)-type cationic transporters, and anionic drug efflux transporters, represented by organic anion transporters (OATs) and organic anion transporting polypeptides (OATPs), are present in humans [1–3]. These transporters play specific roles based on their substrate specificity and localization, and elucidation of their transport properties is important to understand pharmacokinetics. In this chapter, we describe protocols to analyze transporter function using a liposome reconstitution system, with a focus on drug transporters.

Kunihiko Nishino and Mikio Tanabe (eds.), *Bacterial Multidrug Exporters: Methods and Protocols*, Methods in Molecular Biology, vol. 3044, https://doi.org/10.1007/978-1-0716-5316-6_7,
© The Author(s), under exclusive license to Springer Science+Business Media, LLC, part of Springer Nature 2026

Drug transporters have been analyzed using heterologous expression systems in animal cells [4]. Here, the protocol measures the uptake of radiolabeled drugs by cells expressing drug transporters and is simple and easy to use [4, 5]. However, precise control of transport direction and driving force are some challenges of this method. In contrast, the liposome reconstitution method using purified proteins allows the control of the driving force and transport direction, making it possible to measure physiological transport [4]. However, highly hydrophobic drugs often exhibit non-specific adsorption onto nitrocellulose membranes, resulting in a high background signal. Methods using a spin probe and a spin column have been used to overcome these issues. Here, we introduce protocols for measuring the transport activity of P-glycoprotein (Pgp) and MATE1 using a spin probe and spin column [6, 7].

2 Materials

2.1 Materials for Measurements of MATE1 Drug Transport Activity

1. Cork borer (size 2).
2. 1-mL disposable medical plastic syringes (6.5 × 70 mm).
3. Glass test tubes (15 × 105 mm).
4. 500-μL microtubes.
5. Whatman GF/D glass filters.
6. Sephadex G-50 Fine.
7. Lipid buffer: 20 mM MOPS–NaOH (pH 7.0), 0.5 mM dithiothreitol.
8. Reaction buffer: 20 mM Tricine–NaOH (pH 8.0), 100 mM potassium acetate, and 5 mM Mg acetate.
9. Dilution buffer: 20 mM MOPS–NaOH (pH 6.5), 0.1 M potassium acetate, and 5 mM $MgCl_2$.

2.2 Materials for Measurement of P-glycoprotein (Pgp) Drug Transport Activity

2.2.1 Lipid Preparation

1. Nitrogen or argon gas supply.
2. SpeedVac concentrator (Thermo Fisher Scientific).
3. Tip sonicator.
4. Lipid solubilization buffer: 50 mM Tris–SO_4 (pH 7.4), 250 mM sucrose, 20% glycerol (v/v), 1 mM NaN_3, 5 mM 6-aminohexanoic acid, and 2%(w/v) octyl-ß-glucoside.

2.2.2 MTS-Verapamil

1. MTS reaction buffer: 50 mM HEPES/NaOH [pH 8.5], 70% [v/v] methanol.
2. MTS-verapamil solution: 6.25 mM MTS-verapamil, 50 mM HEPES/NaOH, pH 8.5, and 70% methanol.
3. HPLC column for SL-verapamil: XTerra RP column (4.6 × 150 mm; Waters).
4. Mobile phase for HPLC: 4 mM Tricine/NaOH [pH 8.5], 60% (v/v) methanol.

2.2.3 EPR

1. EPR reaction mixture: 50 mM Tris–HCl, pH 7.5, 140 mM NaCl, 50 mM KCl, 10 mM ATP, 10 mM $MgCl_2$, 1 mM PEP, 0.5 mg/mL PK, 50 μM SL-verapamil, and 2.2 μM Pgp.
2. 1.43× concentrated EPR reaction mixture: 71.5 mM Tris–HCl, pH 7.5, at room temperature, 200.2 mM NaCl, and 71.5 mM KCl.
3. ATP/$MgCl_2$ mixture (10 mM each).
4. TPX capillaries (IGC Medical Advances).
5. Microsyringes (10 μL, 1 μL).
6. Bruker EMX X-band EPR spectrometer with a loop-gap resonator (Medical Systems).

3 Methods

3.1 Measurements of MATE1 Drug Transport Activity

3.1.1 Introduction

MATE transporters are distributed in various organisms, from bacteria and plants to humans, and play important roles in the export of toxic drugs [8]. In humans, MATE1 is expressed and localized in the biliary membrane of hepatocytes and excretes cationic drugs into the bile solution in an H^+/drug antiport manner [5]. Human MATE1 (hMATE1) and MATE2 (hMATE2) are also expressed and localized in the brush border membrane of renal tubular cells and excrete cationic drugs via the urine. hMATE1 and hMATE2 recognize and transport various drugs, including tetraethylammonium (TEA), 1-methyl-4-phenylpyridinium, cimetidine, creatinine, and metformin [5, 9]. In this chapter, we describe a protocol to measure the transport activity of purified and reconstituted MATE1 [7].

3.1.2 Lipid Preparation

Although various lipids can be used for transport assays, crude or mixed lipids yield the best results. In this assay, we used Type II-S phosphatidylcholine, which is a crude extract of soybean phospholipids.

1. Place 20 mg of soybean L-α-phosphatidylcholine (Sigma Type II-S) in a glass tube (*see* **Note 1**).
2. Add 2 mL of lipid buffer (20 mM MOPS–NaOH [pH 7.0] and 0.5 mM dithiothreitol) and cover the tube with Parafilm.
3. Disperse lipids via sonication in a bath-type sonicator. Sonication should be continued until the solution becomes opalescent (*see* **Note 2**).
4. Divide into aliquots and store at −80 °C until use.

3.1.3 Spin Column Preparation

Several methods can be used to separate the reconstituted proteoliposomes from the reaction mixture. Specifically, the spin column method is suitable for drug transporters. Multidrug transporter

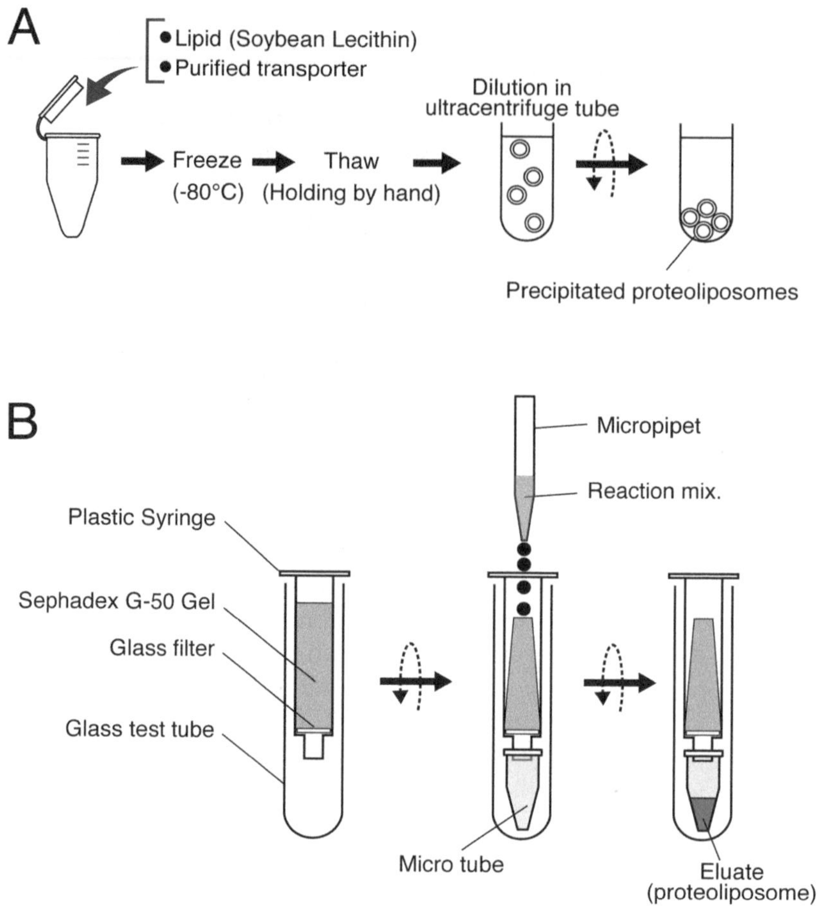

Fig. 1 (**a**) Reconstitution of transporters into proteoliposomes. Mix the purified protein and preformed liposomes in a microtube and freeze at −80 °C. Thaw the mixture by holding it in your hand. Subsequently, dilute the thawed mixture to reconstitute the proteoliposomes and harvest them via ultracentrifugation. (**b**) Illustration of the spin column. Fill a plastic syringe with Sephadex G-50 gel and centrifuge to remove excess solution. After pre-spinning, apply the sample to the spin column and centrifuge immediately

substrates are often adsorbed onto nitrocellulose membranes, thereby increasing the background. The spin column method separates them by gel filtration and is suitable for hydrophobic substrates, such as verapamil, chloroquine, and chloramphenicol [10, 11]. This column efficiently separates proteoliposomes from the reaction mixture (approximately 1/10,000 leakage; Fig. 1). In this section, we describe the preparation of handmade spin columns filled with Sephadex G-50 Fine gel.

1. Add approximately 20 mL of reaction buffer containing (20 mM Tricine–NaOH [pH 8.0], 100 mM potassium acetate, and 5 mM Mg acetate) to 1 g of Sephadex G-50 Fine (Sigma-Aldrich). Keep at room temperature for over 3 h for swelling (*see* **Note 3**).

2. Cut out the Whatman glass GF/D filter with a cork borer (size: 2).
3. Place a glass filter at the bottom of a 1-mL disposable medical plastic syringe (6.5 × 70 mm). The filter should be flat and placed horizontally.
4. Place the plastic syringe in a glass test tube (15 × 105 mm).
5. Fill the plastic syringe with 1 mL of Sephadex G-50 Fine gel (bed volume) using a Pasteur pipette. Take care to ensure no bubbles are present in the column.
6. Wash the column with 1 mL of the same buffer.
7. Spin columns should be wrapped with plastic wrap and placed at room temperature. Use within a day.

3.1.4 Reconstitution of Purified MATE1

This section describes the reconstitution of hMATE1 using the freeze–thaw method [4, 7]. In this method, purified protein solution containing a detergent is mixed with liposome and frozen at −80 °C. Freezing of the protein–liposome mixture destabilizes the micelles, facilitating protein incorporation into liposomes. After liposome reformation via dilution, proteins can be efficiently incorporated into proteoliposomes. Notably, this method is useful for various transporters, including MATE1 [10, 11].

1. Thaw the frozen transporter and lipid solution by holding the tube in your hand and placing it on ice. Do not raise the temperature above approximately 10 °C during thawing.
2. Add 30 μg of purified hMATE1 to an ice-cold microtube containing 50 μL of 10 mg/mL lipid solution (500 μg lipids) and mix thoroughly.
3. Freeze the mixture at −80 °C for 15 min.
4. Thaw the mixture by holding the tube in your hand. Do not significantly increase the temperature; maintain ice temperature.
5. Dilute the mixture over 20-fold by adding the mixture to a centrifuge tube containing an ice-cold dilution buffer (20 mM MOPS–NaOH [pH 6.5], 0.1 M potassium acetate, and 5 mM $MgCl_2$).
6. Centrifuge the diluted mixture for 60 min at 150,000× *g* and 4 °C (*see* **Note 4**).
7. Suspend the precipitate in 200 μL of dilution buffer. Use a 2-mL glass homogenizer (approximately five strokes) and keep it on ice until use (*see* **Note 5**).

3.1.5 Transport Assay

As indicated above, hMATE1 is a bidirectional H^+/drug antiporter. Therefore, drug uptake by liposomes can be measured using an acidic luminal H^+ gradient. To this end, we reconstitute hMATE1 under pH 6.5 and add reconstituted proteoliposomes to a reaction buffer (pH 8.0). Among cationic drugs, TEA is a suitable substrate to measure the transport activity of MATEs. In this section, we describe a protocol for the TEA transport assay using reconstituted hMATE1 by applying an H^+ gradient, in which the vesicle interior is acidic.

1. Add 140 μL of reaction mixture (20 mM Tricine–NaOH [pH 8.0], 100 mM potassium acetate, 5 mM magnesium acetate, and 50 μM ^{14}C-TEA [0.5 MBq/μmol]) to a glass test tube and incubate for 2 min at 30 °C.
2. Place the spin column in a glass test tube (15 × 105 mm) and centrifuge for 2 min at 2000 rpm (750× *g*; *see* **Note 6**).
3. Transfer the pre-spun spin column to a new glass tube containing a 500-μL microtube at the bottom (Fig. 1). The cap should be removed before placing the microtube in the test tube. Keep the column until the reaction is completed (*see* **Note 7**).
4. Add 10 μL of proteoliposomes to the reaction mixture and mix via vortexing.
5. Two minutes after starting the reaction, withdraw 130 μL of reaction mixture and apply to a pre-spun spin column.
6. Centrifuge the spin column for 2 min at 4 °C and 2000 rpm (750× *g*).
7. Collect the eluate and measure the radioactivity using a liquid scintillation counter.

3.2 Measurements of Pgp Drug Transport Activity

3.2.1 Introduction

Pgp is a multidrug transporter belonging to the ABC family that uses the ATP hydrolysis energy to transport drugs out of the cells [1]. Pgp is present in the kidneys, small intestines, liver, and other tissues and plays a key role in drug extrusion from the body. This transporter recognizes diverse molecular structures but prefers cationic and hydrophobic drugs as transport substrates [12, 13]. Many drug transporters are bidirectional and use ion gradients as their driving force [8]. Drug uptake into cells can be measured by controlling the appropriate ionic gradients. ABC transporters, including Pgp, use ATP hydrolysis energy to transport drugs out of the cell. Therefore, drug transport by Pgp is unidirectional, making it difficult to measure drug transport via cellular uptake. Additionally, Pgp substrates are hydrophobic and highly membrane permeable. The membrane filtration method used to measure the transport of hydrophobic substrates is difficult due to non-specific binding to the filter. These features of Pgp and its substrates make it difficult to accurately measure its transport activity.

A protocol to measure the transport activity of Pgp using electron paramagnetic resonance (EPR) is described below. The EPR signal is highly sensitive to changes in the mobility of

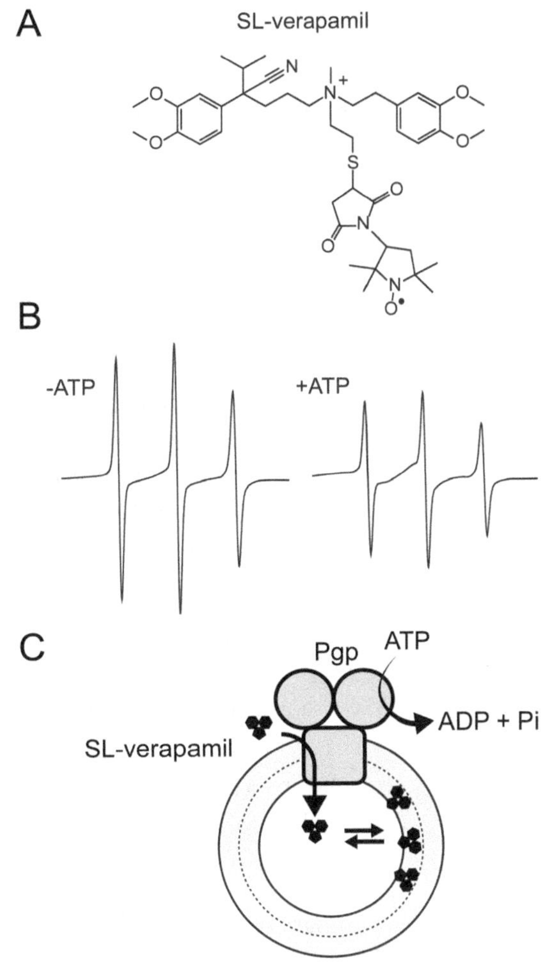

Fig. 2 (**a**) Structure of spin-labeled verapamil (SL-verapamil). (**b**) Electron paramagnetic resonance (EPR) spectra of SL-verapamil in the proteoliposome solution. Mobility of SL-verapamil decreases after the addition of ATP. (**c**) Schematic illustration of SL-verapamil transport into the vesicles. When SL-verapamil is transported to the luminal side of the vesicles, it is adsorbed onto the inner leaflet of the lipid bilayer, resulting in a mobility change in the spin probe

molecules upon binding to lipids [6]. Verapamil, a typical Pgp substrate, is highly hydrophobic. We introduced a spin probe and a fixed positive charge into this molecule to form spin-labeled verapamil (SL-verapamil; Fig. 2). SL-verapamil is a good substrate for Pgp and stimulates ATPase activity with a higher affinity than verapamil. When SL-verapamil is mixed with liposomes, the EPR signal changes due to its binding to lipids. The partition coefficient of SL-verapamil in the lipid membrane can be calculated by titrating the lipid concentration. As the lipid concentration on the luminal side of liposomes is high, the majority of the transported

SL-verapamil is partitioned into the lipid layer facing the lumen. Therefore, the EPR signal of the highly mobile free SL-verapamil decreases during transport. SL-verapamil is not membrane permeable, probably because of its fixed positive charge, thus facilitating accurate transport measurements. Using this method, Pgp was shown to form a ten-fold concentration gradient with SL-verapamil [6].

3.2.2 Purification and Reconstitution of Human Pgp

Pgp purification has been previously described [6, 14]. The concentration of typical purified Pgp is 0.03–0.05 mg/mL in a buffer containing 25 mM Tris–SO_4, pH 7.4, 20% (v/v) glycerol, 50 mM Na_2SO_4, 1 mM NaATP, 1 mM $MgSO_4$, 2 mM ß-mercaptoethanol, 250 mM imidazole, 1.4% (w/v) octyl-ß-glucoside, 0.1% (w/v) lipids, and 1 mM phenylmethylsulfonyl fluoride [14]. Purified Pgp was reconstituted via dialysis against 50 mM Tris–Cl, pH 7.4, 1 mM dithiothreitol (DTT), 1 mM EGTA, and 5 mM 6-aminohexanoic acid. The reconstituted proteoliposomes were suspended in EPR buffer (50 mM Tris–HCl, pH 7.5, 140 mM NaCl, 0.1 mM DTT, and 50 mM KCl) after precipitation via ultracentrifugation.

3.2.3 Lipid Preparation

We use mixed lipids for Pgp reconstitution.

1. Mix 60% (w/w) *Escherichia coli* ether/acetone-precipitated lipid (Avanti Polar Lipids #100600; 20 mg/mL solution in chloroform), 17.5% egg phosphatidylcholine (Sigma-Aldrich #P3556), 10% bovine brain phosphatidylserine (Sigma-Aldrich #B1627), and 12.5% cholesterol (Sigma-Aldrich #C8667) in a glass tube (*see* **Note 8**).
2. Dry under a nitrogen or argon gas stream.
3. Dry under vacuum until the solvent disappears (confirm by weighing). We used SpeedVac (Thermo Fisher Scientific) here.
4. Add lipid solubilization buffer containing 50 mM Tris–SO_4 (pH 7.4), 250 mM sucrose, 20% glycerol (v/v), 1 mM NaN_3, 5 mM 6-aminohexanoic acid, and 2% (w/v) octyl-ß-glucoside to the lipid film.
5. Sonicate the turbid lipid solution with a microtip sonicator until it becomes opalescent under a nitrogen or argon gas stream (*see* **Note 9**).
6. After aliquoting into microtubes, equilibrate with argon gas and store at −80 °C.

3.2.4 Preparation of SL-Verapamil

SL-verapamil can be synthesized by reacting MTS-verapamil (Toronto Research Chemicals) with proxylmaleimide (Merck) [6].

1. Apply 3.2 g of immobilized dithiothreitol (0.229 mEq/g, Calbiochem) to a plastic column, wash with distilled water

until the smell disappears, and equilibrate with the MTS reaction buffer (50 mM HEPES/NaOH [pH 8.5] and 70% [v/v] methanol) at room temperature.

2. Apply 3.2 mL of MTS-verapamil (6.25 mM MTS-verapamil, 50 mM HEPES/NaOH pH 8.5, and 70% methanol) to the column.
3. Add 400 μL of the reaction buffer and wrap the column with aluminum foil. Keep the column at room temperature for 1 h.
4. Add 24 mL of the MTS reaction buffer to elute the reduced MTS-verapamil (SH-verapamil).
5. Add 400 μL of 50 mM maleimide-proxyl (20 μmol) to the eluate and keep at room temperature for 1 h.
6. Concentrate SL-verapamil to approximately 400 μL in an argon gas stream.
7. Purify crude SL-verapamil via high-performance liquid chromatography using the XTerra RP column (4.6 × 150 mm; Waters) with 60% (v/v) methanol and 4 mM Tricine/NaOH, pH 8.5, as the mobile phase.
8. Monitor absorbance at 278 nm and collect the SL-verapamil fractions.
9. Concentrate the SL-verapamil fractions using an argon stream.
10. Add an equal volume of dimethyl sulfoxide and store under −80 °C.

3.2.5 SL-Verapamil Transport Assay

Reactions are performed using an EPR reaction mixture containing 50 mM Tris–HCl, pH 7.5, 140 mM NaCl, 50 mM KCl, 10 mM ATP, 10 mM $MgCl_2$, 1 mM PEP, 0.5 mg/mL PK, 50 μM SL-verapamil, and 2.2 μM Pgp. After initiating the reaction, the mixture is loaded into TPX capillaries (IGC Medical Advances), and EPR spectra are measured at room temperature.

1. Dilute reconstituted Pgp to 3.1 μM with 1.43× concentrated EPR reaction mixture in a microtube.
2. Add 3.5 μL of diluted Pgp to a new microtube and mix with other components to achieve the final concentrations described above. The total volume is 4.75 μL. Use a microsyringe.
3. Add 0.25 μL of 10 mM ATP/10 mM $MgCl_2$ mixture and immediately load the mixture onto the TPX capillary.
4. Measure EPR spectra using the Bruker EMX X-band EPR spectrometer with a loop-gap resonator (Medical Systems) at room temperature.
5. Determine the peak-to-peak intensity of the high-field resonance. This signal indicates the concentration of free SL-verapamil in the reaction mixture.

6. Calculate the free SL-verapamil concentration in the reaction mixture and plot it against the reaction time. Take care to prevent non-specific binding of SL-verapamil to plastic in the absence of liposomes. You may need to estimate the free SL-verapamil concentration by extrapolating the lipid titration curve.

4 Notes

1. Lipids are hygroscopic. Do not use weighing paper.
2. You can use a tip sonicator as required. The turbid solution may contain multilamellar vesicles. Sonicate the lipid solution until the turbidity disappears.
3. Swelling of the Sephadex G-50 gel should be performed at room temperature to avoid microbubble generation. Do not use a cold reaction buffer for this purpose. We recommend swelling the gel 1 day before use.
4. After ultracentrifugation, the pellet becomes transparent.
5. Reconstitution, spin column preparation, and transport assays should be completed in 1 day. Proteoliposomes can be reconstituted during spin column preparation.
6. Use a thick-walled glass test tube (15 × 105 mm) that fits the spin column (Fig. 1). Pre-spinning is essential to effectively separate the proteoliposomes from the reaction mixture. Without pre-spinning, a part of the reaction mixture will be eluted along with the proteoliposomes.
7. Do not keep the pre-spun column. Pre-spin just before use.
8. Lipids are easily solubilized in chloroform and are hygroscopic. Therefore, we recommend weighing these lipids directly in glass tubes.
9. We recommend using a pulsed sonication mode (ca. 30 s cycles) and an ice bath to avoid heat generation. Sonicate the lipid solution until the turbidity disappears. Turbid solutions may contain multilamellar vesicles that are unsuitable for protein reconstitution.

Conflicts of Interest

The authors have no conflicts of interest to declare that are relevant to the content of this chapter.

References

1. Gottesman MM, Pastan I (1993) Biochemistry of multidrug resistance mediated by the multidrug transporter. Annu Rev Biochem 62:385–427
2. Pelis RM, Wright SH (2014) SLC22, SLC44, and SLC47 transporters–organic anion and cation transporters: molecular and cellular properties. Curr Top Membr 73:233–261
3. Lusvarghi S, Robey RW, Gottesman MM, Ambudkar SV (2020) Multidrug transporters: recent insights from cryo-electron microscopy-derived atomic structures and animal models. [version 1; peer review: 2 approved]. F1000Res 9:17
4. Omote H, Moriyama Y (2018) Reconstitution and transport analysis of eukaryotic transporters in the post-genomic era. Methods Mol Biol 1700:343–352
5. Otsuka M, Matsumoto T, Morimoto R et al (2005) A human transporter protein that mediates the final excretion step for toxic organic cations. Proc Natl Acad Sci USA 102:17923–17928
6. Omote H, Al-Shawi MK (2002) A novel electron paramagnetic resonance approach to determine the mechanism of drug transport by P-glycoprotein. J Biol Chem 277:45688–45694
7. Kawasaki T, Matsumoto T, Iwai Y et al (2018) Purification and reconstitution of polyspecific H^+/organic cation antiporter human MATE1. Biochim Biophys Acta Biomembr 1860:2456–2464
8. Moriyama Y, Hiasa M, Matsumoto T, Omote H (2008) Multidrug and toxic compound extrusion (MATE)-type proteins as anchor transporters for the excretion of metabolic waste products and xenobiotics. Xenobiotica 38:1107–1118
9. Tanihara Y, Masuda S, Sato T et al (2007) Substrate specificity of MATE1 and MATE2-K, human multidrug and toxin extrusions/H^+-organic cation antiporters. Biochem Pharmacol 74:359–371
10. Juge N, Moriyama S, Miyaji T et al (2015) *Plasmodium falciparum* chloroquine resistance transporter is a H^+-coupled polyspecific nutrient and drug exporter. Proc Natl Acad Sci USA 112:3356–3361
11. Nagarathinam K, Nakada-Nakura Y, Parthier C et al (2018) Outward open conformation of a Major Facilitator Superfamily multidrug/H^+ antiporter provides insights into switching mechanism. Nat Commun 9:4005
12. Al-Shawi MK, Omote H (2005) The remarkable transport mechanism of P-glycoprotein: a multidrug transporter. J Bioenerg Biomembr 37:489–496
13. Seelig A (2020) P-glycoprotein: one mechanism, many tasks and the consequences for pharmacotherapy of cancers. Front Oncol 10: 576559
14. Figler RA, Omote H, Nakamoto RK, Al-Shawi MK (2000) Use of chemical chaperones in the yeast *Saccharomyces cerevisiae* to enhance heterologous membrane protein expression: high-yield expression and purification of human P-glycoprotein. Arch Biochem Biophys 376: 34–46

Chapter 8

Analytical Ultracentrifugation for Studying the Oligomeric State and Protein–Protein Interactions of Membrane Transporters

Zachary D. Tillett, Renwick C. J. Dobson, and Rachel A. North

Abstract

Analytical ultracentrifugation is still the gold-standard technique for detecting and characterizing the interactions between particles in solution, including membrane transporter proteins. However, because membrane transporter proteins require detergents (and sometimes phospholipids) to solubilize the protein in the aqueous phase, analytical ultracentrifugation is not as simple as for soluble proteins. Here, we present and develop our protocol for carrying out sedimentation velocity-type analytical ultracentrifugation experiments to help understand the mechanism of multi-component tripartite ATP-independent (TRAP) transporter proteins, which are essential for nutrient uptake in bacteria. We describe the expression and purification of TRAP transporters in detergent and exchange them into alternative membrane mimetics, including amphipol and nanodiscs, and we outline the methodology to determine TRAP transporter stoichiometry, define the average number of detergent molecules that are bound to the protein, and perform in-solution studies in native-like membrane environments free from detergents. These methodologies can be applied to other membrane transporters and complexes to answer fundamental questions related to size, shape, stoichiometry, and interactions with other proteins in solution.

Key words Analytical ultracentrifugation, Sedimentation velocity, Oligomeric state, Membrane mimetics, Membrane transporters, Protein, protein interaction, Tripartite ATP-independent periplasmic (TRAP) transporters

1 Introduction

Because cells are separated from their immediate environment by a lipid bilayer (plasma membrane), nature has evolved proteins that embed into the membrane to interact with the surroundings. This includes, for example, receptors that bind extracellular signal molecules and send the signal into the cell (e.g., quorum sensing in bacteria), or integral membrane transporters that catalyze the movement of molecules across the membrane (e.g., secondary-active transporters).

Kunihiko Nishino and Mikio Tanabe (eds.), *Bacterial Multidrug Exporters: Methods and Protocols*, Methods in Molecular Biology, vol. 3044, https://doi.org/10.1007/978-1-0716-5316-6_8,
© The Author(s), under exclusive license to Springer Science+Business Media, LLC, part of Springer Nature 2026

Secondary-active transporters move molecules across membranes by harnessing the free energy associated with electrochemical ion gradients to drive transport [1–4]. No matter the flavor of the membrane protein that one is interested in, they tend to be difficult to study in vitro because they need to be removed from their native phospholipid membrane environment and solubilized in an alternative lipid-like environment, which is usually a detergent.

Tripartite ATP-independent periplasmic (TRAP) transporters are a unique class of secondary-active transporters found only in bacteria and archaea [5, 6]. Unlike other secondary transporters, TRAP transporters cannot receive their substrates directly and rely on a secreted substrate-binding protein in a manner analogous to ATP-binding cassette (ABC) importers. In lieu of utilizing energy-demanding ATP hydrolysis like that of ABC importers, TRAP transporters use energetically favorable sodium ion gradients, coupling the import of specific carboxylate-containing nutrients to the movement of sodium ions down their concentration gradient [7].

TRAP transporter systems comprise three subunits: a soluble substrate-binding "P-subunit" known as SiaP, a membrane-bound complex consisting of a smaller "Q-subunit" containing four transmembrane helices, and a larger "M-subunit" containing 12 transmembrane helices [8], known collectively as SiaQM. For a small number of TRAP transporter systems, the Q- and M-subunits are fused together as a single polypeptide [6], resulting in an additional transmembrane arm helix connecting the Q- and M-subunits [9]. Like many other heteromeric transporter systems and complexes, the stoichiometry of the TRAP transporter complex in the membrane is critical for understanding protein function and mechanism, which can be elucidated in vitro using analytical ultracentrifugation.

Our interest in TRAP transporters is in part to define their mechanism of action [8–13] but also because they transport molecules that are essential for bacterial pathogenesis. TRAP transporters are used by commensal and pathogenic bacteria to import sialic acids, which are amino sugars that the bacteria use for energy, cell wall precursors, and immune evasion. For several pathogens, including *Haemophilus influenzae*, sialic acid import has been demonstrated to be essential for effective colonization and persistence [14, 15]. We have previously focused on other sialic acid transporter families [16, 17] and on the enzymes that manipulate the catabolism of sialic acids in bacteria [18–29]. Furthermore, we have designed inhibitors to sialic acid transporters from other transporter families [30]. Here, we focus on the sialic acid TRAP transporter from *H. influenzae*, which we recently studied [9].

Analytical ultracentrifugation (AUC) is still the gold-standard technique to quantitatively characterize the mass, shape, and interactions of macromolecules and their complexes in the solution environment, including solubilized membrane proteins. AUC combines an ultracentrifuge and an optical system (usually UV/Vis absorbance) to track the sedimentation and diffusion behavior of molecules under biologically relevant buffer conditions (e.g., pH, reducing potential, and ionic strength) [31]. Unlike size exclusion chromatography, which is often used to establish and characterize interactions, AUC is a first-principles method; that is, the experiment can be precisely modeled using established biophysical theory to give the size distribution, shape, and molecular weight of the molecules in solution. If the system under study involves a protein–protein interaction, the interaction dynamics of the system can be determined, including the dissociation constant (K_D) and, in some cases, the on and off rates (k_{on} and k_{off}) [32].

AUC data is collected in real time using the UV/Vis absorbance, Rayleigh interference, or fluorescence emission optical systems [31, 32]. When studying membrane proteins, all three methods of detection are useful. Because membrane proteins tend to have a higher abundance of aromatic residues, they are usually easily tracked using absorbance at 280 nm. However, most detergents do not absorb at this wavelength, and here the Rayleigh interference system for detection becomes important. The fluorescence optical system, if available, offers superb sensitivity and selectivity, enabling reliable measurements down to picomolar concentrations within complex solutions [33, 34]. When using this system to detect protein–protein interactions, one of the interacting partners must be extrinsically labeled with a fluorescent probe that is compatible with the excitation laser (488 nm), which may introduce confounding factors to the experiment. Furthermore, care must be taken when operating at very low protein concentrations to avoid sample sticking to the surface of the ultracentrifuge cell. The addition of a non-labeled "carrier" protein (e.g., bovine serum albumin, or kappa casein) to the sample buffer helps prevent this [32].

Software and computing advances over recent decades have driven the expansion of AUC usage, accessibility, and analysis options. Two popular software packages for AUC data processing are SedFit [35] and UltraScan [36, 37], but there are other excellent packages. Each software package has its own workflow and modeling method, although the results for identical data analyzed between the two packages tend to be similar. GUSSI is an additional data visualization and post-processing tool that is primarily aimed at supplementing the tools included in SedFit [38]. Results from UltraScan can also be imported and processed in GUSSI with a few extra steps. Conveniently, GUSSI has implemented an analysis method that differentiates the signal coming from protein and detergent molecules [39]. This tool enables users to define the mass and oligomeric state of membrane proteins in detergent micelles in solution [40–42], which has previously been difficult because the signal from the protein and detergent components could not be easily separated.

Using TRAP transporters as an example, we outline methods for determining the stoichiometry of a membrane protein complex in detergent and alternative membrane mimetics and define the average number of detergent molecules that are bound to the protein.

2 Materials

Here, we describe the materials required and methods for expressing and purifying a fused TRAP transporter from *H. influenzae* [9] known as *Hi*SiaQM for sedimentation velocity-type analytical ultracentrifugation experiments.

The described methods include: (1) the classical solubilization and purification of the *Hi*SiaQM membrane transporter in detergent micelles; (2) detergent-exchange of *Hi*SiaQM into amphipol, short amphipathic polymers that adsorb onto transmembrane regions of membrane proteins to keep them water-soluble; or (3) incorporation of *Hi*SiaQM into nanodiscs, discoidal fragments of lipid bilayer that are made water-soluble by two encircling amphipathic helical protein belts termed membrane scaffold proteins.

These and other membrane mimetics are compatible with analytical ultracentrifugation studies, and depending on the research question, researchers may pick some or all the described methods for purifying their protein of interest for analytical ultracentrifugation studies. Because membrane proteins are inherently unstable outside of their native lipid membrane environment and some have specific lipid requirements that modulate their function, it is often important to conduct functional experiments in more than one type of membrane mimetic.

2.1 TRAP Transporter Expression and Purification

1. *Escherichia coli* TOP10 cells.
2. LB agar: 2 g tryptone, 1 g yeast extract, 2 g NaCl, and 3 g agar per 200 mL medium.
3. Phosphate buffer: 0.17 M KH_2PO4 and 0.72 M K_2HPO_4.
4. Terrific broth (TB): 24 g yeast extract, 20 g tryptone, 4 mL glycerol, and 100 mL phosphate buffer per 1 L of medium (*see* **Note 1**).
5. 100 mg/mL ampicillin stock solution.
6. 20% L-arabinose stock solution.
7. Lauryl maltose neopentyl glycol (L-MNG), Anagrade (Anatrace).

8. Phosphate-buffered saline: 80 g NaCl, 2 g KCl, 14.4 g Na_2HPO_4, 2.4 g KH_2PO_4, and 10× stock solution.
9. 100 mM phenylmethylsulfonyl fluoride (PMSF) stock solution.
10. Cell resuspension buffer: 1× phosphate-buffered saline, pH 7.4, 0.5 mg/mL lysozyme, and 0.1 mM PMSF.
11. Membrane resuspension buffer: 1 × phosphate-buffered saline, pH 7.4, 6% glycerol, 5 mM dithiothreitol (DTT), 0.1 mM PMSF, and 2% w/v L-MNG.
12. Buffer A: 50 mM Tris–HCl, pH 8.0, 150 mM NaCl, and 0.002% (w/v) L-MNG.
13. Buffer B: 50 mM Tris–HCl, pH 8.0, 150 mM NaCl, 500 mM imidazole, and 0.002% (w/v) L-MNG.
14. Chromatography: Ni-NTA affinity column and a HiLoad 16/60 Superdex 200 column.

2.2 Membrane Scaffold Protein Expression and Purification

1. *E. coli* BL21-Gold (DE3) cells.
2. Terrific broth (TB): 24 g yeast extract, 20 g tryptone, 4 mL glycerol, 100 mL phosphate buffer per 1 L of medium (*see* **Note 1**).
3. 50 mg/mL kanamycin stock solution.
4. 1 M Isopropyl ß-D-1-thiogalactopyranoside (IPTG) stock solution.
5. 100 mM PMSF stock solution.
6. Phosphate-buffered saline: 80 g NaCl, 2 g KCl, 14.4 g Na_2HPO_4, 2.4 g KH_2PO_4, 10 × stock solution.
7. Ni-NTA agarose, Qiagen.
8. Membrane Scaffold Protein (MSP) Wash Buffer A: 40 mM Tris–HCl, pH 8.0, 300 mM NaCl, 1% Triton X-100.
9. MSP Wash Buffer B: 40 mM Tris–HCl, pH 8.0, 300 mM NaCl, 20 mM Na-cholate, 20 mM imidazole.
10. MSP Wash Buffer C: 40 mM Tris–HCl, pH 8.0, 300 mM NaCl, 50 mM imidazole.
11. MSP Elution buffer: 40 mM Tris–HCl, pH 8.0, 300 mM NaCl, 400 mM imidazole.
12. MSP Storage buffer: 50 mM Tris–HCl, pH 8.0, 150 mM NaCl.
13. Tobacco etch virus (TEV) protease.
14. Chromatography: Econo-column (Bio-Rad), Ni-NTA affinity column, and a HiLoad 16/60 Superdex 200 column.

2.3 Amphipol Reconstitution

1. 100 mg/mL Amphipol A8–35 stock solution, Anatrace.
2. Bio-Beads™ SM-2 resin, Bio-Rad Laboratories.
3. Amphipol reconstitution buffer: 50 mM Tris–HCl, pH 8.0, 150 mM NaCl.
4. Chromatography: 10/300 Superdex 200 Increase column.

2.4 Nanodisc Reconstitution

1. Powdered *E. coli* polar lipid extract, Avanti.
2. SEC-purified MSP1D1 (*see* Subheadings 2.2 and 3.2).
3. Bio-Beads™ SM-2 resin, Bio-Rad Laboratories.
4. Ni-NTA agarose, Qiagen.
5. Nanodisc reconstitution buffer: 50 mM Tris–HCl, pH 8.0, 150 mM NaCl.
6. Nanodisc elution buffer: 50 mM Tris–HCl, pH 8.0, 150 mM NaCl, 250 mM imidazole.
7. Chromatography: 10/300 Superdex 200 Increase column.

2.5 Sedimentation Velocity Analytical Ultracentrifugation

1. Beckman Coulter Optima, XLI or XLA analytical ultracentrifuge instrument (*see* **Note 2**).
2. Epon 2-channel 12 mm centerpiece AUC cell with sapphire windows for Rayleigh interference experiments, or quartz windows for absorbance or fluorescence experiments.
3. An-50/60 Ti rotor (Beckman Coulter).
4. Narrow-diameter gel-loading tips.
5. SedFit, SedNterp, and GUSSI software (*see* **Note 3**).

3 Methods

3.1 TRAP Transporter Expression and Purification

3.1.1 Protein Expression

1. Transform *E. coli* TOP10 cells with the recombinant pBad-HisA plasmid containing the sequence encoding the TRAP transporter (*see* **Notes 4** and **5**).
2. Plate cells evenly on LB agar plates containing 100 μg/mL ampicillin to select for transformants.
3. Incubate the plate for 14–16 h at 37 °C until small, distinct colonies appear, which correspond to transformed TOP10 cells containing the plasmid of interest.
4. Inoculate a single colony into 5 mL TB medium supplemented with 5 μL ampicillin stock solution (100 μg/mL final concentration) and incubate for 8 h at 37 °C, 200 rpm to prepare a starter culture.
5. Inoculate a 1/50 dilution of the starter culture into fresh TB medium containing ampicillin (100 μg/mL final concentration) and incubate overnight at 30 °C, 200 rpm for ~14 h.

6. Prepare large-scale cultures (*see* **Note 6**) of TB medium supplemented with ampicillin (100 μg/mL final concentration). Inoculate with a 1/50 dilution of the overnight culture, and incubate at 37 °C, 200 rpm until an optical density at 600 nm (OD_{600}) of 1.7–1.8 is reached.
7. Induce recombinant protein expression by adding L-arabinose to a final concentration of 0.2% and continue incubation for 3 h at 37 °C and 200 rpm.
8. Harvest cells by centrifugation at 8000 g for 10 min at 4 °C. Discard the supernatant and snap-freeze the pellet for storage at −80 °C until required.

3.1.2 Membrane Preparation

1. Thaw cells at room temperature and homogenize in 4× w/v cell resuspension buffer.
2. Lyse by ultrasonication at 70% amplitude in 0.5 s on, 0.5 s off cycles for 30 min.
3. Centrifuge the cell lysate at 18,000× *g* for 30 min at 4 °C to remove unbroken cell debris.
4. Carefully transfer the supernatant to a clean tube and repeat the centrifugation under the same conditions. Discard the resulting pellets.
5. Harvest membranes by ultracentrifugation at 240,000× *g* for 2 h at 4 °C, carefully discard the supernatant, and snap-freeze the resulting membrane pellet for storage at −80 °C until required.

3.1.3 Protein Purification

1. Resuspend the membranes in membrane resuspension buffer and solubilize with 2% (w/v) L-MNG for 2 h at 4 °C with gentle agitation using a rotary shaker.
2. Ultracentrifuge at 160,000× *g* for 1 h at 4 °C to remove insoluble material; retain the clarified supernatant and discard the resulting pellets.
3. Equilibrate the Ni-NTA affinity column by washing with three column volumes of Buffer A.
4. Apply the supernatant containing solubilized TRAP transporter protein from **step 2** into the pre-equilibrated Ni-NTA affinity column.
5. Wash the loaded Ni-NTA affinity column with 20 column volumes of Buffer A.
6. Elute bound TRAP transporter protein with ten column volumes of Buffer B and collect elution in fractions of approximately 1 mL each.
7. Use SDS–PAGE to visually assess the purity of peak fractions and pool those corresponding to >90% purity.

8. Concentrate pooled fractions using 100 kDa molecular weight cutoff (MWCO) spin concentrators to 1 mL and perform SEC on a HiLoad 16/60 Superdex 200 column equilibrated with three column volumes of Buffer A. Collect 1 mL elution fractions.
9. Confirm the purity of peak fractions by SDS–PAGE. Pool the pure fractions, concentrate to an appropriate concentration (e.g., 1 mg/mL) for subsequent experiments, or flash-freeze in liquid nitrogen for storage at −80 °C until use.

3.2 Membrane Scaffold Protein Expression and Purification

3.2.1 Expression

1. Transform *E. coli* BL21-Gold (DE3) cells with the recombinant pET28a plasmid containing the sequence encoding MSP1D1 (*see* **Note 7**).
2. Plate cells evenly on LB agar plates containing 50 μg/mL kanamycin to select for transformants.
3. Incubate the plate for 14–16 h at 37 °C until small, distinct colonies appear, which correspond to transformed BL21-Gold (DE3) cells containing the plasmid of interest.
4. Inoculate a single colony into 5 mL TB medium supplemented with 5 μL kanamycin stock solution (50 μg/mL final concentration) and incubate for 8 h at 37 °C, 200 rpm to prepare a starter culture.
5. Inoculate a 1/50 dilution of the starter culture into fresh TB medium containing kanamycin (50 μg/mL final concentration) and incubate overnight at 30 °C, 200 rpm for ~14 h.
6. Prepare large-scale cultures (*see* **Note 8**) of TB medium supplemented with kanamycin (50 μg/mL final concentration). Inoculate with a 1/50 dilution of the overnight culture, and incubate at 37 °C, 200 rpm until an OD_{600} of 1.7–1.8 is reached.
7. Induce recombinant protein expression by adding IPTG to a final concentration of 1 mM and continue incubation for 3 h at 37 °C and 200 rpm.
8. Harvest cells by centrifugation at 8000 g for 10 min at 4 °C. Discard the supernatant and snap-freeze the pellet for storage at −80 °C until required.

3.2.2 Protein Purification

1. Thaw cells at room temperature, weigh, and homogenize in 4× w/v PBS, then add 1% Triton X-100 and 1 mM PMSF.
2. Lyse by ultrasonication at 70% amplitude in 0.5 s on, 0.5 s off cycles for 5 min.
3. Pellet unbroken cell debris by centrifugation at 20,000 g for 30 min at 4 °C and pour the supernatant into a glass beaker. Centrifuge the cell lysate again at 20,000× *g* for 30 min at 4 °C

to remove unbroken cell debris and pour the supernatant into a glass beaker.

4. Add Ni-NTA agarose resin (approximately 1 mL resin/g of initial cell pellet) to the supernatant, supplement with imidazole to a final concentration of 10 mM, and incubate for 3 h at 4 °C with gentle agitation using a magnetic stirrer to allow binding.
5. Pour the resin-lysate mixture into an Econo-column and discard the flow-through. Wash the column sequentially with 6 column volumes of MSP wash Buffer A, 6 column volumes of MSP wash Buffer B, and 20 column volumes of MSP wash Buffer C.
6. Elute bound MSP1D1 with three column volumes of MSP elution buffer. Add TEV protease to remove the N-terminal His-tag at a ratio of 1:5 (MSP1D1:TEV protease in mg/mL) and dialyze overnight in MSP storage buffer at 4 °C.
7. To remove TEV and any uncleaved MSP1D1, add Ni-NTA agarose resin (approximately 1 mL/40 mg of MSP1D1) to the protein solution in an Econo-column. Seal and incubate for 1 h at 4 °C with gentle agitation on a rotary shaker.
8. Unseal the column and collect the flow-through. Wash the resin with three column volumes of MSP storage buffer, collecting and combining the flow-through fractions containing cleaved MSP1D1 protein.
9. If necessary, concentrate to 5–10 mg/mL using 30 kDa MWCO and check purity by SDS–PAGE (*see* **Note 9**).
10. Aliquot and flash-freeze in liquid nitrogen for storage at −80 °C until use.

3.3 Amphipol Reconstitution

1. Incubate SEC-purified TRAP transporter protein with Amphipol A8–35 at a 1:5 mass ratio for 2 h on ice.
2. Add Bio-Beads SM-2 resin (100 mg/mL) and incubate overnight at 4 °C with gentle agitation on a rotary shaker to remove detergent.
3. Remove Bio-Beads by centrifugation at 10,000× *g* for 10 min at 4 °C and collect the supernatant.
4. Wash the beads with 500 μL of amphipol reconstitution buffer, repeat centrifugation, and combine supernatants.
5. Concentrate the amphipol-reconstituted TRAP transporter using 100 kDa MWCO spin concentrators to 250 μL.
6. Centrifuge at 13,000× *g* for 20 min at 4 °C to remove aggregates.
7. Perform SEC on a 10/300 Superdex 200 Increase column equilibrated with three column volumes of amphipol

reconstitution buffer to remove free amphipol and assess protein monodispersity. Collect ~500 μL fractions.

8. Pool peak fractions and concentrate using 100 kDa MWCO spin concentrators to ~1 mg/mL for downstream experiments or flash-freeze in liquid nitrogen for storage at −80 °C.

3.4 Nanodisc Reconstitution

1. Prepare lipid extract by solubilizing powdered *E. coli* polar lipid extract in reconstitution buffer to a final concentration of 20 mM (*see* **Note 10**).
2. Prepare a 300 μL final volume of SEC-purified SiaQM, purified MSP1D1, and solubilized *E. coli* polar lipid extract at a 1:5:100 ratio of (SiaQM:MSP1D1:lipid). Incubate the mixture for 30 min at room temperature.
3. Add 50 mg Bio-Beads SM-2 resin to the SiaQM:MSP1D:lipid mixture to remove detergent and promote nanodisc formation. Incubate for 2 h at room temperature, then overnight at 4 °C with gentle agitation on a rotary shaker.
4. Remove Bio-Beads by centrifugation at 10,000× *g* for 10 min at 4 °C and collect the supernatant.
5. Wash the beads with 500 μL of nanodisc reconstitution buffer, repeat centrifugation, and combine supernatants.
6. Centrifuge the sample at 13,000× *g* for a further 20 min at 4 °C to remove aggregates.
7. Add Ni-NTA agarose resin (approximately 1 mL of resin per mg of SiaQM) and incubate for 1 h at 4 °C with gentle agitation on a rotary shaker to remove empty nanodiscs.
8. Wash the resin with five column volumes of nanodisc reconstitution buffer.
9. Elute nanodisc-reconstituted SiaQM with two column volumes of nanodisc elution buffer.
10. Concentrate to 250 μL using 100 kDa MWCO spin concentrators, and centrifuge again at 13,000× *g* for 20 min at 4 °C to remove aggregates.
11. Perform SEC on a 10/300 Superdex 200 Increase column equilibrated with three column volumes of nanodisc reconstitution buffer to assess protein monodispersity. Collect ~500 μL fractions.
12. Verify peak fractions by SDS–PAGE (*see* **Note 11**), pool them, and concentrate using 100 kDa MWCO spin concentrators to ~2 mg/mL for subsequent AUC experiments.

3.5 Sedimentation Velocity Analytical Ultracentrifugation of TRAP Transporter to Determine the Stoichiometry and Number of Detergent Molecules Solubilizing the Protein

3.5.1 Sedimentation Velocity Analytical Ultracentrifugation Sample Preparation and Experiment Run

1. Prepare ~500 μL of both sample and matching buffer. The buffer must be detergent-free but otherwise identical to the sample buffer.
2. Using a UV–Vis spectrometer, ensure the sample has an absorbance of ~0.5 at 280 nm for a 1 cm pathlength, or select an alternative wavelength that yields a similar absorbance (*see* **Note 12**).
3. Assemble a sapphire window AUC cell with a 12 mm Epon centerpiece for use by torquing the top screw to 120 ft-lbs (*see* **Note 13**).
4. Place the cell on its side with the screw ring facing you and the filling holes at the top. The sample sector corresponds to the filling hole on the right-hand side.
5. Using a narrow-diameter gel-loading tip, add 400 μL of sample to the sample sector of the AUC cell.
6. Load 400 μL of reference buffer into the reference sector using narrow-diameter gel-loading tips.
7. To seal the cell after loading, place a fill-hole gasket into the reference and sample filling holes (either PFTE or Neoprene fill-hole gaskets), and insert a socket-fill hole screw into each of the filling holes and tighten to finger-tight.
8. Repeat **steps 3–6** for any additional samples.
9. Prepare a Beckman counterbalance by adding the appropriate weights to ensure the mass is within 200 mg of the loaded experimental cell.
10. Place the counterbalance into the final position of the AUC rotor and load all cells symmetrically to ensure a balanced rotor.
11. Insert the rotor into the analytical ultracentrifuge and allow it to reach thermal equilibrium for 1 h at 20 °C (*see* **Note 14**), or a temperature suitable for your system.
12. Typically, we run experiments at 42,000–50,000 rpm, but if the particles are large (e.g., >300 kDa), use a slower speed (*see* **Note 15**).
13. Ensure that you are collecting interference data and either absorbance or intensity data (*see* **Note 16**).
14. Start the run by using the instrument control software to set the desired speed.

3.5.2 Sedimentation Velocity Analytical Ultracentrifugation Data Analysis; SedFit and GUSSI

For detailed information on required parameters and software options, *see* **Notes 17** and **18**.

1. Load the raw absorbance data (.ra* files) into SedFit and select the c(s) model.

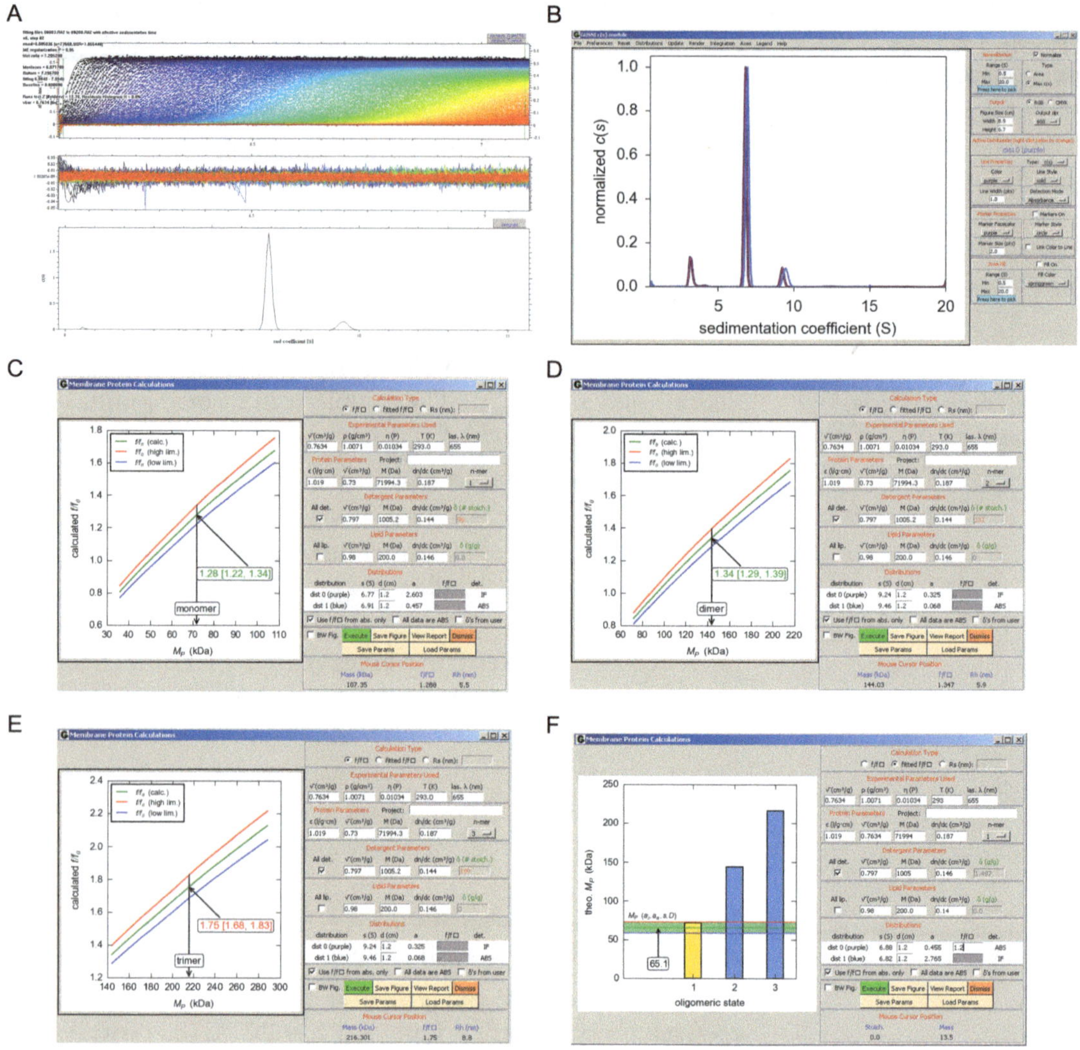

Fig. 1 Analysis of *Hi*SiaQM using SedFit and GUSSI. Explanation for each pane is in the text where referenced

2. Set the meniscus and limits of the data to be included by adjusting the green and red bars in the graphical interface.
3. Add the relevant parameters to the "Parameters box"—protein partial specific volume, buffer density, and buffer viscosity.
4. Fit the data to the *c(s)* model with baseline correction and floating the frictional ratio (example fit for *Hi*SiaQM is shown in Fig. 1a).
5. Load the distribution into GUSSI by selecting *GUSSI c(s) Plot* from the plot tab, then save the distribution data.
6. Repeat the SedFit analysis for the interference data (.ip* files) using the same parameters.

Table 1
Oligomeric state parameters used in the analytical ultracentrifugation studies

Parameters	*Hi*SiaQM in L-MNG
Buffer density (g/mL)	1.0071
Buffer viscosity (poise)	0.01034
Temperature (K)	293.15
Laser wavelength (nm)	655.0
Partial specific volume (P, mL/g)	0.7634
Mass extinction coefficient (P, l/g.cm)	1.019
Molecular mass (P, Da)	71,994.3
Refractive index increment (P, mL/g)	0.187
Partial specific volume (D, mL/g)	0.797
Molecular mass (D, Da)	1005.2 Da
Refractive index increment (D, mL/g)	0.146

P protein, *D* detergent

7. Load the *c(s)* distribution into GUSSI, and also load the saved absorbance distribution into the same GUSSI session.
8. The window in GUSSI should resemble Fig. 1b, where distributions are normalized to the ~7 S peak (for *Hi*SiaQM).
9. Open the *Memb. Protein Cals.* Module in the integration tab. Select the range corresponding to the peak of interest (here, this is ~7S).
10. Add the parameters for the buffer, protein, and detergent (Table 1). Set *n-mer* = *1* to test the hypothesis that the ~7 S peak represents a monomeric protein complex (in this case 1 SiaQ and 1 SiaM) with bound detergent.
11. Click *Execute* to perform the calculation (*see* Fig. 1c). A monomeric particle with a frictional ratio of 1.26 and 96 L-MNG molecules is sensible in this case.
12. Similar analyses can be performed for the peak at ~9.5 S, to test alternative oligomeric states, such as a dimeric protein complex (2 SiaQ and 2 SiaM subunits, Fig. 1d) or a trimeric protein complex (3 SiaQ and 3 SiaM subunits, Fig. 1e). In this case, the results indicate that this peak likely represents a dimer rather than a trimer.
13. To validate the integrated peak (e.g., the ~7 S monomeric complex), compare with fitted frictional ratios from SedFit or UltraScan by selecting *fitted* f/f_0 in the module and entering

the frictional ratio obtained from the *c(s)* fit. The resulting analysis (Fig. 1f) reveals that the mass of the protein is 65.1 kDa, consistent with a 1:1 SiaQ:SiaM complex (theoretical mass = 72.0 kDa).

3.6 Experiment 2: Sedimentation Velocity Analytical Ultracentrifugation of the TRAP Transporter in Nanodiscs

The experiments below can be conducted on membrane proteins solvated in detergents, amphipol, or nanodiscs. Here, we outline an experiment using *Hi*SiaQM inserted into a nanodisc, as described in Subheading 3.4.

3.6.1 Sedimentation Velocity Analytical Ultracentrifugation Sample Preparation

1. Immediately prior to the AUC experiment, perform a size exclusion chromatography experiment to separate empty nanodiscs from those containing membrane proteins, as described in Subheading 3.4, **step 11**. For *Hi*SiaQM, multiple peaks are evident from the size exclusion profile (Fig. 2, colored regions), each of which can be analyzed by sedimentation velocity.
2. Prepare ~500 μL of sample and matching buffer (here, 20 mM Tris–HCl, pH 7.4, 100 mM NaCl, and 0.5 mM EDTA).
3. The sample should, ideally, have an absorbance of 0.3–0.6 OD at 280 nm for a 1 cm pathlength, or choose a wavelength where the absorbance is within this range.

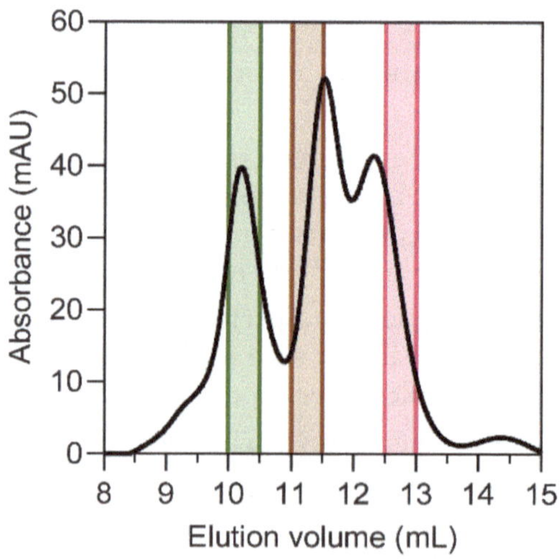

Fig. 2 Solubilization of *Hi*SiaQM in nanodiscs. The size exclusion chromatogram following nanodisc reconstitution with a 1:4:80 ratio of *Hi*SiaQM:MSP:lipid identifies the presence of multiple species. The three fractions across the elution profile (shown by the colored bars) are analyzed with sedimentation velocity AUC in the following figures. (Reproduced from Currie et al. (2024) with permission from eLife [9])

4. Use a properly cleaned and torqued AUC cell, typically a two-sector, 1.2 cm pathlength cell with sapphire or quartz windows (*see* **Note 13**).
5. Place the cell on its side with the screw ring facing you and the filling holes at the top. The sample sector corresponds to the filling hole on the right-hand side.
6. Add 380–400 μL of sample to the sample sector using narrow-diameter gel-loading tips.
7. With the cell on its side and the screw ring facing you, the reference sector is the filling hole on the left-hand side.
8. If collecting in absorbance mode, the volume of reference buffer added to the reference sector should be ~20 μL more than that of the sample volume (380–400 μL), which allows precise identification of the meniscus position. If using intensity mode, you should load the same volume of sample or buffer into the reference or B sector.
9. To seal the cell after loading, place fill-hole gaskets (PFTE or Neoprene) into both the sample and reference filling holes.
10. Insert a socket-fill hole screw into each of the filling holes and tighten finger-tight, taking care not to cross-thread or over-tighten the screws.
11. Place the cell into the rotor (either an An-50 or An-60 Ti Rotor, Beckman Coulter).
12. Ensure there is a counterbalance or loaded cell in the final rotor position (either 4 or 8) with an appropriately adjusted weight (depends on the cell type).
13. Align the etched notches of each cell with the alignment marks on the rotor.
14. Place the rotor into the analytical ultracentrifuge and let the rotor and samples reach thermal equilibrium (*see* **Note 14**).
15. Typically, experiments are performed at 42,000–50,000 rpm, but if the particles are large (e.g., >300 kDa), a slower speed may be more appropriate (*see* **Note 15**).
16. Start the experiment by setting the desired speed and number of scans/duration in the control software (Beckman Coulter or UltraScan). Collect data in either absorbance or intensity mode at the desired wavelength.

3.6.2 Sedimentation Velocity Analytical Ultracentrifugation Data Analysis

1. Import experimental data into UltraScan by selecting the Utilities tab and then "Import Experimental Data" (this step will be done automatically if running the Optima AUC through UltraScan).
2. Define the buffer by either manually entering the density and viscosity, or using the predefined buffer components and

inputting the buffer composition in the "Buffer Management Utility" tool.

3. Define the partial specific volume of your analyte(s) either manually or by inputting the amino acid sequence in the "Analyte Management Utility" tool.
4. Define your solutions by selecting the correct analyte and buffer for each experimental setup.
5. Edit the experimental data by selecting the meniscus and the end of the data range. For interference optics experiments, the air gap region must also be defined.
6. Scans collected after pelleting the sample can be excluded from the data. If desired, reduce the total number of scans to expedite subsequent data processing steps.
7. Perform a time derivative analysis of the data. Note the high and low S values (x-axis) where the data returns to baseline. The default plotting range may require adjustment to accurately determine the high S-value. For a lower limit, 1 S is usually sufficient.
8. Using the limits noted in **step 7** as the S-value range, perform an initial 2DSA analysis of the data through either a LIMS server or the UltraScan GUI, fitting time-invariant noise. The default f/f_0 range of 1–4 is usually appropriate for most protein solutions.
9. Examine the results from **step 8**. If the residuals appear random, proceed to the next step. If a diagonal pattern is observed, adjust the S and f/f_0 limits based on the discrete distribution from the initial model and repeat **step 8** until satisfactory residuals are obtained.
10. Fit the meniscus by submitting an additional 2DSA analysis request, using the same limits used in **step 8**. Ensure you are fitting both time- and radially invariant noise and select the option to fit the meniscus.
11. Once this analysis is complete, use the "fit meniscus" utility to verify the results and define the final meniscus position. If the plotted results do not fit the expected r.m.s.d. curve, return to **step 8** and recheck the meniscus position.
12. Perform a 2DSA iterative analysis using the same limits as in **steps 8** and **10**. Fit both time- and radially invariant noise and select Iterative Refinement, with the refinement level set to ten iterations.
13. Visualize the results using the "FE model viewer" in UltraScan and save the outputs to the database.
14. From this point, the analysis trajectory will vary depending on the properties of the analyte distribution. If a paucidisperse

(i.e., containing few discreet analytes) solution is obtained, parsimonious regularization using the genetic algorithm method is recommended. For all other distributions, it is advisable to proceed with a 2DSA analysis in conjunction with a 50 or 100 Monte-Carlo analysis.

15. To perform a genetic algorithm refinement of the 2SDA-IT solution, use the "initialise genetic algorithm" utility in UltraScan.
16. Here, either manually or automatically define your solutes by drawing bins around the pseudo-3D distributions.
17. Save the bins and use them to initialize the genetic algorithm on the LIMS server (Fig. 3a). The results from the genetic algorithm analysis can be visualized in the compare pseudo-3D distributions (Fig. 3b).
18. To perform a Monte-Carlo analysis, submit an additional 2DSA request on the LIMS server. In this case, do not fit radially or time-invariant noise; instead, apply the noise corrections from the 2DSA iterative analysis step and select either 50 or 100 Monte-Carlo iterations.
19. Depending on the Monte-Carlo results, a subsequent genetic algorithm analysis may be beneficial. Automatic definition of solutes is disabled when initializing a genetic algorithm from Monte-Carlo data, and they must be manually defined by drawing bins around clusters of analyte points.

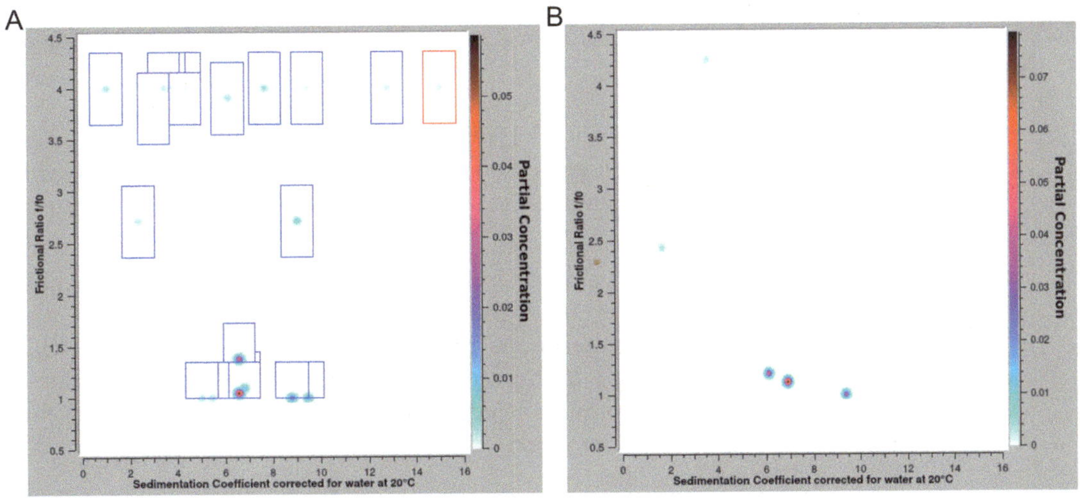

Fig. 3 Genetic algorithm analysis of *Hi*SiaQM in nanodiscs. (**a**) Initialize the genetic algorithm utility, showing the results of the final 2DSA iterative distribution of the largely monomeric sample (brown curve in Fig. 4) with the solute bins to be saved for genetic algorithm analysis. (**b**) The results of the genetic algorithm analysis showing parsimonious regularization of the model to a minimum number of analytes

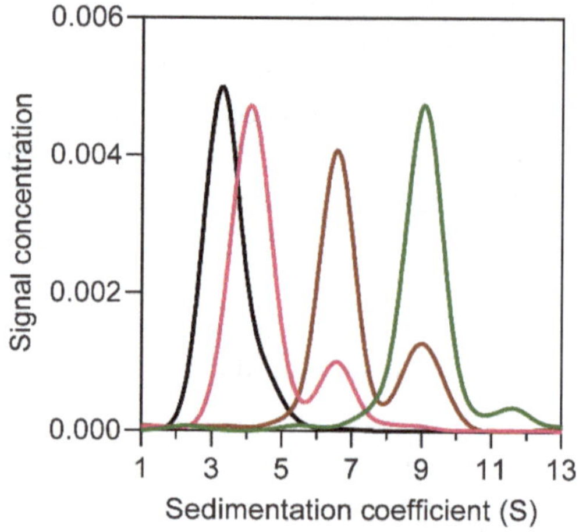

Fig. 4 Sedimentation velocity analysis of *Hi*SiaQM in nanodiscs. Sedimentation velocity AUC analysis of nanodisc solubilized *Hi*SiaQM (pink, brown, and green from the SEC trace in Fig. 2) and empty nanodiscs (black) shows that later eluted fractions gave a main species consistent with empty nanodiscs (pink, 4.0 S), middle eluted fractions gave a main species consistent with a monomer incorporated (brown, 6.8 S), and earlier eluted fractions gave a larger 9.0 S main species (green). The larger species at 9.0 S is most likely a dimer of *Hi*SiaQM in a nanodisc, as it has been shown to exist as a dimer in both detergent and amphipol. (Reproduced from Currie et al. (2024) with permission from eLife [9])

20. Compare datasets (e.g., different peaks from an SEC trace) using the "combine discrete distributions" utility. Figure 4 shows the results from the genetic algorithm analysis of *Hi*SiaQM SEC fractions (Fig. 2), along with an empty nanodisc control.

4 Notes

1. When preparing 1 L of TB medium, add 900 mL of deionized water to the yeast extract, tryptone, and glycerol, and sterilize by autoclaving. Once cool, add 100 mL of sterile phosphate buffer.
2. Beckman Coulter Optima XLI or XLA analytical ultracentrifuge instrument. Although not detailed in this protocol, if the intended experiment requires fluorescence detection (which can be used to detect protein–protein interactions), one of these instruments needs to be fitted with an AVIV AUC fluorescence detection system (usually the XLA or XLI instruments).
3. These software programs are free for academic users and run on a Windows PC. Alternatively, you can use UltraScan or the UltraScan in A Box Server (these are also free for academic users and run on a Linux PC) and GUSSI.

4. The gene sequence encoding a TRAP transporter can be cloned into pBad-HisA by GeneArt, using XhoI and EcoRI restriction sites so that a hexahistidine affinity tag is fused at the 5′ end.
5. A small proportion of TRAP transporter Q- and M-subunits are fused into a single polypeptide, but most are non-fused and natively expressed as separate polypeptides. For non-fused TRAP transporter systems, both genes encoding the Q- and M-subunits should be inserted into the expression plasmid with the native intergenic region intact, as it contains a Shine-Dalgarno ribosomal binding site, which enables both subunits to be recombinantly expressed in a single plasmid.
6. Large-scale cultures were performed in baffled 5 L borosilicate conical flasks, with 1 L of culture in each. To obtain a suitable amount of protein for future experiments, we recommend culturing approximately 12 L of the TRAP transporter for purification. Of course, this depends on the yield.
7. Choose an MSP that will form a nanodisc with an appropriately sized scaffolding diameter for your membrane protein of interest.
8. Large-scale cultures were performed in baffled 5 L borosilicate conical flasks, with 1 L of culture in each. We recommend culturing 2 L of MSP1D1 for each purification. Following Ni-affinity purification and cleavage of MSP1D1, you can expect to obtain approximately 80–100 mg of protein per 1 L of culture.
9. If MSP1D1 is not pure after Ni-affinity purification and cleavage, then SEC can be used as a final purification step.
10. Choose a lipid or lipid combination for nanodisc reconstitution that is appropriate for your membrane protein of choice.
11. Bands on the SDS–PAGE gel corresponding to your MSP and membrane protein of choice should be of a similar intensity, indicating a successful and stable nanodisc reconstitution.
12. The AUC cell has a 1.2 cm pathlength; therefore, setting the sample absorbance to ~0.5 at 280 nm using a 1 cm cuvette ensures the detector operates within its linear range during the experiment. If a wavelength other than 280 nm is used, one will need to determine the extinction coefficient of their protein at the selected wavelength for accurate data analysis.
13. Detailed instructions for loading AUC cells are provided in the Beckman Coulter Analytical Ultracentrifuge User Manual. Additional guidance and training materials can be found on the UltraScan website: https://resources.aucsolutions.com/, including the AUC User Guide (https://resources.aucsolutions.com/AUCuserGuideVolume-1-Hardware.pdf).

14. The rotor should be incubated at the required temperature of the run for at least 1 h before starting the experiment. If the protein is known to be somewhat unstable, the experiment can be conducted at lower temperatures (e.g., 4 °C or 10 °C), but 20 °C is standard.
15. Select a rotor speed and run time that ensure complete sedimentation of the analyte. Simulation tools such as SViMULATE [43], SedFit, or the UltraScan Finite Element Model [44] can aid in determining suitable parameters. If uncertain, an overnight run (≥12 h) is recommended. Scans collected after complete sedimentation can be excluded during data processing.
16. The Optima instrument will automatically record both intensity and absorbance data if the absorbance optics are used. In contrast, the XLI can record data in either absorbance or intensity mode, but not both. If using intensity mode on the XLI, be sure that the OD for the reference sector does not exceed 0.3 to avoid resetting the gain settings. We regularly collect intensity data, and this can be easily converted to pseudoabsorbance in either SedFit or UltraScan.
17. Before analysis, information on the buffer (density and viscosity), protein (partial specific volume, molecular mass, molar attenuation coefficient, and refractive index increment), and detergent of choice (partial specific volume, molecular mass, and refractive index increment) is required. Buffer properties and the partial specific volume of the protein can be calculated using UltraScan or SedNterp or can be determined experimentally using a densitometer/viscometer (e.g., Anton Paar). Parameter values for common detergents such as DDM and L-MNG are available in the literature (Table 1).
18. If using UltraScan, the results from Two-Dimensional Spectrum Analysis (2DSA), Parametrically Constrained Spectrum Analysis (PCSA), or genetic algorithm distributions can be manually loaded into GUSSI. To do this, export the distributions from UltraScan and copy-paste them individually into GUSSI from a.csv file. The workflow for analyzing these distributions is like that used for a c(s) distribution; however, as UltraScan distributions are automatically converted to $S_{(20,W)}$, the buffer details in GUSSI must be entered as water at 20 °C.

Acknowledgments

R.C.J.D. and R.A.N. acknowledge funding support from the Marsden Fund, managed by Royal Society Te Apārangi (contract UOC1506), Canterbury Medical Research Fund (contract CMRF 08), and the Biomolecular Interaction Centre (UC).

Conflicts of Interest The authors have no conflicts of interest to declare that are relevant to the content of this chapter.

References

1. Drew D, North RA, Nagarathinam K, Tanabe M (2021) Structures and general transport mechanisms by the major facilitator superfamily (MFS). Chem Rev 121(9):5289–5335
2. Pao SS, Paulsen IT, Saier MH (1998) Major facilitator superfamily. Microbiol Mol Biol Rev 62(1):1–34
3. Reizer J, Reizer A, Saier MH (1994) A functional superfamily of sodium/solute symporters. BBA-Rev Biomembranes 1197(2): 133–166
4. Ren QH, Paulsen IT (2005) Comparative analyses of fundamental differences in membrane transport capabilities in prokaryotes and eukaryotes. PLoS Comput Biol 1(3):190–201
5. Rosa LT, Bianconi ME, Thomas GH, Kelly DJ (2018) Tripartite ATP-independent periplasmic (TRAP) transporters and tripartite tricarboxylate transporters (TTT): from uptake to pathogenicity. Front Cell Infect Microbiol Rev 8:33
6. Kelly DJ, Thomas GH (2001) The tripartite ATP-independent periplasmic (TRAP) transporters of bacteria and archaea. FEMS Microbiol Rev 25(4):405–424
7. Vetting MW et al (2015) Experimental strategies for functional annotation and metabolism discovery: targeted screening of solute binding proteins and unbiased panning of metabolomes. Biochemistry 54(3):909–931
8. Davies JS et al (2023) Structure and mechanism of a tripartite ATP-independent periplasmic TRAP transporter. Nat Commun 14(1): 1120
9. Currie MJ et al (2024) Structural and biophysical analysis of a *Haemophilus influenzae* tripartite ATP-independent periplasmic (TRAP) transporter. elife 12:RP92307
10. Davies JS et al (2024) TRAPs: the 'elevator-with-an-operator' mechanism. Trends Biochem Sci 49(2):134–144
11. Goyal P et al (2024) Molecular determinants of Neu5Ac binding to a tripartite ATP independent periplasmic (TRAP) transporter. elife 13: RP98158
12. Davies JS et al (2021) Selective nutrient transport in bacteria: multicomponent transporter systems reign supreme. Front Mol Biosci 8: 699222
13. Newton-Vesty MC et al (2025) Structural basis of isethionate transport by a TRAP transporter from a sulfate-reducing bacterium. Structure. https://doi.org/10.1016/j.str.2025.10.011
14. Severi E et al (2005) Sialic acid transport in *Haemophilus influenzae* is essential for lipopolysaccharide sialylation and serum resistance and is dependent on a novel tripartite ATP-independent periplasmic transporter. Mol Microbiol 58(4):1173–1185
15. Jenkins GA et al (2010) Sialic acid mediated transcriptional modulation of a highly conserved sialometabolism gene cluster in *Haemophilus influenzae* and its effect on virulence. BMC Microbiol 10:48
16. North RA et al (2018) The sodium sialic acid symporter from *Staphylococcus aureus* has altered substrate specificity. Front Chem 6:233
17. Wahlgren WY et al (2018) Substrate-bound outward-open structure of a Na^+-coupled sialic acid symporter reveals a new Na^+ site. Nat Commun 9(1):1753
18. Coombes D et al (2020) The basis for non-canonical ROK family function in the N-acetylmannosamine kinase from the pathogen *Staphylococcus aureus*. J Biol Chem 295(10):3301–3315
19. Davies JS et al (2019) Functional and solution structure studies of amino sugar deacetylase and deaminase enzymes from Staphylococcus aureus. FEBS Lett 593(1):52–66
20. Currie MJ et al (2021) *N*-Acetylmannosamine-6-phosphate 2-epimerase uses a novel substrate-assisted mechanism to catalyze amino sugar epimerization. J Biol Chem 297(4):101113
21. Arif T et al (2021) Synthesis of *N*-acetylmannosamine-6-phosphate derivatives to investigate the mechanism of *N*-acetylmannosamine-6-phosphate 2-epimerase. Carbohydr Res 510: 108445
22. Horne CR et al (2021) Mechanism of NanR gene repression and allosteric induction of bacterial sialic acid metabolism. Nat Commun 12(1):1988
23. Manjunath L et al (2018) Crystal structures and kinetic analyses of *N*-acetylmannosamine-6-phosphate 2-epimerases from *Fusobacterium*

nucleatum and *Vibrio cholerae*. Acta Crystallogr F Struct Biol Commun 74(Pt 7):431–440
24. Caing-Carlsson R et al (2017) Crystal structure of *N*-acetylmannosamine kinase from *Fusobacterium nucleatum*. Acta Crystallogr F Struct Biol Commun 73(Pt 6):356–362
25. North RA et al (2014) Cloning, expression, purification, crystallization and preliminary X-ray diffraction analysis of *N*-acetylmannosamine kinase from methicillin-resistant *Staphylococcus aureus*. Acta Crystallogr F Struct Biol Commun 70(Pt 5):643–649
26. North RA et al (2014) Cloning, expression, purification, crystallization and preliminary X-ray diffraction analysis of *N*-acetylmannosamine-6-phosphate 2-epimerase from methicillin-resistant *Staphylococcus aureus*. Acta Crystallogr F Struct Biol Commun 70 (Pt 5):650–655
27. North RA et al (2013) Cloning, expression, purification, crystallization and preliminary X-ray diffraction studies of *N*-acetylneuraminate lyase from methicillin-resistant *Staphylococcus aureus*. Acta Crystallogr Sect F Struct Biol Cryst Commun 69(Pt 3):306–312
28. North RA et al (2016) Structure and inhibition of *N*-acetylneuraminate lyase from methicillin-resistant *Staphylococcus aureus*. FEBS Lett 590(23):4414–4428
29. Bouchet V et al (2003) Host-derived sialic acid is incorporated into *Haemophilus influenzae* lipopolysaccharide and is a major virulence factor in experimental otitis media. Proc Natl Acad Sci USA 100(15):8898–8903
30. Bozzola T et al (2022) Sialic acid derivatives inhibit SiaT transporters and delay bacterial growth. ACS Chem Biol 17(7):1890–1900
31. Cole JL, Lary JW, Moody TP, Laue TM (2008) Analytical ultracentrifugation: sedimentation velocity and sedimentation equilibrium. Method Cell Biol 84:143–179
32. Demeler B (2010) Methods for the design and analysis of sedimentation velocity and sedimentation equilibrium experiments with proteins. Curr Protoc Protein Sci 7:7 13 1-7
33. MacGregor IK, Anderson AL, Laue TM (2004) Fluorescence detection for the XLI analytical ultracentrifuge. Biophys Chem 108(1–3):165–185
34. Zhao H, Mayer ML, Schuck P (2014) Analysis of protein interactions with picomolar binding affinity by fluorescence-detected sedimentation velocity. Anal Chem 86(6):3181–3187
35. Schuck P, To SC, Zhao H (2023) An automated interface for sedimentation velocity analysis in SEDFIT. PLoS Comput Biol 19(9):e1011454
36. Cao W, Demeler B (2005) Modeling analytical ultracentrifugation experiments with an adaptive space-time finite element solution of the Lamm equation. Biophys J 89(3):1589–1602
37. Demeler B (2005) UltraScan a comprehensive data analysis software package for analytical ultracentrifugation experiments. In: Scott DJ, Harding SE, Rowe AJ (eds) Modern analytical ultracentrifugation: techniques and methods. Royal Society of Chemistry (UK), pp 210–229
38. Brautigam CA (2015) Calculations and publication-quality illustrations for analytical ultracentrifugation data. Methods Enzymol 562:109–133
39. Ebel C (2011) Sedimentation velocity to characterize surfactants and solubilized membrane proteins. Methods 54(1):56–66
40. Le Roy A et al (2015) AUC and small-angle scattering for membrane proteins. Methods Enzymol 562:257–286
41. le Maire M et al (2008) Gel chromatography and analytical ultracentrifugation to determine the extent of detergent binding and aggregation, and Stokes radius of membrane proteins using sarcoplasmic reticulum Ca2+-ATPase as an example. Nat Protoc 3(11):1782–1795
42. Salvay AG, Santamaria M, le Maire M, Ebel C (2007) Analytical ultracentrifugation sedimentation velocity for the characterization of detergent-solubilized membrane proteins Ca++-ATPase and ExbB. J Biol Phys 33(5–6):399–419
43. Brautigam CA (2023) SViMULATE: a computer program facilitating interactive, multimode simulation of analytical ultracentrifugation data. Eur Biophys J 52(4–5):293–302
44. Demeler B (2024) Methods for the design and analysis of analytical ultracentrifugation experiments. Curr Protoc 4(2):e974

Part III

Computational and Simulation Approaches

Chapter 9

Molecular Dynamics Simulation Analyses of the Conformational Change of an *Escherichia coli* Multidrug Exporter, MdfA

Tohru Terada

Abstract

Molecular dynamics (MD) simulations are a powerful tool to investigate the dynamics and the conformational distributions of proteins. My collaborators and I applied them to study the proton transport-coupled conformational change of an *Escherichia coli* multidrug exporter, MdfA. In this chapter, I first describe the theoretical basis for studying the dynamics and the conformational distributions of a protein using MD simulations. Second, I describe the methods to construct the simulation system, in which the protein is embedded in a solvated lipid bilayer, and those employed to conduct the MD simulations and to analyze the results. Finally, I describe the problems that remain unsolved and present a plan to further study the transport mechanism of MdfA using MD methods.

Key words Molecular dynamics, Potential energy function, Newton's equations of motion, Statistical ensemble, Free energy, Conformational distribution, Conformational change

1 Introduction

An *Escherichia coli* multidrug exporter, MdfA, is a member of the Major Facilitator Superfamily (MFS) and exchanges small molecules, including antibiotics, in the cytosol with H^+ in the periplasm [1–3]. The tertiary structure of MdfA was first determined in the inward-open (or inward-facing) conformation (I_o) by Heng et al. [4], and subsequently that in the outward-open (or outward-facing) conformation (O_o) was determined by Tanabe's group (Fig. 1) [6]. Similarly with typical MFS transporters, the structure of MdfA is composed of 12 transmembrane (TM) helices (TM1–TM12) and is divided into two structural domains termed the N- and C-lobes, each of which is composed of six TM helices (N-lobe: TM1–TM6; C-lobe: TM7–TM12) [4, 6]. The structural comparison of O_o and I_o using the DynDom server [7] indicates no structural changes in N- and C-lobes and a rotational movement

Kunihiko Nishino and Mikio Tanabe (eds.), *Bacterial Multidrug Exporters: Methods and Protocols*, Methods in Molecular Biology, vol. 3044, https://doi.org/10.1007/978-1-0716-5316-6_9,
© The Author(s), under exclusive license to Springer Science+Business Media, LLC, part of Springer Nature 2026

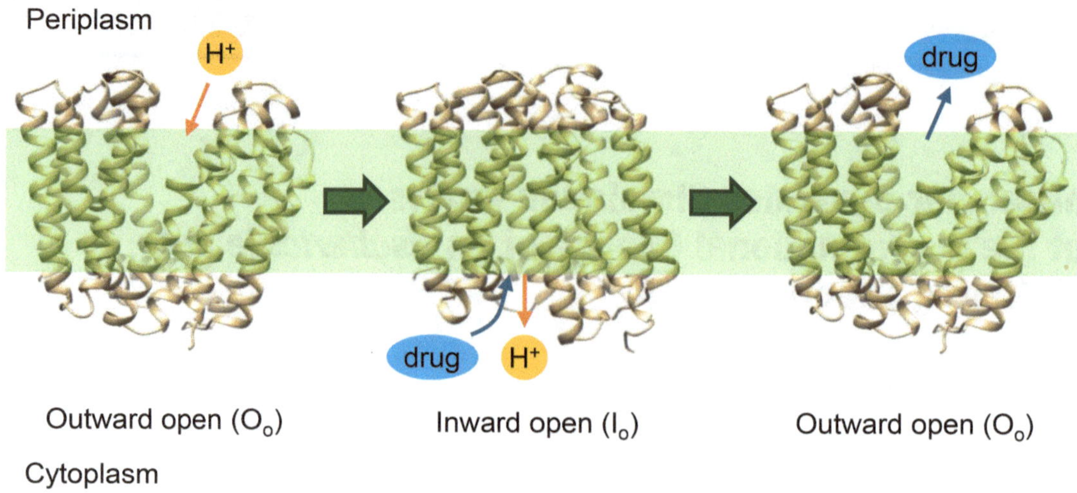

Fig. 1 The alternating-access model for substance transport by MdfA. The tertiary structures of MdfA in the outward-open (O_o) and the inward-open (I_o) conformations are shown in a ribbon representation. (All structural images were created with UCSF Chimera [5])

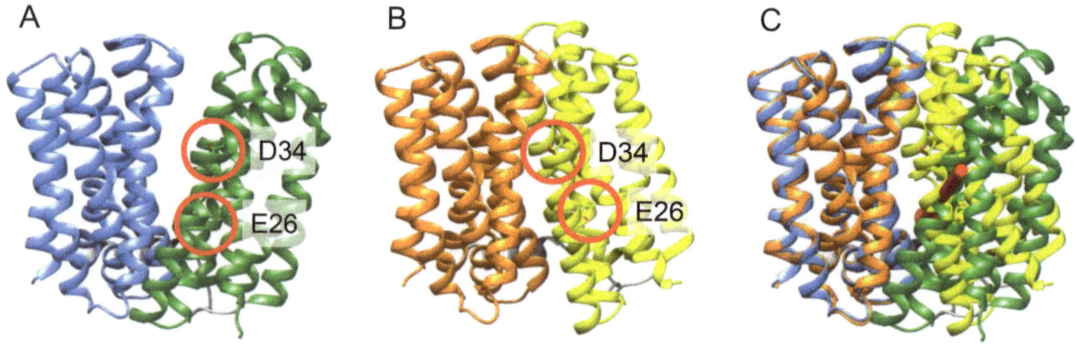

Fig. 2 Comparison between the O_o and I_o structures. (**A**) The O_o structure. The N-lobe and C-lobe are colored green and blue, respectively. E26 and D34 are shown in a stick representation and indicated by red circles. (**B**) The I_o structure. The N-lobe and C-lobe are colored yellow and orange, respectively. E26 and D34 are shown in a stick representation and indicated by red circles. (**C**) Superposition of the two structures. The rotation axis is indicated by a red arrow

of N-lobe around an axis going through the center of TM5 and the helix between the N- and C-lobes (Fig. 2). Based on these structures, the transport mechanism of MdfA is explained by the alternating-access model, as are those of the other MFS transporters, in which a transporter transports a substance by cycling between the two conformations (*see* Fig. 1) [6, 8].

The alternating-access model implies that proton binding to a protein residue induces a conformational change from O_o to I_o. There are two acidic residues, E26 and D34, in the middle of TM1 of MdfA. Because TM helices are usually composed of hydrophobic residues except at the terminals of the helices, the two residues have attracted the attention of experimental researchers. It has been shown that the neutralizing mutations of the two residues abolish the H^+/drug antiport activity [9, 10]. Therefore, the two residues have been considered as candidates for the proton-binding site important for the transport activity of MdfA.

The experiments have thus clarified the tertiary structures and the important residues for the transport activity. However, they have limitations to clarify the process of the change between O_o and I_o and the mechanism of how the protonation of the acidic residues induces the conformational change. I therefore took a theoretical approach to solve these problems. I performed molecular dynamics (MD) simulations for the O_o and the I_o structures of MdfA in the four possible protonation states of E26 and D34 and analyzed the conformational changes. As a result, my collaborators and I found that the protonation on D34 of MdfA in O_o as well as in I_o induces the transition to the occluded conformation, which has not yet been identified by the experiments. Based on these results, the process of the conformational change coupled with the proton transport is illustrated as follows (Fig. 3) [6]:

(a) MdfA waits for a proton from the periplasm in O_o.

(b) A proton binds to D34, which induces the transition to the occluded conformation.

(c) The proton is dissociated from D34 and diffuses into the cytoplasm, which induces the transition to I_o.

In this chapter, I first outline the theory of MD. Then, I describe the methods used in this study. It should be noted that the methods described here are generally applicable to study the dynamics and the conformational distributions of membrane proteins. Finally, I describe a plan to further study the transport mechanism of MdfA using MD methods.

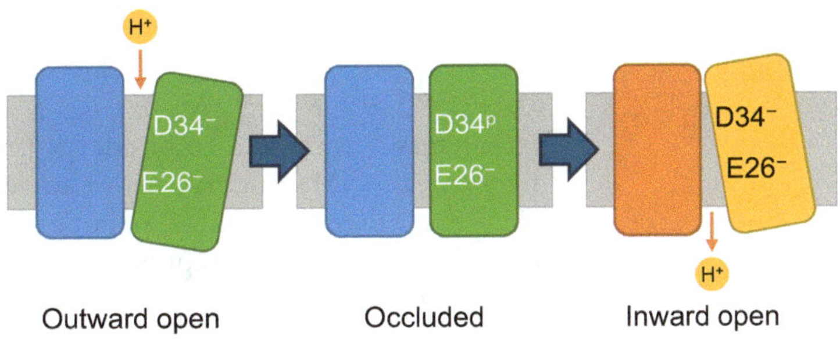

Fig. 3 The process of the proton transport-coupled conformational change of MdfA suggested by MD simulations

2 Theory

MD methods simulate the time evolution of the coordinates and the velocities of the atoms in a molecular system. In this section, I describe the theoretical basis for studying the dynamics and the conformational distributions of a protein using MD simulations.

2.1 Potential Energy Function

MD methods are based on the potential energy function and Newton's equations of motion. The potential energy function $E(\mathbf{r})$ is given by the sum of interaction energy terms [11]:

$$E(\mathbf{r}) = E_{\text{bond}}(\mathbf{r}) + E_{\text{angle}}(\mathbf{r}) + E_{\text{UB}}(\mathbf{r}) + E_{\text{dihedral}}(\mathbf{r}) + E_{\text{improper}}(\mathbf{r}) + E_{\text{cmap}}(\mathbf{r}) + E_{\text{electrostatic}}(\mathbf{r}) + E_{\text{van der Waals}}(\mathbf{r}), \tag{1}$$

where $\mathbf{r}$ represents the coordinates of the atoms. Each energy term is given by the following equations:

$$E_{\text{bond}}(\mathbf{r}) = \sum_b K_b \left(r_b - r_b^0\right)^2, \tag{2}$$

$$E_{\text{angle}}(\mathbf{r}) = \sum_a K_a \left(\theta_a - \theta_a^0\right)^2, \tag{3}$$

$$E_{\text{UB}}(\mathbf{r}) = \sum_u K_u \left(r_u - r_u^0\right)^2, \tag{4}$$

$$E_{\text{dihedral}}(\mathbf{r}) = \sum_d K_d [1 + \cos(n_d \phi_d - \delta_d)], \tag{5}$$

$$E_{\text{improper}}(\mathbf{r}) = \sum_m K_m \left(\psi_m - \psi_m^0\right)^2, \tag{6}$$

$$E_{\text{electrostatic}}(\mathbf{r}) = \sum_{j>i} \frac{q_i q_j}{r_{ij}}, \tag{7}$$

$$E_{\text{van der Waals}}(\mathbf{r}) = \sum_{j>i} \varepsilon_{ij} \left[\left(\frac{\sigma_{ij}}{r_{ij}}\right)^{12} - \left(\frac{\sigma_{ij}}{r_{ij}}\right)^{6} \right]. \tag{8}$$

K_b, r_b^0, K_a, θ_a^0, K_u, r_u^0, K_d, n_d, δ_d, K_m, ψ_m^0, q_i, ε_{ij}, and σ_{ij} are the parameters determined to fit energy profiles obtained by quantum mechanical (QM) calculations or to reproduce experimental data. They are called force-field parameters. $E_{\text{bond}}(\mathbf{r})$, $E_{\text{angle}}(\mathbf{r})$, $E_{\text{dihedral}}(\mathbf{r})$, and $E_{\text{improper}}(\mathbf{r})$ represent the bond, angle, proper dihedral angle, and improper dihedral angle interactions that depend on bond r_b, angle θ_a, proper dihedral angle ϕ_d, and improper dihedral angle ψ_m, respectively. $E_{\text{UB}}(\mathbf{r})$ represents the Urey–Bradley interaction that depends on the distance r_u between the atoms at both ends of an angle. $E_{\text{electrostatic}}(\mathbf{r})$ and $E_{\text{van der Waals}}(\mathbf{r})$ represent electrostatic and van der Waals interactions between atoms separated by distance r_{ij}. The sums are taken over all the atom pairs except those involved in the bond and the angle interactions. $E_{\text{cmap}}(\mathbf{r})$ is a correction term applied to the energy terms of backbone dihedral angles ϕ and ψ to accurately reproduce the $\phi - \psi$ potential energy map obtained by high-level QM calculations. The force-field parameters for standard amino acids, nucleotides, and common molecules, including water, ions, sugars, and lipids, have been determined and are distributed along with MD simulation software packages, such as Amber [12] and CHARMM [13]. The energy terms are slightly different between Amber and CHARMM, and those described above are those of CHARMM.

2.2 Numerical Solution of Newton's Equations of Motion

Newton's equations of motion are given by

$$m_i \frac{d^2\mathbf{r}_i}{dt^2} = \mathbf{F}_i(\mathbf{r}), \tag{9}$$

where m_i and $\mathbf{r}_i$ are the mass and the coordinates of atom i of a molecular system, and $\mathbf{F}_i(\mathbf{r})$ is the force acting on atom i and depends on the coordinates of all the atoms in the system. The force is obtained by calculating the gradient of the potential energy function:

$$\mathbf{F}_i(\mathbf{r}) = -\frac{\partial E(\mathbf{r})}{\partial \mathbf{r}_i} \tag{10}$$

By using the relation $\mathbf{v}_i = d\mathbf{r}_i/dt$, where $\mathbf{v}_i$ is the velocity of atom i, the above equations are converted into the first-order differential equations:

$$\begin{cases} \dfrac{d\mathbf{r}_i}{dt} = \mathbf{v}_i \\ m_i \dfrac{d\mathbf{v}_i}{dt} = \mathbf{F}_i(\mathbf{r}) \end{cases} \tag{11}$$

The time evolution of the coordinates and the velocities of the atoms in a molecular system can be simulated by solving these equations. However, Newton's equations of motion cannot be solved analytically for systems composed of three or more bodies. Therefore, they must be solved numerically when they are applied to a molecular system composed of three or more atoms. The velocity Verlet method is widely used to numerically solve Newton's equations of motion [14]. It approximately calculates the coordinates and the velocities at $t + \Delta t$ from those at t as follows:

1. The velocities at $t + \Delta t/2$ are calculated using the forces at t as

$$\mathbf{v}_i\left(t + \frac{\Delta t}{2}\right) = \mathbf{v}_i(t) + \frac{\mathbf{F}_i(\mathbf{r}(t))}{m_i}\frac{\Delta t}{2}. \tag{12}$$

2. The coordinates at $t + \Delta t$ are calculated using the velocities at $t + \Delta t/2$ as

$$\mathbf{r}_i(t + \Delta t) = \mathbf{r}_i + \mathbf{v}_i\left(t + \frac{\Delta t}{2}\right)\Delta t. \tag{13}$$

3. The forces at $t + \Delta t$ are calculated, and the velocities at $t + \Delta t$ are calculated as

$$\mathbf{v}_i(t + \Delta t) = \mathbf{v}_i\left(t + \frac{\Delta t}{2}\right) + \frac{\mathbf{F}_i(\mathbf{r}(t + \Delta t))}{m_i}\frac{\Delta t}{2}. \tag{14}$$

These steps are repeated until the time reaches the desired time. Because the error of the approximation is proportional to Δt^2, it is necessary to use small Δt for an accurate and stable simulation. If Newton's equations of motion were solved analytically, the total energy, which is the sum of the kinetic and the potential energies, would be conserved. When they are solved numerically, the total energy deviates from the initial value. The accuracy of the numerical solution can be measured by the degree of conservation of the total energy. Generally, values from 1/20 to 1/10 of the period of the fastest vibration are recommended for Δt to compromise the accuracy and efficiency (*see* **Note 1**).

2.3 Generation of Statistical Ensembles

In addition to tracking the time evolution of coordinates and velocities of the atoms in a molecular system, MD simulations can be used to generate statistical ensembles. In this subsection, I explain the basics of the microcanonical, canonical, and isothermal–isobaric ensembles.

2.3.1 Microcanonical Ensemble

As mentioned above, the total energy is kept nearly constant during a simulation conducted by numerically solving Newton's equations of motion. This means that the trajectory of the simulation contains only states (i.e., pairs of coordinates and velocities) having the specific total energy. The states in the trajectory are regarded as samples from an *NVE* ensemble, in which the number of atoms, the volume, and the total energy of the system are constant. This ensemble is also called the microcanonical ensemble. In this ensemble, the temperature of each sample depends on the potential energy, because the sum of the kinetic energy and the potential energy is constant.

2.3.2 Canonical Ensemble

However, in the real system, the temperature is kept constant. The statistical ensemble in thermal equilibrium with an external heat bath is called the canonical ensemble (and is also called the *NVT* ensemble). In this ensemble, the probability of existence of a state follows the canonical distribution, given by

$$p(\mathbf{r}, \mathbf{v}) = \frac{1}{Z}\exp\left\{-\frac{1}{RT}[K(\mathbf{v}) + E(\mathbf{r})]\right\}, \tag{15}$$

where R is the gas constant, T is the temperature, $K(\mathbf{v})$ is the kinetic energy given by

$$K(\mathbf{v}) = \sum_i \frac{m_i |\mathbf{v}_i|^2}{2}, \tag{16}$$

and Z is the partition function given by

$$Z = \int d\mathbf{r} d\mathbf{v} \exp\left\{ -\frac{1}{RT} [K(\mathbf{v}) + E(\mathbf{r})] \right\}. \tag{17}$$

The temperature is regulated by coupling a thermostat to the system. Consequently, a series of states following the canonical distribution is generated. Such MD simulations are called constant-temperature MD simulations. Various methods, including the Nosé–Hoover method [15], Langevin dynamics [16], and the velocity rescaling method [17], have been proposed for constant-temperature MD simulations (*see* **Note 2**).

An experimentally observed value of a real system is the weighted average of the corresponding physical quantity by the probability of existence of each state. In general, the average of an arbitrary physical quantity $A(\mathbf{r}, \mathbf{v})$ is given by

$$\langle A \rangle = \int d\mathbf{r} d\mathbf{v} A(\mathbf{r}, \mathbf{v}) p(\mathbf{r}, \mathbf{v}) = \frac{\int d\mathbf{r} d\mathbf{v} A(\mathbf{r}, \mathbf{v}) \exp\left\{ -\frac{1}{RT} [K(\mathbf{v}) + E(\mathbf{r})] \right\}}{Z}. \tag{18}$$

Because the states generated by a constant-temperature MD simulation are regarded as samples from a canonical ensemble, the average over time coincides with that given by Eq. (18), if the simulation is conducted for an infinitely long time:

$$\lim_{T \to \infty} \frac{1}{T} \sum_{t=1}^{T} A(\mathbf{r}(t), \mathbf{v}(t)) = \langle A \rangle. \tag{19}$$

Thus, the value obtained by averaging a physical quantity over the time of an MD simulation is comparable with the value observed in an experiment.

Suppose that a protein changes its conformation along a reaction coordinate $\xi(\mathbf{r})$. Let $A(\mathbf{r}) = \delta(\xi' - \xi(\mathbf{r}))$, the average gives the probability of the existence of the conformation having the reaction coordinate value of ξ'. The substitution of $A(\mathbf{r})$ in Eq. (18) gives

$$\begin{aligned} P(\xi') &= \int d\mathbf{r} d\mathbf{v} \delta(\xi' - \xi(\mathbf{r})) p(\mathbf{r}, \mathbf{v}) \\ &= \frac{\int d\mathbf{r} (\xi' - \xi(\mathbf{r})) \exp\left\{ -\frac{E(\mathbf{r})}{RT} \right\} \int d\mathbf{v} \exp\left\{ -\frac{K(\mathbf{v})}{RT} \right\}}{Z}. \end{aligned} \tag{20}$$

Similarly, the substitution of $A(\mathbf{r})$ in Eq. (19) gives

$$\lim_{T\to\infty}\frac{1}{T}\sum_{t=1}^{T}\delta(\xi' - \xi(\mathbf{r}(t))) = p(\xi'), \tag{21}$$

which corresponds to the probability of appearance of conformation ξ' during the MD simulation. The Helmholtz free energy of a system is associated with the partition function as

$$F = -RT\ln Z. \tag{22}$$

When the range of the integral of the partition function is limited to a subspace, the value obtained by substituting Z in Eq. (22) with the integral over the subspace corresponds to the free energy of the subspace. Consequently, the free energy of conformation ξ' is given by

$$F(\xi') = -RT\ln\int d\mathbf{r}(\xi' - \xi(\mathbf{r}))\exp\left\{-\frac{E(\mathbf{r})}{RT}\right\}\int d\mathbf{v}\exp\left\{-\frac{K(\mathbf{v})}{RT}\right\} = -RT\ln P(\xi') + F. \tag{23}$$

This relation is rewritten as

$$\Delta F(\xi') = F(\xi') - F = -RT\ln P(\xi'). \tag{24}$$

Thus, the relative free energy of conformation ξ' is associated with the probability of appearance of conformation ξ' in the MD simulation. This relation indicates that the conformation that frequently appears in an MD simulation is stable with a low free energy. Therefore, an MD simulation can be used to find more stable conformations and to relax or refine a structural model.

2.3.3 Isothermal–Isobaric Ensemble

In a real system, the pressure is also kept constant, along with the temperature. The corresponding statistical ensemble is the isothermal–isobaric ensemble (and is also called the *NPT* ensemble). The state in this ensemble follows the isothermal–isobaric distribution, given by

$$p(\mathbf{r}, \mathbf{v}, V) = \frac{1}{Q}\exp\left\{-\frac{1}{RT}[K(\mathbf{v}) + E(\mathbf{r}) + PV]\right\}, \tag{25}$$

where Q is the partition function given by

$$Q = \int d\mathbf{r}d\mathbf{v}dV\exp\left\{-\frac{1}{RT}[K(\mathbf{v}) + E(\mathbf{r}) + PV]\right\}. \tag{26}$$

The Gibbs free energy is associated with the isothermal–isobaric ensemble. It is calculated from the partition function by

$$G = -RT\ln Q. \tag{27}$$

The pressure regulation is usually applied in combination with the periodic boundary condition. Under the periodic boundary condition, the system is formed in a rectangular parallelepiped shape called "cell," and the cell is infinitely replicated throughout space. As a result, an atom going out from a cell through a face enters the cell through the opposite face, which ensures that the number of atoms in the cell is constant. The instantaneous pressure of a system is given by

$$P = \frac{2K(\mathbf{v})}{3V} + \frac{1}{3V}\sum_{j>i} \mathbf{r}_{ij} \cdot \mathbf{F}_{ij}, \tag{28}$$

where $\mathbf{r}_{ij} = \mathbf{r}_i - \mathbf{r}_j$ and $\mathbf{F}_{ij} = \mathbf{F}_i(\mathbf{r}) - \mathbf{F}_j(\mathbf{r})$ (*see* **Note 3**). To regulate the pressure of a system, a barostat is coupled with the system. The Parrinello–Rahman method [18, 19], the Langevin piston method [20], and the stochastic cell rescaling method [21] are used for this purpose (*see* **Note 2**). In these methods, the cell volume changes depending on the difference between the pressure of the system and the reference pressure. When the pressure is less than the reference pressure, the cell is shrunken. Conversely, when the pressure is greater than the reference pressure, the cell is enlarged. MD simulations keeping the pressure and the temperature constant are called constant-pressure, constant-temperature MD simulations. States generated by a constant-pressure, constant-temperature MD simulation are samples from the isothermal–isobaric ensemble. Equations similar to Eqs. (18)–(24) hold for the isothermal–isobaric ensemble, although the Helmholtz free energy F is replaced by the Gibbs free energy G in the equations.

3 Methods

MD simulations for MdfA were conducted under different conditions of the conformation of the initial structure (O_o or I_o) and the protonation states of E26 and D34 (protonated or deprotonated). Therefore, eight runs, i.e., $O_o(E26^-/D34^-)$, $O_o(E26^p/D34^-)$, $O_o(E26^-/D34^p)$, $O_o(E26^p/D34^p)$, $I_o(E26^-/D34^-)$, $I_o(E26^p/D34^-)$, $I_o(E26^-/D34^p)$, and $I_o(E26^p/D34^p)$ were conducted in total, where superscript—and superscript p denote the deprotonated and the protonated states, respectively. In these MD simulations, each structural model of MdfA is embedded in a solvated lipid bilayer. Its surroundings are quite different from those in the protein crystal in which the structure was determined. Moreover, E26 and/or D34 were protonated in six runs, though both of them are considered to be deprotonated under the conditions of the structure determination. Therefore, it is important to make the protein structure adapted to the simulation environment. An MD simulation conducted for this purpose is called an equilibration run. Position restraints are often imposed on the protein atoms to avoid a large deviation from the initial structure, and the restraints are gradually weakened during the equilibration run. Then, a long MD simulation called a production run is conducted without restraints. As explained in the previous section, the protein conformation will be changed to a more stable one during an MD simulation. The conformation that appears the most frequently in the MD simulation is the most stable. Therefore, I analyzed the conformational distributions in the trajectories of the production runs to find the stable conformations under the respective conditions. In the following subsections, I explain the detailed methods of the MD simulations for MdfA.

3.1 Preparation of Protein Structure

The coordinates of the structures of MdfA were obtained from the website of the Protein Data Bank (PDB). The identification codes of the structures in O_o and I_o are 6GV1 and 4ZOW, respectively. From the coordinates of 6GV1, the antibody, which binds to the cytoplasmic side of the O_o structure, and a sulfate ion were removed. From the coordinates of 4ZOW, chloramphenicol, which binds to the cavity between N- and C-lobes as a transport substrate, was removed. Note that crystallographic water molecules should not be removed. The protein coordinates registered in the PDB are determined with respect to their crystal lattices. Therefore, the orientation and the position of a protein in a lipid bilayer must be predicted based on the principle that hydrophobic residues should be embedded in a lipid bilayer. The orientations of proteins in membranes (OPM) database provides the predicted results of representative membrane proteins (*see* **Note 4**) [22]. The OPM database also provides the coordinates of a protein that are rotated and translated so that the lipid bilayer embedding the protein is centered at $Z = 0$ and parallel to the X-Y plane (Fig. 4), following the custom of the MD simulations of membrane proteins. I obtained the transformed coordinates for the PDB entry of 4ZP0, which has essentially the same structure as 4ZOW but binds a different transport substrate. The coordinates of 6GV1 and 4ZOW were rotated and translated to align the transformed coordinates of 4ZP0. The coordinates of the Cα atoms of residues 203–400 (i.e., C-lobe) of 6GV1 and those of residues 10–400 of 4ZOW were used in the structure alignments.

3.2 Embedding Protein Structure in a Lipid Bilayer

I used the CHARMM-GUI server to generate the initial coordinate file and the parameter/topology files necessary to conduct MD simulations (*see* **Note 5**) [23]. To embed a protein structure in a lipid bilayer, the user chooses "Bilayer Builder" in "Membrane Builder" from the menu of "Input generator" [24]. Using the Bilayer Builder, a system is constructed in the following order:

(a) Upload a protein structure file in the PDB format.

(b) Modify the structure.

(c) Determine the system size.

(d) Select lipids.

(e) Add ions.

(f) Download the generated files.

In **step 2**, the user can cap the terminals, change the protonation states, add disulfide bonds, replace amino acids, and add

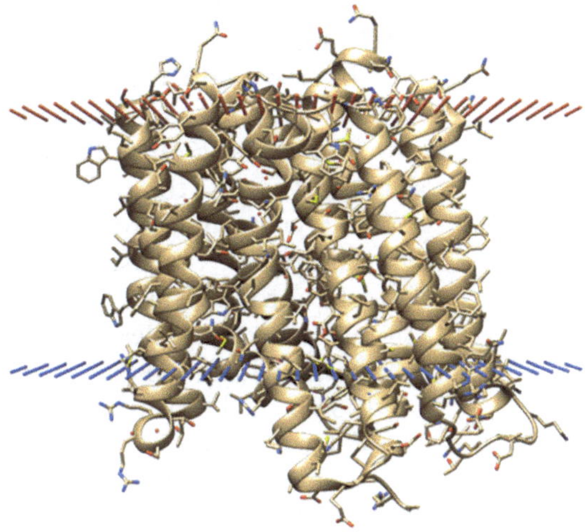

Fig. 4 The coordinates downloaded from the OPM database for 4ZPO. The backbone is shown in a ribbon representation, and the atoms are shown in a stick representation. The hydrocarbon chains of the lipid bilayer are located in the region between red and blue points. The red and blue points indicate the boundaries with the periplasm and the cytoplasm, respectively

posttranslational modifications. If the protein binds small molecules or contains non-standard amino acids, their parameter and topology files are uploaded in this step. In the MD simulations of MdfA, the N- and C-terminals were capped with an acetyl and an *N*-methyl group, respectively. Because Gln131 was replaced with Arg in 4ZOW [4], the Arg residue was replaced with the wild-type residue, Gln.

In **step 3**, the user gives the size of the system. For water thickness, the default value of 22.5 Å is appropriate for most cases. In the CHARMM-GUI server, the X length of the system is set to the same value as the Y length. The user determines the length, referring to the X extent and the Y extent of the protein shown on the web page. From these values, the widths of the protein along the X and the Y axes are calculated. I think that the larger width plus 32 Å is appropriate for the X and the Y lengths of the system (*see* **Note 6**). In the case of 4ZOW, the X extent is −28 Å to 30 Å, and the Y extent is −29 Å to 26 Å. Because the larger width is 58 Å, the X and the Y lengths of the system should be 90 Å or larger.

In **step 4**, the user specifies the composition of the lipids in the lipid bilayer. When the composition of the lipids in the lipid bilayer where the protein functions is known, it should be used. Although the lipid composition is not known for MdfA, I used 1-palmitoyl-2-oleoyl-*sn*-glycero-3-phosphatidylethanolamine (POPE) for the lipid, because phosphatidylethanolamine amounts to 70% of the lipids in the cell membrane of *Escherichia coli*, and the combination of palmitic acid (16:0) and oleic acid (18:1) is frequently used in the MD simulations of membrane proteins (*see* **Note 7**).

In **step 5**, the user specifies the types and the concentration of ions to be added to the system. The default choice of K^+ and Cl^- for the ions and the default concentration of 0.15 M are appropriate. The numbers of the cation and the anion are adjusted so that the net charge of the system is zero (*see* **Note 8**).

In **step 6**, the user can download an archive containing the initial coordinate file, the parameter/topology files, and the MD simulation setting files. The server can generate these files in several formats, including those of Amber, CHARMM, Gromacs, NAMD, and so on [25]. The user can readily start the MD simulations using these files.

The system for the I_O structure was composed of one protein, 233 POPE, 41 or 42 K^+, 47 or 48 Cl^-, and about 14,700 water molecules. The system size was 92.01 Å × 92.01 Å × 98.11 Å. The system for the O_O structure was constructed in essentially the same manner. It was composed of one protein, 223 POPE, 40 or 41 K^+, 44 or 45 Cl^-, and about 14,000 water molecules. The system size was 90.76 Å × 90.76 Å × 96.98 Å (Fig. 5). The CHARMM36 force-field parameters [26] were used for protein, lipid, and ions, and the TIP3P model [27] was used for water (*see* **Note 9**).

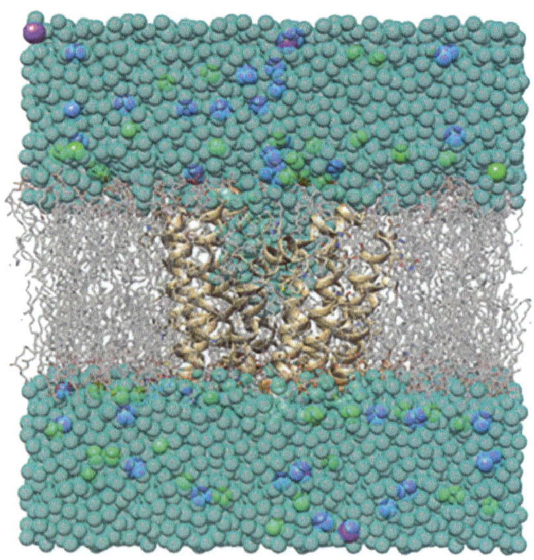

Fig. 5 Initial coordinates for the O_O(E26$^-$/D34$^-$) run. The protein backbone is shown in a ribbon representation. Protein and lipid atoms are shown in a stick representation. Oxygen atoms of water molecules and potassium and chloride ions are shown as spheres. The backbone ribbon and carbon atoms of the protein are colored brown, and the carbon atoms of lipids are colored gray. Oxygen atoms of water molecules are colored cyan, and those of protein and lipids are colored red. Nitrogen, phosphorus, and sulfur atoms are colored blue, orange, and yellow, respectively. Potassium and chloride ions are colored violet and green, respectively

3.3 MD Simulations

MD simulations were conducted using the setting files generated by the CHARMM-GUI server. Each system was first energy-minimized and was equilibrated at the temperature of 303.15 K and the pressure of 1×10^5 Pa. The production runs were performed for 1.6 μs or 3.2 μs for the MD simulations starting from the O_o structure and 1.0 μs for the MD simulations starting from the I_o structure. The Nosé–Hoover and the Parrilello–Rahman methods were used to regulate the temperature and the pressure, respectively. To conduct the MD simulation efficiently, the LINCS algorithm [28, 29] was used to constrain the bond lengths involving hydrogen atoms, which allowed the use of the time step Δt of 2 fs (*see* **Note 1**). Electrostatic interactions were calculated with the particle mesh Ewald method (*see* **Note 8**) [30, 31]. All the MD simulations were conducted with Gromacs 5.0.5 [32].

3.4 Analyses of Trajectories

An MD simulation saves the coordinates in a trajectory file at regular intervals. In the MD simulations of MdfA, I saved the coordinates every 10 ps. Therefore, the trajectory file of the 1-μs simulation contains 100,000 snapshots of the coordinates. I first analyzed the time evolutions of the deviations from the reference structures during the MD simulations (*see* **Note 10**). The deviations are measured by the root mean square deviations (RMSDs) given by

$$\mathrm{RMSD}(t) = \sqrt{\frac{1}{N}\sum_{i=1}^{N}\left|\mathbf{U}\mathbf{r}_i(t) + \mathbf{d} - \mathbf{r}_i^0\right|^2}, \qquad (29)$$

where $\mathbf{r}_i(t)$ is the coordinate of the i-th atom at time t in the N atoms used for the calculation of RMSDs, $\mathbf{r}_i^0$ is the coordinate of the corresponding atom in the reference structure, and the rotation matrix $\mathbf{U}$ and the translation vector $\mathbf{d}$ are determined to minimize RMSD(t). The Cα atoms were used to calculate the RMSDs, and the crystal structures in O_o and I_o were used as the reference structures. Figure 6 shows the plots of the time evolutions of the RMSDs. When the MD simulations started from the O_o structure, RMSDs from the O_o crystal structure are usually smaller than those from the I_o crystal structure. However, the plot for the O_o(E26$^-$/D34^P) run shows that the RMSDs from the I_o crystal structure became smaller than those from the O_o crystal structure in a short time, indicating that a conformational transition occurred upon protonation of D34. To investigate it more precisely, I calculated the conformational distributions as a function of the minimum distance between the Cα atoms in the periplasmic halves of TM5 and TM8, d_1, and the minimum distance between the Cα atoms in the cytoplasmic halves of TM5 and TM8, d_2, which were used as the reaction coordinates describing the transition between O_o and I_o (Fig. 7). As shown in Fig. 8, the conformation with small

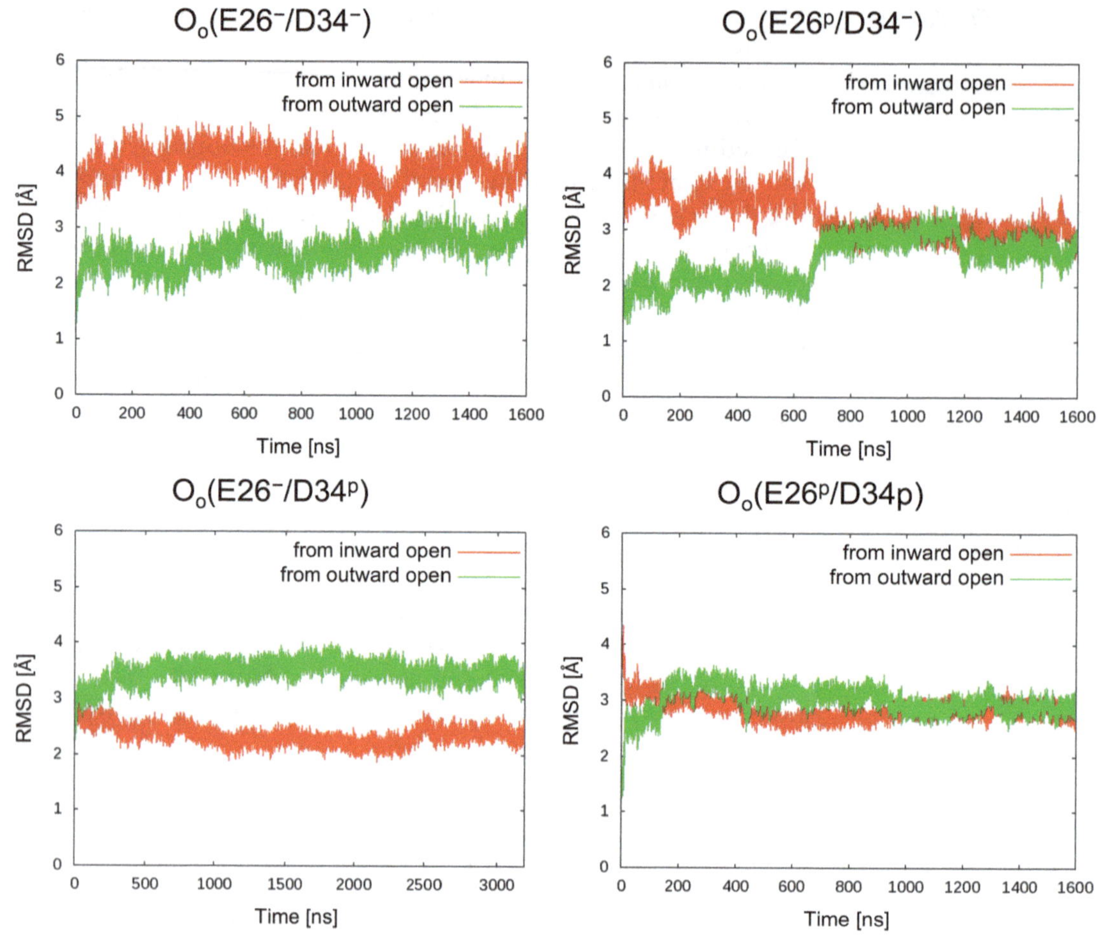

Fig. 6 Plots of the time evolutions of RMSDs calculated for the MD simulations starting from the O_o structure. Red and green lines indicate RMSDs from the I_o and O_o crystal structures

d_1 and d_2 is highly populated when D34 is protonated, irrespective of the initial conformation. The final structures of the $O_o(E26^-/D34^p)$ and the $I_o(E26^-/D34^p)$ runs closely agree with each other (Fig. 9), indicating that the conformations of the two runs converged to the identical conformation. Because both the periplasmic and the cytoplasmic sides are closed in this conformation, we termed it "the occluded conformation." Based on these results, we concluded that the protonation of D34 induces the transition from O_o to the occluded conformation [6].

In the MD simulation for $I_o(E26^-/D34^p)$, the transition to the occluded conformation was also observed (*see* Fig. 8). However, the proposed transport mechanism (*see* Fig. 1) demands the transition from the occluded conformation to I_o. Therefore, we speculated that the deprotonation of D34 would induce this transition. This speculation was supported by the flat free energy profile between the occluded conformation and I_o calculated using the

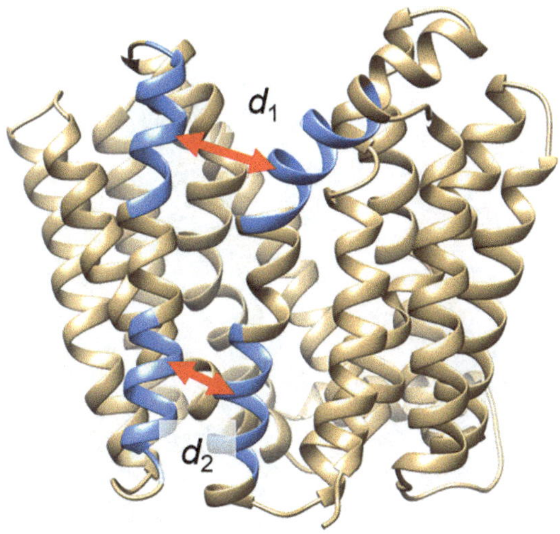

Fig. 7 Definition of distances d_1 and d_2. d_1 is the minimum distance between the Cα atoms of the residues in the periplasmic halves of TM5 and TM8. d_2 is the minimum distance between the Cα atoms of the residues in the cytoplasmic halves of TM5 and TM8. The residues used for the distance calculations are colored blue

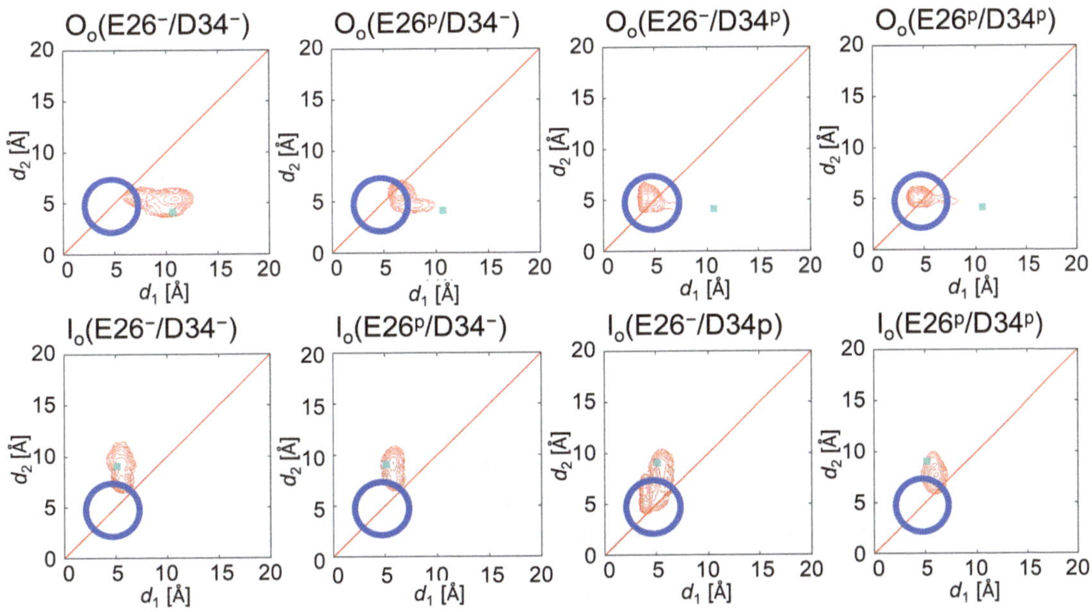

Fig. 8 Plots of conformational distributions as a function of d_1 and d_2 calculated from the trajectories of the eight MD simulation runs. A blue circle is placed at the same position in each plot. Cyan squares indicate the corresponding distances in the initial structures

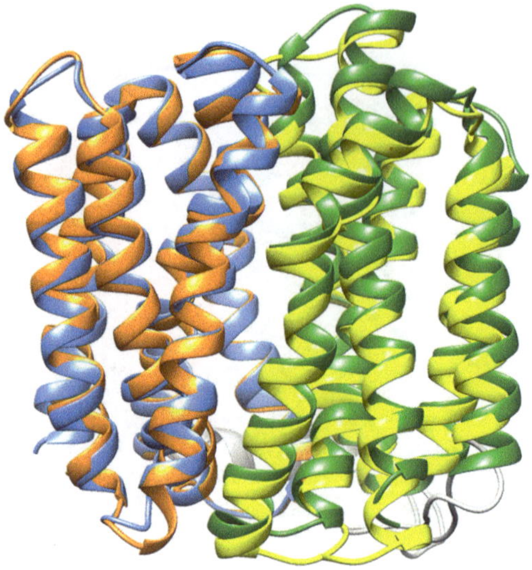

Fig. 9 Superposition of the final structures of the $O_o(E26^-/D34^p)$ and the $I_o(E26^-/D34^p)$ runs. N-lobe and C-lobe of the structure of the $O_o(E26^-/D34^p)$ run are colored green and blue, respectively. The N-lobe and C-lobe of the structure of the $I_o(E26^-/D34^p)$ run are colored yellow and orange, respectively

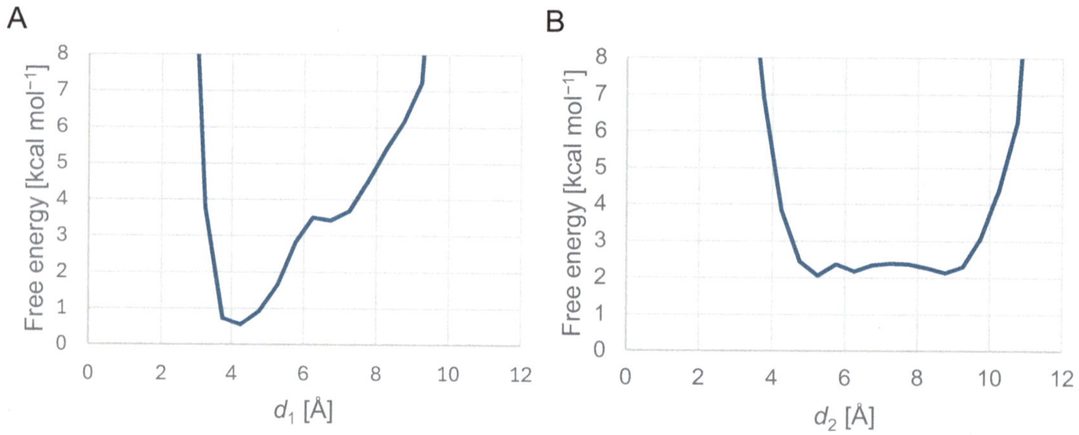

Fig. 10 Free energy profiles as a function of d_1 in the $O_0(E26^-/D34^p)$ MD run (**A**) and d_2 in the $I_0(E26^-/D34^p)$ MD run (**B**). In the O_o crystal structure, d_1 and d_2 are 10.64 Å and 4.15 Å, respectively. In the I_o crystal structure, d_1 and d_2 are 5.12 Å and 9.07 Å, respectively

trajectory of the $I_o(E26^-/D34^p)$ run, suggesting that the transition from the occluded conformation to I_o is similarly possible to the transition from I_o to the occluded conformation (Fig. 10) (*see* **Note 11**). However, the transition from the occluded conformation to I_o was not observed in an MD simulation that started from the occluded conformation after D34 was deprotonated. Therefore, the transition from the occluded conformation to I_o occurs more slowly than the time scale of the MD simulation. Enhanced sampling methods are useful to study such slow processes. I am now applying one of these methods, the replica exchange umbrella sampling method [33], to clarify the mechanism of the transition from the occluded conformation to I_o as well as that of the drug export.

4 Notes

1. The fastest motion in a protein is the stretching vibration of the bonds involving hydrogen atoms. The period of the vibration is about 10 fs. Therefore, 0.5–1.0 fs is recommended for Δt. The LINCS algorithm [28, 29] constrains the bond lengths involving hydrogen atoms. The second fastest motion in a protein is the stretching vibration of the C–C bonds, etc. Because the period of the vibration of these bonds is about 20 fs, 2.0 fs can be used for Δt.
2. Berendsen's weak coupling methods [34] are also used to regulate the temperature and the pressure, especially in equilibration runs, because of their stability, though they do not produce states following the canonical or the isothermal–isobaric distribution.
3. Because the instantaneous pressure P shows large fluctuations during MD simulations, P sometimes takes negative values.
4. The URL of the OPM database is https://opm.phar.umich.edu/. When the protein of interest is not registered in the OPM database, the orientation and the position of the protein in a lipid bilayer can be predicted with the PPM server at https://opm.phar.umich.edu/ppm_server
5. The URL of the CHARMM-GUI server is https://charmm-gui.org/. Academic and government users can use the server for free, but commercial users must contact the developer before they use it.
6. The buffer size of 32 Å corresponds to four lipid layers, because the area occupied by a lipid molecule is approximated by a square of 8 Å × 8 Å. When the protein is at the center of the system, there are two lipid layers between the protein and the boundary of the system.
7. For proteins from eukaryotes, 1-palmitoyl-2-oleoyl-*sn*-glycero-3-phosphatidylcholine (POPC) is used.
8. This requirement (the net charge of the system is zero) is imposed when the electrostatic interactions are calculated without a cutoff by using the particle mesh Ewald method [30, 31].
9. Use of the CHARMM36m force-field parameters [35] is now recommended.

10. The gmx rms program of Gromacs or the cpptraj program of Amber can be used to calculate RMSDs from a reference structure.

11. Free energy profiles were calculated by using Eq. (24). The free energy profile between O_o and the occluded conformation, calculated using the trajectory of the $O_o(E26^-/D34^P)$ run, shows a steep slope from O_o to the occluded conformation. This indicates that the transition from O_o to the occluded conformation occurs quickly, but the transition from the occluded conformation to O_o does not occur.

Conflicts of Interest

The authors have no conflicts of interest to declare that are relevant to the content of this chapter.

References

1. Edgar R, Bibi E (1997) MdfA, an Escherichia coli multidrug resistance protein with an extraordinarily broad spectrum of drug recognition. J Bacteriol 179(7):2274–2280. https://doi.org/10.1128/jb.179.7.2274-2280.1997
2. Sigal N, Cohen-Karni D, Siemion S, Bibi E (2006) MdfA from *Escherichia coli*, a model protein for studying secondary multidrug transport. J Mol Microbiol Biotechnol 11(6): 308–317. https://doi.org/10.1159/000095633
3. Yardeni EH, Zomot E, Bibi E (2018) The fascinating but mysterious mechanistic aspects of multidrug transport by MdfA from Escherichia coli. Res Microbiol 169(7):455–460. https://doi.org/10.1016/j.resmic.2017.09.004
4. Heng J et al (2015) Substrate-bound structure of the E. coli multidrug resistance transporter MdfA. Cell Res 25(9):1060–1073. https://doi.org/10.1038/cr.2015.94
5. Pettersen EF et al (2004) UCSF Chimera–a visualization system for exploratory research and analysis. J Comput Chem 25(13): 1605–1612. https://doi.org/10.1002/jcc.20084
6. Nagarathinam K et al (2018) Outward open conformation of a major facilitator superfamily multidrug/H+ antiporter provides insights into switching mechanism. Nat Commun 9(1):4005. https://doi.org/10.1038/s41467-018-06306-x
7. Veevers R, Hayward S (2019) Methodological improvements for the analysis of domain movements in large biomolecular complexes. Biophys Physicobiol 16:328–336. https://doi.org/10.2142/biophysico.16.0_328
8. Drew D, North RA, Nagarathinam K, Tanabe M (2021) Structures and general transport mechanisms by the major facilitator superfamily (MFS). Chem Rev 121(9):5289–5335. https://doi.org/10.1021/acs.chemrev.0c00983
9. Fluman N, Ryan Christopher M, Whitelegge Julian P, Bibi E (2012) Dissection of mechanistic principles of a secondary multidrug efflux protein. Mol Cell 47(5):777–787. https://doi.org/10.1016/j.molcel.2012.06.018
10. Sigal N, Fluman N, Siemion S, Bibi E (2009) The secondary multidrug/proton antiporter MdfA tolerates displacements of an essential negatively charged side chain*. J Biol Chem 284(11):6966–6971. https://doi.org/10.1074/jbc.M808877200
11. Mackerell AD (2004) Empirical force fields for biological macromolecules: overview and issues. J Comput Chem 25(13):1584–1604. https://doi.org/10.1002/jcc.20082
12. Case DA et al (2005) The Amber biomolecular simulation programs. J Comput Chem 26(16): 1668–1688. https://doi.org/10.1002/jcc.20290
13. Brooks BR et al (2009) CHARMM: the biomolecular simulation program. J Comput Chem 30(10):1545–1614. https://doi.org/10.1002/jcc.21287

14. Swope WC, Andersen HC, Berens PH, Wilson KR (1982) A computer simulation method for the calculation of equilibrium constants for the formation of physical clusters of molecules: application to small water clusters. J Chem Phys 76(1):637–649. https://doi.org/10.1063/1.442716
15. Hoover WG (1985) Canonical dynamics: equilibrium phase-space distributions. Phys Rev A 31(3):1695–1697. https://doi.org/10.1103/PhysRevA.31.1695
16. Pastor RW, Brooks BR, Szabo A (1988) An analysis of the accuracy of Langevin and molecular dynamics algorithms. Mol Phys 65(6): 1409–1419. https://doi.org/10.1080/00268978800101881
17. Bussi G, Donadio D, Parrinello M (2007) Canonical sampling through velocity rescaling. J Chem Phys 126(1). https://doi.org/10.1063/1.2408420
18. Parrinello M, Rahman A (1981) Polymorphic transitions in single crystals: a new molecular dynamics method. J Appl Phys 52(12): 7182–7190. https://doi.org/10.1063/1.328693
19. Nose S, Klein ML (1983) Constant pressure molecular dynamics for molecular systems. Mol Phys 50(5):1055–1076. https://doi.org/10.1080/00268978300102851
20. Feller SE, Zhang YH, Pastor RW, Brooks BR (1995) Constant pressure molecular dynamics simulation: the Langevin piston method. J Chem Phys 103(11):4613–4621. https://doi.org/10.1063/1.470648
21. Bernetti M, Bussi G (2020) Pressure control using stochastic cell rescaling. J Chem Phys 153(11):114107. https://doi.org/10.1063/5.0020514
22. Lomize MA et al (2012) OPM database and PPM web server: resources for positioning of proteins in membranes. Nucleic Acids Res 40 (D1):D370–D376. https://doi.org/10.1093/nar/gkr703
23. Jo S, Kim T, Iyer VG, Im W (2008) CHARMM-GUI: a web-based graphical user interface for CHARMM. J Comput Chem 29(11):1859–1865. https://doi.org/10.1002/jcc.20945
24. Wu EL et al (2014) CHARMM-GUI membrane builder toward realistic biological membrane simulations. J Comput Chem 35(27): 1997–2004. https://doi.org/10.1002/jcc.23702
25. Lee J et al (2016) CHARMM-GUI input generator for NAMD, GROMACS, AMBER, OpenMM, and CHARMM/OpenMM simulations using the CHARMM36 additive force field. J Chem Theory Comput 12(1): 405–413. https://doi.org/10.1021/acs.jctc.5b00935
26. Best RB et al (2012) Optimization of the additive CHARMM all-atom protein force field targeting improved sampling of the backbone ϕ, ψ and side-chain χ_1 and χ_2 dihedral angles. J Chem Theory Comput 8(9):3257–3273. https://doi.org/10.1021/ct300400x
27. Jorgensen WL et al (1983) Comparison of simple potential functions for simulating liquid water. J Chem Phys 79(2):926–935. https://doi.org/10.1063/1.445869
28. Hess B, Bekker H, Berendsen HJC, Fraaije J (1997) LINCS: a linear constraint solver for molecular simulations. J Comput Chem 18(12):1463–1472. https://doi.org/10.1002/(sici)1096-987x(199709)18:12<1463::aid-jcc4>3.3.co;2-l
29. Hess B (2008) P-LINCS: a parallel linear constraint solver for molecular simulation. J Chem Theory Comput 4(1):116–122. https://doi.org/10.1021/ct700200b
30. Darden T, York D, Pedersen L (1993) Particle mesh Ewald: an N·log(N) method for Ewald sums in large systems. J Chem Phys 98(12): 10089–10092. https://doi.org/10.1063/1.464397
31. Essmann U et al (1995) A smooth particle mesh Ewald method. J Chem Phys 103(19): 8577–8593. https://doi.org/10.1063/1.470117
32. Hess B, Kutzner C, van der Spoel D, Lindahl E (2008) GROMACS 4: algorithms for highly efficient, load-balanced, and scalable molecular simulation. J Chem Theory Comput 4(3): 435–447. https://doi.org/10.1021/ct700301q
33. Sugita Y, Kitao A, Okamoto Y (2000) Multidimensional replica-exchange method for free-energy calculations. J Chem Phys 113(15): 6042–6051. https://doi.org/10.1063/1.1308516
34. Berendsen HJC et al (1984) Molecular dynamics with coupling to an external bath. J Chem Phys 81(8):3684–3690. https://doi.org/10.1063/1.448118
35. Huang J et al (2017) CHARMM36m: an improved force field for folded and intrinsically disordered proteins. Nat Methods 14(1): 71–73. https://doi.org/10.1038/nmeth.4067

Chapter 10

Computer Simulations of Multidrug Efflux Pumps of the Resistance–Nodulation–Cell Division Superfamily

Mohd Athar, Silvia Gervasoni, Giuliano Malloci, Paolo Ruggerone, and Attilio V. Vargiu

Abstract

Bacterial efflux pumps of the resistance–nodulation–cell division (RND) superfamily are major contributors to intrinsic and acquired multidrug resistance in Gram-negative bacteria. Structural and biochemical studies on AcrAB-TolC (*E. coli*) and MexAB-OprM (*P. aeruginosa*) pumps performed in the past decade have elucidated the molecular basis of allosteric transport by these complex machineries. Alongside these efforts, studies have also characterized the broad substrate specificity (polyspecificity) and molecular determinants that dictate the fate of substrates, inhibitors, and "avoiders" (molecules not actively transported) of RND transporters. Computational approaches have contributed to elucidating the structural and dynamic details of the functioning mechanisms of these transporters, including the interactions with substrates and inhibitors. In this chapter, we will discuss and illustrate several protocols developed in our lab on molecular docking, homology modeling, all-atom molecular dynamics simulations, and binding free energy estimation that have provided valuable insights into substrate recognition, extrusion mechanisms, optimization of efflux avoidance, and potential inhibition strategies.

Key words Efflux pumps, Gram-negative bacteria, Membrane proteins, RND transporters, MD simulations, Molecular docking, Homology modeling, Free energy calculations, MM/GBSA, Fragment-based drug design

1 Introduction

The increasing tolerance of bacteria to multiple antibiotics, known as multidrug resistance, is a significant threat to public health [1] and is recognized as a priority and not a deferrable task by the scientific community [2–5]. Gram-negative bacteria are a major concern as they encompass most clinical strains exhibiting multidrug resistance, including extreme or total resistance, with few or

Scripts/Tutorials: https://doi.org/10.5281/zenodo.16720686

Mohd Athar and Silvia Gervasoni contributed equally with all other contributors.

Kunihiko Nishino and Mikio Tanabe (eds.), *Bacterial Multidrug Exporters: Methods and Protocols*, Methods in Molecular Biology, vol. 3044, https://doi.org/10.1007/978-1-0716-5316-6_10,
© The Author(s), under exclusive license to Springer Science+Business Media, LLC, part of Springer Nature 2026

no effective antibiotics available [6, 7]. Mechanisms contributing to this phenomenon, such as the action of multidrug efflux pumps of the resistance–nodulation–cell division (RND) superfamily, play a crucial role in both intrinsic and acquired resistance among Gram-negative pathogens [8–16].

These efflux pumps are intricate assemblies comprised at least three units (Fig. 1a), spanning the entire periplasmic space [15, 17–26]. They consist of: (i) a RND transporter embedded in the inner membrane, acting as the engine of the pump [27–34] and responsible for drug recognition and selectivity [11, 35–44]; (ii) a channel protein embedded in the outer membrane (Outer Membrane Factor, OMF), facilitating the passage of toxic molecules into the outer medium [45–49], and (iii) an assembly of membrane fusion proteins (MFPs) that bridge the aforementioned components, enabling trans-envelope efflux of noxious compounds [25, 48, 50–56]. A fourth component, named AcrZ, was identified in the AcrAB-TolC efflux pump [20, 57, 58]. Figure 1 shows the general scheme of an RND efflux pump.

Over the past years, structural biologists have been actively elucidating key properties of the individual components of RND efflux pumps. As a result, a significant number of structures have been determined for the major efflux pumps found in *Escherichia coli* (AcrAB(Z)-TolC), *Pseudomonas aeruginosa* (MexAB-OprM), and *Acinetobacter baumannii* (AdeABC). In these systems, AcrB, MexB, and AdeB serve as the transporters, AcrA, MexA, and AdeA function as the MFPs, and TolC, OprM, and AdeC act as the OMFs [20, 22, 28–32, 46, 51–53, 59, 60, 62–87].

The availability of structural information, coupled with advances in simulation techniques [61, 88–100] and increased computational power [101–104] has fostered numerous computational studies aimed at understanding the mechanistic aspects of compounds' extrusion by these pumps. Molecular Dynamics (MD) simulations of such large complexes are challenging from a computational point of view. One strategy to handle these huge complexes is to use simplified descriptions of the system, such as coarse-grained methods [105–108]. However, given the substantial size of these molecular complexes, most in silico investigations have been limited to studying only individual components [42, 61, 109–131]. Since RND transporters act as the engine of these pumps, it is not surprising that a significant portion of studies have focused on the functional dynamics of these proteins and their detailed interactions with substrates and inhibitors. Among these, AcrB of *E. coli* has been extensively studied, as it can recognize a vast array of unrelated compounds ranging from organic solvents to high molecular-mass antibiotics and detergents. For this reason, AcrB has been serving as a paradigm system for multidrug transporters [25, 33, 82].

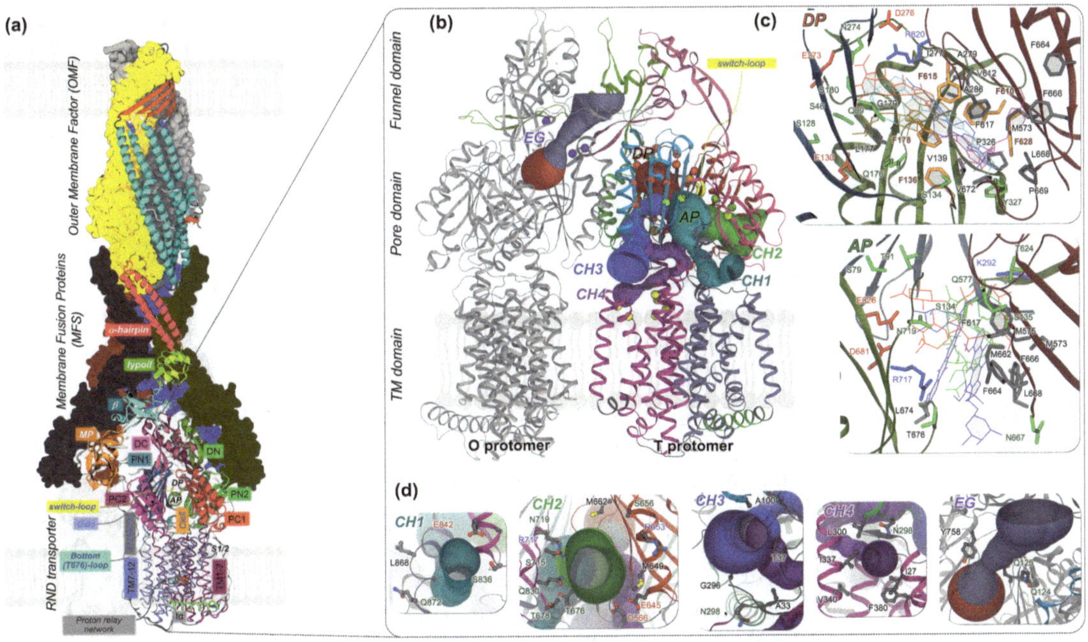

Fig. 1 (**a**) Structure of a fully assembled RND efflux pump. AcrAB-TolC of *E. coli* is shown as an exemplification (PDB ID: 5O66 [59]). For each component, one monomer is shown in cartoons and the others using molecular surfaces of different colors (one AcrA monomer is transparent, while the L protomer of AcrB is hidden for the sake of clarity). AcrB and AcrA domains are indicated by black and white bold italic labels, respectively. AcrB transmembrane and periplasmic domains are shown in ribbons and cartoons, respectively. Multifunctional sites are shown as semitransparent colored spheres and black italic labels. Additional key elements are indicated by colored arrows. (**b**) Detailed view of the RND transporter AcrB of *E. coli* (PDB ID: 4DX5) [60], with an indication of four putatively functional entry channels (CH1–CH4, shown as solid channels colored differently within the T protomer) and of the exit gate (EG) opening toward the Funnel Domain in the O protomer. Cα atoms of residues lining the multifunctional sites AP, DP (HT), and S1/2 are shown as green, red (orange), and yellow spheres, respectively. (**c**) Residues lining the DP and the AP are shown as sticks colored by residue type (apolar, polar, negatively charged, and positively charged residues in gray, green, red, and blue, respectively). Phenylalanines lining the HT are colored orange. Selected antibiotics and inhibitors co-crystallized within each site are also shown as sticks colored by molecule. (**d**) Residues lining the entrances of transport channels CH1–4 and the bottleneck of the EG are shown as sticks colored by atom type. (Adapted from [61])

This chapter outlines the computational methods adopted in our lab for studying RND efflux pumps, with a focus on AcrB and homologous transporters. We describe the primary tools and protocols employed to model the features of these proteins, including homology modeling, molecular docking, and the insertion of the protein–ligand complexes into a lipid membrane environment, ensuring a realistic representation of the physiological conditions. Then, we consider the equilibration of the system followed by MD simulations coupled with estimation of the binding free energy, fragment-based mapping, and analysis of receptor–ligand contacts. The combined use of these computational techniques provides molecular insights into the structural and functional aspects of AcrB and other RND transporters, which are crucial for understanding their roles in multidrug resistance and for designing effective therapeutic strategies.

2 Theory

In this section, we briefly discuss the theoretical background of the computational methods used to model and simulate bacterial RND transporters.

2.1 Homology Modeling

Homology modeling (*a.k.a.* comparative modeling) has long been a key computational approach for predicting protein structures in the absence of experimental determination [132–134]. The method relies on the principle that proteins sharing significant amino acid sequence similarity are likely to possess similar structures and functions. Thus, the initial step in any molecular modeling calculation involves a reasonably accurate starting structure of a homologous protein of the target under investigation. Utilizing an available template structure (homolog), homology modeling predicts the three-dimensional arrangement of a target protein. For effective prediction, a high degree of sequence similarity between the target (new protein sequences without experimental structure) and the template (homolog) is essential. Several new methods have emerged in protein structure prediction, including AlphaFold by Google DeepMind and Neuralplexer by Iambic Therapeutics [135]. These methods leverage artificial intelligence-based methods to accelerate the process, relying on a priori information (existing knowledge) to make predictions [136, 137].

Among other widely used standalone popular homology modeling tools, MODELLER employs a unique strategy to generate the target protein's structure by computing positional restraints derived from the template's 3D structure [138, 139]. These restraints ensure that the spatial arrangement of atoms between corresponding residues in the target and template remains consistent. Additionally, stereochemical restraints, such as bond lengths, angles, and non-bonded atom contacts, obtained from molecular mechanics force fields [139–141] are incorporated to refine the models. Subsequently, models are generated by minimizing violations of these restraints. In Subheading 4, we will describe how to use MODELLER to build a 3D model for an RND transporter.

2.2 Molecular Docking

Molecular docking is a computational technique that allows to (i) predict the bound conformation (pose) of a (macro)molecule onto another macromolecule and (ii) the binding affinity of the corresponding complex, starting from their unbound conformations. This chapter specifically explores the binding between a small

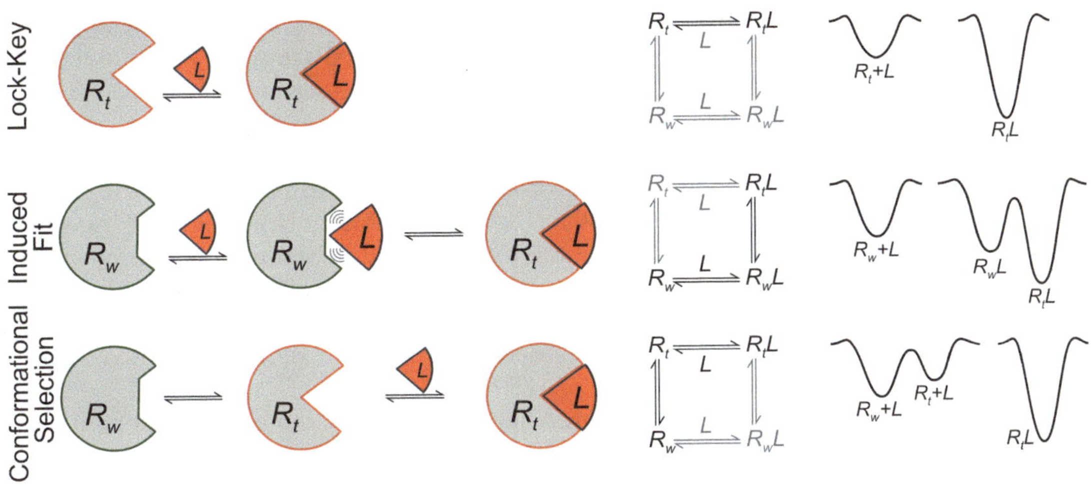

Fig. 2 Schematic illustrations of the three protein–ligand binding models: (upper row) Lock-and-key; (middle row) Induced fit; and (lower row) Conformational selection. (Reproduced with permission from [161])

molecule and a macromolecular target (RND transporter, e.g, AcrB). The orientation and interaction between a small molecule (ligand) and a receptor is primarily driven by non-covalent interactions [142]. The subject of molecular docking is particularly relevant due to its applications in medicinal chemistry, such as computer-aided-drug-design, virtual screening of vast chemical libraries for potential drug candidates (hit identification), and lead optimization [143–145]. The traditional lock-and-key model, where the receptor acts as a lock and the ligand as a key, has been the foundation of molecular docking. However, experimental observations suggest that both ligand and receptor can undergo conformational changes upon binding. This has led to the evolution of the lock-and-key model to the more accurate "induced-fit" and "conformational selection" models (Fig. 2). Modern docking software attempts to account for this flexibility to achieve more realistic predictions of protein–ligand interactions [146].

Ideally, a docking simulation should consider the flexibility of both the ligand and the protein. However, a fully flexible approach is computationally expensive and often impractical for large-scale studies. Therefore, most docking simulations employ a semi-flexible approach, where the protein structure is typically kept rigid, or only specific parts of the active site are allowed to rotate [147]. Ligand flexibility is often addressed by generating multiple conformations either prior to docking, using conformational sampling or MD, or dynamically on-the fly during the docking process.

While semi-flexible docking offers a balance between accuracy and computational efficiency, achieving a more rigorous sampling of protein conformations is desired. One alternative approach to achieve this involves utilizing pre-generated ensembles of protein structures [148, 149]. These ensembles can be obtained from methods like conventional or enhanced-sampling MD simulations [150, 151], or by other techniques such as generative machine learning (ML) based methods like Boltz [137, 152, 153]. Alternatively, external servers like CABS-flex [154] can be employed to explore the protein's conformational flexibility. This strategy has a lower computational cost compared to fully flexible docking simulations, while potentially improving the binding mode predictions. Similarly, conformational flexibility of the ligand can be accounted by pre-generating of ensembles through MD simulations [155, 156] or using dedicated software such as RDKit or OpenEye's OMEGA [157–159].

Each docking method contains two essential components: the search algorithm and the scoring function. The former is responsible for exploring different ligand conformations and orientations (poses) whereas the latter provides a score (that can be interpreted as a free energy of binding) to rank docking poses. Each pose is assigned a value of the free energy of binding by means of a "master equation" including several terms contributing to the scoring [142]. Different docking programs utilize different search algorithms and scoring functions.

Earlier, molecular docking has proven to be a powerful tool for studying how various compounds interact with MDR transporters like AcrB and MexB [114–116, 160]. This chapter delve into AutoDock and related program such as Vina (one of the popular molecular docking program) [144, 146]. AutoDock evaluates molecular potentials within a defined grid of points (grid-based search space), capturing ligand-receptor interaction during docking. The program allows flexibility of selected rotatable bonds of ligand, although it does not explicitly consider the flexibility of the protein backbone.

2.3 Molecular Dynamics Simulations

MD simulations are commonly used to inspire and complement experiments, allowing to investigate the dynamics of (macro)-molecules with very high spatial accuracy [61, 91, 92, 95]. While a comprehensive atomic-level description should ideally incorporate quantum-mechanical (QM) properties, the computational cost often leads researchers to simplify simulations by focusing on classical mechanics.

Despite this simplification, the accurate modeling of dynamic systems, such as proteins, requires accounting for a range of interactions, which are incorporated within a model-potential function referred to as the force fields. Typically, a force fields for biomolecular simulations consist of sums of pairwise interaction terms, which can be categorized into bonded (bond stretching, angle bending, and proper/improper dihedral angles) and non-bonded interactions (Coulomb potential for electrostatic interactions and a 6–12 Lennard–Jones potential for Van der Waal forces).

The functional form (Eq. 1) reports one of the force field potentials most used in biomolecular simulations.

$$E_{\text{tot}} = \sum\nolimits_{\text{bonds}} k_r (r - r_{\text{eq}})^2 + \sum\nolimits_{\text{angles}} k_\theta (\theta - \theta_{\text{eq}})^2 + \sum\nolimits_{\text{dihedrals}} k_\phi (1 + \cos[n\phi - \gamma]) + \sum\nolimits_{\text{improper}} k_\omega (\omega - \omega_{\text{eq}})^2 + \sum\nolimits_{i<j}^{vdw} \varepsilon_{ij} \left[\left(\frac{r_{\text{m}}}{r_{ij}} \right)^{12} - 2 \left(\frac{r_{\text{m}}}{r_{ij}} \right)^{6} \right] + \sum\nolimits_{i<j}^{\text{elect}} \frac{q_i q_j}{4\pi\varepsilon_0 r_{ij}}$$

Although new ML force fields have been recently developed [162], and this field of research is very active nowadays, the above functional form is still adopted by most of the force fields [163–169], often implicitly account changes in polarization of the electron density with the inclusion of additional terms [170, 171].

2.3.1 Force Fields for Small Molecules

While the widely used force fields are generally considered reliable [163], the parametrization of generic ligands remains a challenging task that requires ad hoc protocols [172–174]. Generation of force fields for small molecules involves a combination of chemical characterization and classical to quantum calculations [175–177]. In our lab, we typically use the General Amber Force Field (GAFF2) for modeling small molecules [155, 156, 178, 179]. Recently, we released a comprehensive database of antimicrobial compounds (AB-DB) [155, 156], which includes all-atom force field parameters along with physicochemical descriptors derived from quantum-mechanical calculations and extensive classical MD trajectories. Our goal is to provide researchers with valuable resources for studying antimicrobial compounds using computational approaches [180, 181] like we employed in our recent investigations, aiding in the development of novel therapeutic strategies against drug-resistant pathogens [41, 182].

2.3.2 Force Fields for Phospholipid Membranes

The phospholipid membrane is a critical component [183, 184], as it serves as both a matrix for embedding membrane proteins (such as RND transporters and MFP components) and a barrier that regulates cellular transport. The hydrophobic effect governs the assembly of phospholipids into a bilayer structure, which plays essential roles in cellular function [185]. Computationally, MD simulations offer atomic-level resolution to study membrane behavior, although coarse-grained or united-atoms models are also available [186], [187–190]. Unlike force fields for proteins and nucleic acids [191–195], those for simulating membrane dynamics have been a more recent development [196, 197] [189]. One of the force fields developed for phospholipid bilayer dynamics is Lipid21 [169], integrated into the AMBER22 package [198].

Tools like CHARMM-GUI are often used to embed proteins into the phospholipid membrane [199–204]. This is essential for setting up simulations of RND transporters paying particular attention to the presence of lipids within the protein structure.

2.4 Post-Processing of Molecular Docking and MD Simulations

2.4.1 Analysis of Membrane Structure and Dynamics

The structural properties of the lipid bilayer, as derived from MD simulations provide indications into the accuracy of force field parameters. While experimental data on lipid bilayer structure often carry inherent uncertainties [205], it serve as essential references for validating the setup procedures adopted during system configuration [206–210]. An easily accessible structural indicator is the area per lipid (ApL), revealing the surface area occupied by each lipid. Additionally, membrane stability can be assessed by parameters such as volume per lipid and membrane thickness. Another useful property is the electron density profile [211–213].

Calculating ApL, particularly in simulations involving a complete lipid bilayer, is straightforward and can be achieved using software tools as mentioned in ref. [214–216] or custom scripts, leveraging the simulation box dimensions [169, 217]. However, incorporating a membrane protein into the lipid bilayer complicates ApL and other property calculations, given the altered area or volume due to the presence of the protein.

Tools such as Grid-MAT-MD [215] compute ApL for protein–membrane systems, and other methods estimate the area occupied by the protein [218, 219]. Therefore, while ApL is typically a primary check in membrane protein simulations, it is crucial to validate results at every stage. Anomalous ApL values (and other properties) may signal underlying issues with the system or errors in accounting for the protein's dynamics.

2.4.2 Free Energy Calculations and the MM/GBSA Method

While traditional docking scoring functions primarily estimate the enthalpic contribution from intermolecular interactions, a more accurate depiction of the calculation of binding affinity requires considering additional factors. This includes changes in intramolecular interactions upon binding, solvent effects, and entropy. Free energy calculations aim to address these limitations by computing the binding free energy based on the principles of statistical thermodynamics, although the calculation of the conformational entropy remains challenging [220–224].

Free energy estimations often rely on extensive MD or Monte Carlo (MC) simulations of the docked protein–ligand complex; as such, they are computationally expensive compared to traditional docking scoring functions. However, the reward lies in potentially more reliable and quantitative results. Free energy calculations incorporate both energetic (potential and solvation energies) and entropic (protein–ligand flexibility and solvent effects) contributions, offering a more comprehensive picture of binding affinity

[225–227]. Despite the challenges, binding free energy prediction through these methods have supported basic research and drug discovery efforts.

The binding free energy (ΔG_{bind}) of a compound or ligand [B] to the protein or receptor [A] is a thermodynamic quantity describing the affinity between A and B, thus related to the ratio between the concentration [AB] of the complex and the product of those of the partners [A] and [B] [226, 228] as shown in Eq. (2).

$$\Delta G_{\text{bind}} = -\mathrm{R}T \ln K_{\mathrm{d}} \tag{1}$$

$$K_{\mathrm{d}} = {}^{[AB]}/_{[A][B]} \tag{2}$$

where K_d is the equilibrium constant, R is the gas constant, and T is the temperature.

Several computational methods can be employed to calculate free energy differences between bound and unbound states. These methods can be broadly categorized into three main types: (i) alchemical calculation like free energy perturbation (FEP), thermodynamic integration (TI), and Bennet's acceptance ratio (BAR), which gradually transform the ligand from unbound state to bound states within the protein environment, calculating free energy changes at each step [229, 230]; (ii) path sampling methods, including the adaptive biasing force algorithm, temperature acceleration, umbrella sampling, metadynamics, and replica exchange, enhance sampling along the binding pathway [231–234]; (iii) end-point methods like Molecular Mechanics Poisson–Boltzmann Surface Area (MM-PBSA) and Molecular Mechanics Generalized Born Surface Area (MM-GBSA) are less expensive calculations to estimate the free energy differences between bound and unbound states [235–238]. For a more in-depth exploration of the theoretical background of free energy calculations, readers can refer [223, 226, 227, 239, 240].

The latter class of methods, which is of relevance to us, treats solvent implicitly and is widely used for estimating binding affinities of substrates and inhibitors of RND transporters like AcrB and MexB [115, 116]. To compute the difference in free energy between two states representing the bound ([AB]) and unbound ([A] + [B]) states of the system, the total free energy of binding is broken down into gas-phase energies, solvation-free energies, and entropic contributions [236] (Fig. 3). In the MM-G(P)BSA framework, the binding free energy for each compound is determined as:

$$\Delta G_{\mathrm{b}} = G_{\text{com}} - \left(G_{\text{rec}} + G_{\text{lig}}\right) \tag{3}$$

where G_{com}, G_{rec}, and G_{lig} represent the absolute free energies of the complex, receptor, and ligand, respectively, averaged over the equilibrium trajectory of the complex (in a single trajectory approach). Each term in the equation above can be decomposed as:

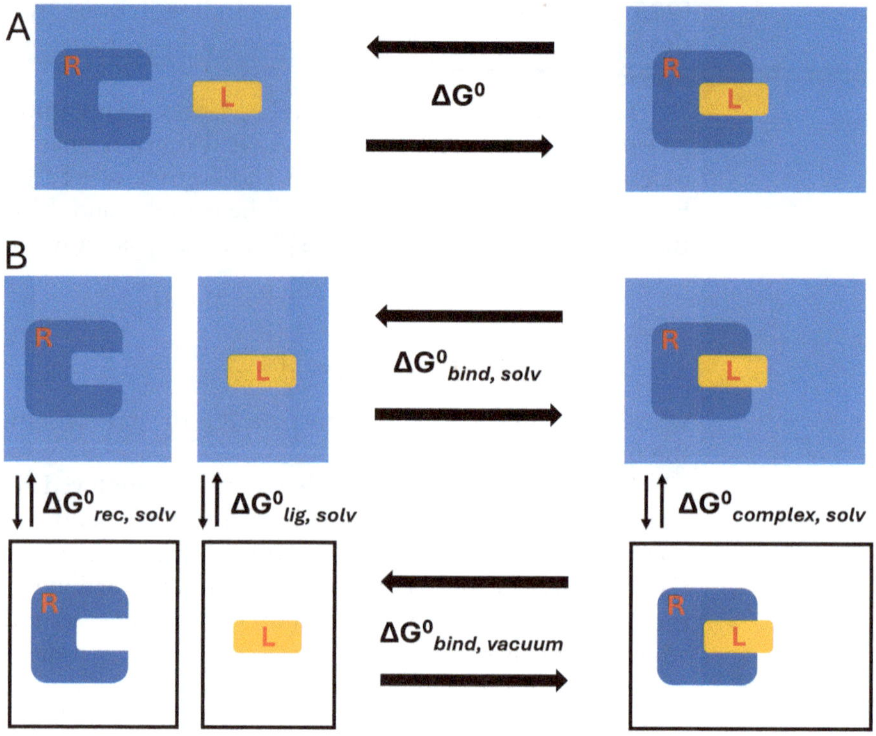

Fig. 3 Scheme of the ideal mechanism to calculate the binding free energy for a receptor (R)–ligand (L) system in the solution, and (B) scheme of the cycle used in MM-G(P)BSA. (Reproduced with permission from [109])

$$\Delta G = \Delta E_{\mathrm{MM}} + \Delta G_{\mathrm{solv}} - T\Delta S_{\mathrm{conf}} \quad (4)$$

where ΔE_{MM} represents the molecular mechanics energy difference, including contributions from bonded (e.g., bond, angle, torsion) and non-bonded terms (e.g., van der Waals, electrostatics). ΔG_{solv} is the solvation-free energy difference, which is split into a polar component and a non-polar component. $T\Delta S_{\mathrm{conf}}$ accounts for the solute conformational entropy difference.

The solvation-free energies (ΔG_{solv}) in Eq. 4, is expressed as the sum of polar ($\Delta G_{\mathrm{solv},\ p}$) [evaluated with MM-G(P)BSA approach] and a non-polar contribution ($\Delta G_{\mathrm{solv},\ np}$) (proportional to the difference in the solvent-exposed surface area) (Eq. 5).

$$\Delta G_{\mathrm{solv}} = \Delta G_{\mathrm{solv},p} + \Delta G_{\mathrm{solv},np} \quad (5)$$

The $T\Delta S_{\mathrm{conf}}$ in Eq. 4 is the solute entropy contribution, which consists of rototranslational and vibrational components, calculated through classical statistical mechanics and normal-mode analysis, respectively [241].

For a comprehensive understanding, readers are referred to excellent reviews on this topic [235–238]. An important feature of MM/GBSA calculations is the possibility to decompose the binding free energy into per-residue, or pairwise-residue, contributions, serving as an alternative to the alanine scanning [242]. A detailed example of binding free energy computation with per-residue decomposition can be found in Subheading 4.4.7 (*see* **Note 1** for additional details).

2.4.3 Fragment-Based Mapping Analysis

We utilized fragment-based protein mapping to identify chemotype-specific multifunctional sites (MFSs) within the putative binding sites (including the Access and Deep Pockets) of AcrB/AcrD from *E. coli* and MexB/MexY from *P. aeruginosa* [123, 131, 243]. This analysis reveals the preferential binding of various fragments to all hotspots of each transporter and assesses the accessibility of sites and channels for substrate and inhibitor recognition. The analysis can be performed either on a single structure or on an ensemble of structures, e.g., representative conformational clusters derived from MD simulations. Such analysis provides insights into the physicochemical characteristics of the site based on probes of diverse types, sizes, shapes, and polarities.

3 Materials

The protocols described in the following *Methods* section require the use of the software and tools described below:

1. MODELLER. It is one of the most popular open-source programs for homology or comparative modeling of protein three-dimensional structures [138, 244, 245]. The program provides additional features such as loop modeling, optimization of protein structure models, as well as tools for clustering and comparing protein structures [246]. MODELLER is available for download at http://www.salilab.org/modeller/
2. MolProbity. It is a free web-based tool used for validating the 3D structures of macromolecules, such as proteins, nucleic acids, and their complexes. It helps researchers assess the accuracy and reliability of their homology models or experimental structures (http://molprobity.biochem.duke.edu).
3. SAVES. This web service offers a collection of programs to check and validate protein structures, during and after model refinement. It is available at http://services.mbi.ucla.edu/SAVES/.
4. AutoDock Vina. It is one of the most popular fast open-source software designed to perform molecular docking runs [147, 247]. It is available at https://github.com/ccsb-scripps/AutoDock-Vina. For more detailed instructions, readers can refer to the manual at https://autodock-vina.readthedocs.io/en/latest/installation.html.

5. AutoDock-GPU. AutoDock-GPU is GPU version derived from AutoDock that utilises OpenCL and CUDA architectures. The program is about 50 times faster than the original code and can be downloaded from the link: https://github.com/ccsb-scripps/AutoDock-GPU [248] AutoDock-GPU requires pre-computed receptor grids generated by the AutoGrid program (*see* Subheading 4.3).
6. GNINA: GNINA is an extension of AutoDock Vina that incorporates deep learning-based scoring functions to improve pose prediction accuracy. It is available at https://github.com/gnina/gnina.
7. AMBER22. AMBER is one of the most up-to-date MD simulation packages. For non-commercial use, it is free and can be downloaded at https://ambermd.org/GetAmber.php. Built on AmberTools23, it integrates seamlessly with Amber22.8. VMD. VMD (http://www.ks.uiuc.edu/Research/vmd) is an open-source software designed for both visualizing and manipulating molecular structures. It offers a comprehensive toolbox for analysis, customization, and *Tk/Tcl* console [249]. Its key features include building and customizing membranes, solvation, and investigating properties of lipid bilayers, such as membrane thickness, lipid tilt angles, and area per lipid [198, 250].
8. FTMap: The FTMap server [251], accessible at https://ftmap.bu.edu/, employs the FTSite algorithm to identify potentially druggable binding sites using fragment probes. It identifies hotspots by analyzing consensus clusters derived from 16 standard solvent probes exhibiting diverse properties. These probes characterize sites for various interaction types, including hydrophilic interactions facilitated by hydrogen bond donor and acceptor molecules, along with hydrophobic (aliphatic) and aromatic interactions.

4 Methods

This section presents the protocols and best practices used in our lab for studying RND superfamily membrane proteins. We assume that readers have access to the programs described in the Materials section and are familiar with the basics of the UNIX command line and shell scripting.

4.1 Homology Modeling of Protein Structures

If the structure of a protein has not been resolved yet by experimental means (e.g., for the RND transporter MexY), homology modeling allows users to build a reliable 3D model in silico based on the structure of a homologous protein (template). In the following, we describe the steps to be done (*see* Fig. 4) to create such a

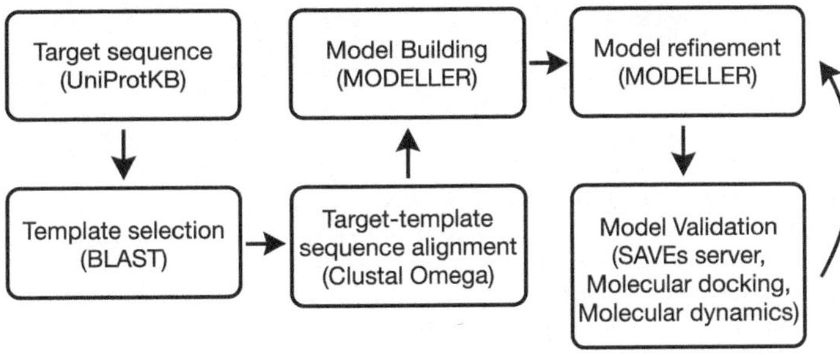

Fig. 4 Workflow of the homology modeling process, comprising target and template selection, alignment, homology modeling, and model refinement. (Reproduced with permission from [109])

homology model structure of a target protein using the software MODELLER [244]. The program can be quickly installed through conda (conda install salilab::modeller) available at Modeller:: Anaconda.org.

4.1.1 Sequence Download

The first step in homology modeling is to obtain the amino acid sequences of both the target protein and its homologous sequences (which can be potential templates). Various databases, such as UniProtKB (http://www.uniprot.org/help/uniprotkb), can provide these sequences. Tools like the Basic Local Alignment Search Tool (BLAST) assist in identifying suitable templates based on sequence similarity and other criteria [252–254]. Figure 5 shows a snapshot of the BLAST available on the NCBI–protein BLAST webpage (https://blast.ncbi.nlm.nih.gov/). It is essential to ensure that the selected template shares significant sequence similarity, maximum query coverage, the lowest number of gaps, the lowest E-value, and structural homology with the target protein.

4.1.2 Model Building

Once a suitable template protein is identified, its 3D structure can be downloaded in PDB format and the homology modeling process begins. Subsequently, the query target sequence and template are aligned. Here, we provide a simple example for the generation of 25 different models of a generic system. More advanced options can be found in the official manual (https://salilab.org/modeller/manual). The first step is to align the target and the template sequences (this can be done through ClustalW). To accomplish this, we create an alignment file, alignment.ali, in which we define the sequence and the PDB structure to be used as a template.

```
>P1;target_code
sequence:target_name:::::::0.00: 0.00
FASTA sequence of the target
*
>P1;template_code
```

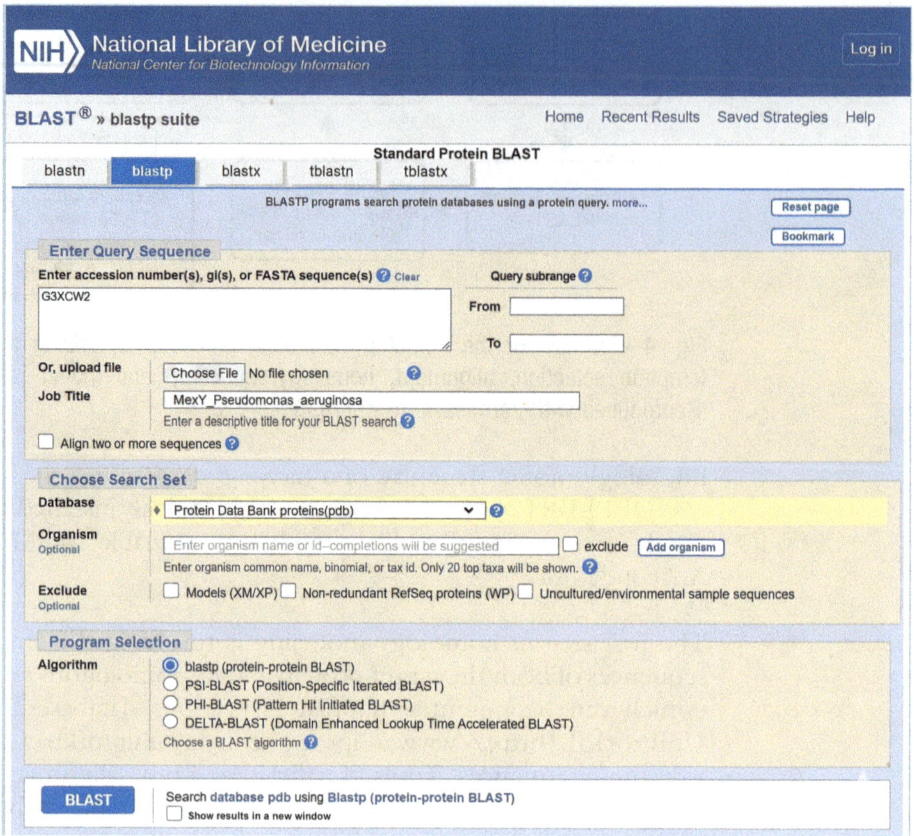

Fig. 5 An example for searching a reliable template for the MexY transporter of *Pseudomonas aeruginosa* using a UniProt ID through protein BLAST in the Protein Data Bank database (https:/blast.ncbi.nlm.nih.gov/)

```
structureX:template_name::::::0:0
*
```

Subsequently, we run the actual homology modeling that will produce the 3D structure of the target protein by using spatial restraints from the template structure (*see* **Note 2**). The reader can easily complete the scheme properly, according to the system under study, by adding the sequence in FASTA format of the target protein and the name of the template (*see* **Note 3** for the explanation of the flags utilized in the script). The example described above is used to build a homology model for a protein using only one template. However, if two or more useful templates are available, the alignment.ali file can be modified to use all these structures during the modeling.

Once the alignment file is ready, there is only one more step to do: build the input file of MODELLER. Several input files are available with different options. Here, we present the basic input, which we call model-generate.py, based on model-default.py of the official MODELLER package. The first lines in the input file typically refer to important information, and usually the user should not change them. The next lines can be easily changed following the example, in which 25 models of the target protein (identified by the target_name in alignment.ali) will be produced using the template_name.pdb as a reference for the 3D structure. The reader should be aware that the values of the *knowns* and *sequence* variables in model-generate.py must be the same as in the alignment file; otherwise, MODELLER would not be able to find the information needed for the run. The models are then automatically scored according to the DOPE (Discrete Optimized Protein Energy) score [255].

```
...
a = automodel(env,
alnfile = 'alignment.ali',
knowns = 'template_name',
sequence = 'target_name',
assess_methods=assess.DOPE)
a.starting_model= 1
a.ending_model = 25
a.auto_align()
a.make()
```

Finally, the homology modeling can be launched by typing in the shell the command:

```
$ modXX model-generate.py
```

where XX is the software version of the MODELLER program installed.

Users can access sample files of the homology modeling of AdeB using AcrB as a template at the link: https://doi.org/10.5281/zenodo.16720685

Model Selection and Refinement

Following model generation, the resulting homology models of 25 files with names, for instance, target_name.B999900XX.pdb, are generated and can be visually inspected. The top-ranking models, based on the DOPE score, are chosen for further validation. DOPE is a statistical potential that estimates the energy of a model based on atomic distances; lower scores indicate more likely, energetically favorable structures [255]. At this stage, it is always good practice to visually inspect the structures with a program like VMD or PyMOL (*see* **Note 4**). In this exercise, we choose the top five models according to the DOPE score and use them for the validation [255]. Tools like MolProbity and SAVES are routinely utilized to assess the quality of the models, identifying potential issues such as steric clashes and backbone angles. In cases of low or moderate validation scores, additional refinement sometimes is necessary, which includes loop refinement and minimization (in vacuum) to optimize the structures (*see* **Note 5**). Another tool we sometimes used to validate predicted models is listed at the link https://rcsb.org/docs/additional-resources/structure-validation-and-quality.

4.1.3 Fragment-Based Mapping Analysis of RND Transporter Binding Sites

In this section, we demonstrate the usage of the FTMap server for probing hotspots. The structure of the RND transporter can be uploaded at the FTMap server [251]. After uploading the structure to the web server, backend calculations are performed to identify hotspot regions for the probes based on the compatibility of sizes, shapes, and polarities of different protein regions. The resulting output, shown in Fig. 6 (depicting a protein with clusters of different chemotypes), enables the identification of consensus sites (CSs) and multifunctional sites (MFSs). CS represents regions where clusters of the same probe type overlap, while MFSs encompass sites with all four probe functionalities: H-bond acceptor, H-bond donor, aliphatic, and aromatic. These sites, particularly relevant for polyspecific RND transporters, are characterized by organic ligands exhibiting various interactions within 5 Å of any residue within the binding site and channels of interest.

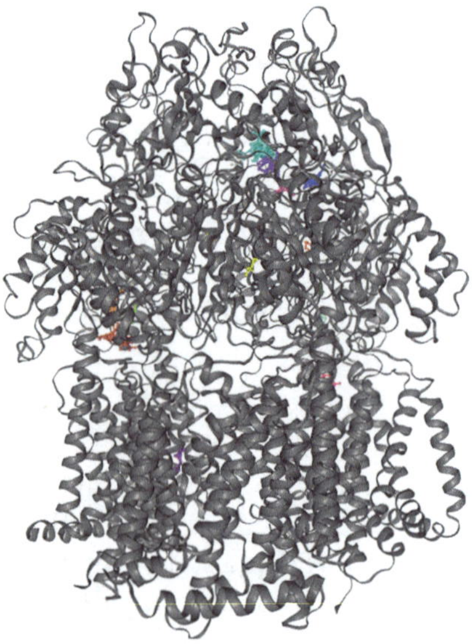

Fig. 6 FTMap analysis for the fragment mapping of the AcrB [PDB ID: 4DX5]. Different colors (in stick representation) represent various chemical groups (chemotypes), highlighting potential hotspots for drug binding/recognition

4.2 Force Field Parameters for Substrates and Inhibitors of RND Transporter

After obtaining a reliable structure of the RND transporter under investigation, the next goal is to obtain a 3D structure of ligands of interest. These can be obtained from various databases such as AB-DB (https://www.dsf.unica.it/~gmalloci/abdb/), PubChem (https://pubchem.ncbi.nlm.nih.gov), ChemSpider (http://www.chemspider.com), or DrugBank (http://www.drugbank.ca). For example, the 3D structure data file (in SDF format) for numerous antimicrobial compounds can be found in PubChem [256]. If the desired ligand structure is not available, the 3D conformation can be obtained starting from the 2D sketch and then converting it to 3D using tools like Open Babel [257], RDKit https://www.rdkit.org, VEGA [258] or ChemAxon Marvin suite of programs https://chemaxon.com/marvin. The latter can also be used to predict the most probable protonation/charge state at physiological pH (7.4) and to determine the distribution of tautomers.

The structure of the ligand, with the desired protonation state and tautomeric form, can be used as input for QM-based calculations. Typically, we use the Gaussian16 [259] package and Density Functional Theory (DFT) [260, 261] with the widely used hybrid B3LYP functional [262], which combines exact exchange with local and gradient-corrected exchange and correlation terms [263, 264]. To represent molecular orbitals and electron density, we typically use the standard Pople-type 6-31G(d,p) Gaussian basis set [265]. First, we perform a geometry refinement to obtain the ground-state optimized structure of the ligand, employing the Polarizable Continuum Model [266] to simulate the effect of a water solvent environment. This helps mimic the solvation effects and prevents the formation of spurious intramolecular hydrogen bonds. To verify that the optimized geometry corresponds to a global minimum on the potential energy surface, a full vibrational analysis is performed. The absence of imaginary frequencies confirms that the optimized structure represents a stable minimum.

Atomic partial charges are derived from a single point energy calculation in vacuum (B3LYP/6-31G(d,p)), performed on the optimized geometry of the ligand. We use the Merz-Singh-Kollman Scheme [267] to construct a grid of points around the molecule, ensuring proper coverage of the electrostatic potential (ESP). Then, the two-step restrained electrostatic potential (RESP) method [268] implemented in the Antechamber module of AMBER [269] is used to assign atomic partial charges, while maintaining the overall electric dipole moment of the molecule. Finally, the GAFF2 force field [178] is assigned to the ligand. The all-atom MD simulation is performed in explicit solvent, typically using the OPC water model, which provides improved accuracy in reproducing water properties [270]. Counter ions (K^+/Cl^- or Na^+/Cl^-) are added to neutralize the system and achieve a physiological ionic strength (typically 0.1–0.15 M; *see* **Note 6**).

Please refer to online tutorials and resources (e.g., http://ambermd.org/tutorials/basic/tutorial4b) for detailed instructions on setting up and running MD simulations of small molecules in explicit solvent using AMBER.

4.3 Molecular Docking

In this section, we describe our general docking protocol in which an ensemble of ligand conformations is docked onto an ensemble of receptor conformations. This approach, known as "ensemble docking," is adopted due to the inherent flexibility of RND transporters, which undergo conformational changes during compound transport. The aim here is not merely to provide a tutorial on using AutoDock Vina, but rather to demonstrate its application for docking antibiotics/inhibitors onto RND efflux transporters.

To initiate a docking experiment from scratch, one first needs to prepare the starting structures. This preparation involves adding any missing hydrogens and assigning protonation states to both the receptor and the ligand. For ligand, assigning the correct protonation state is crucial, and predicting this accurately remains a challenge for most programs [271]. We typically rely on the protonation state predicted by the MarvinSketch ChemAxon program [272] or the Epik utility of Maestro Schrodinger [273]. Following this, the conformation of the ligand can be generated by using the genConf.py utility of the RDKit tool (RDKit Conformer Generation Tool).

For receptor preparation, it is possible to utilize servers like PDB2PQR [274] (https://server.poissonboltzmann.org/pdb2pqr/) or MolProbity (https://molprobity.biochem.duke.edu). For the protein structures with poor experimental resolution, it is often advisable to follow the steps mentioned in Subheading 4.1. Subsequently, solvent molecules (i.e., water) and non-interacting ions must be removed from experimental structures, which can be easily done manually or using some visualization tool (VMD, PyMol, UCSF Chimera, etc.). Once the structures are prepared, all available receptor conformations must be aligned. Assuming different conformations of both receptors (rec_i.pdb, $i = 1, 2, \ldots,$ R) and ligand (lig_j.pdb, $j = 1, 2, \ldots,$ L) are available in standard *.pdb* format, they must be converted into *.pdbqt* format for use with AutoDock and related programs. The *.pdbqt* format is a modified PDB file that includes information about molecular flexibility, such as rotatable bonds. As an example, the whole procedure to run an AutoDock docking calculation of a compound on an RND transporter via command-line tools is reported below, following and integrating the instructions at the link: https://autodock-vina.readthedocs.io/en/latest/docking_basic.html.

```
MGL_DIR="$path/mgltools_x86_64Linux2_1.5.7/MGLToolsPckgs/
AutoDockTools/Utilities24" # Ensure this environment variable
is set
```

```
# prepare the receptor for docking calculations
$MGL_DIR/prepare_receptor -r receptor.pdb -o receptor.pdbqt
# convert pdb or cif or any other format of the ligand into
mol2 extension using obabel; then, prepare the ligand
$ obabel -ixxx ligand.xxx -omol2 -O ligand.mol2
$MGL_DIR/prepare_ligand.py -i ligand.mol2 -o ligand.pdbqt
# generate affinity maps for AutoDock FF
$MGL_DIR//prepare_gpf.py -l ligand.pdbqt -r receptor.pdbqt -p
npts="125 125 90" -p gridcenter="0 0 0" -p spacing="0.375" -o
ligand_receptor.gpf
$path/autogrid4 -p ligand_receptor.gpf -l ligand_receptor.glg
# run the docking and score using the AutoDock4 forcefield
$ vina --ligand ligand.pdbqt --maps receptor --scoring ad4 --
exhaustiveness 512 --out ligand_ad4_out.pdbqt
```

If hydrogen atoms are absent in the protein structure file (receptor.pdb) they can be added using the *-A* "hydrogens" flag. The *.pdbqt* files are similar to the original pdb files but contain an additional column reporting the Gasteiger type charges for each atom, and the information on rotatable bonds. If enough ligand conformations have been sampled externally, for instance, through methods like MD simulations, we can proceed with rigid docking. In this scenario, we deactivate all active torsions using the tool prepare_ligand.py with the *-Z* option:

```
$MGL_DIR/prepare_ligand.py -Z -i ligand.mol2 -o ligand.pdbqt
```

All the available options of the Python scripts *prepare_ligand.py* and *prepare_receptor.py* are listed by simply typing the commands without arguments. With the structures prepared, the user can proceed to run the docking (vina v1.2.5), which can be easily installed in Linux OSs using conda (conda install conda-forge:: vina).

Before running AutoDock Vina, it is necessary to prepare the docking configuration file, containing all the parameters essential for the docking calculations. The available options can be listed simply by typing:

```
$ vina or vina -help
```

A typical configuration file (config.txt) for a docking search in a rectangular box covering the entire AcrB protein, with its z-axis aligned with the membrane bilayer normal, and centered at the origin, looks like the following:

```
scoring = ad4 # ad4 scoring function
ligand = ligand.pdbqt # ligand pdbqt file
maps = ligand_receptor # prefix of maps files
```

```
npts 125 125 90 # num.grid points in xyz
gridfld ligand_receptor.maps.fld # grid_data_file
gridcenter 0 0 0 # xyz-coordinates or auto
spacing 0.375 # spacing(A)
exhaustiveness = 512 # value of exhaustiveness parameter
num_modes = 50 # maximum number of output poses
out = ligand_ad4_out.pdbqt # output file with poses
```

It includes mandatory fields like receptor file, ligand file, and search space dimensions, along with optional parameters like *num_modes* (default 9) and *exhaustiveness* (default 8, higher values are preferred but require higher computational resources) (*see* **Note 7** for setting gridbox dimensions). Once the configuration file is set up, the docking program can be executed by typing:

```
$ vina --config config.txt
```

The output models, along with their corresponding binding affinities, are written to the *ligand_ad4_out.pdbqt* file.

For result analysis, one can load the docked.pdbqt file into visualization software like PyMol or UCSF Chimera. If using VMD, it is necessary to first split the *docked.pdbqt* file into single pdbqt files using vina_split.

```
$ vina_split -input docked.pdbqt
```

Alternatively, the docked.pdbqt file can be converted into a pdb file using a combination of shell commands:

```
$ grep -v BRANCH docked.pdbqt | grep -v ROOT | grep -v TORS >
docked.pdb
```

To facilitate the complete ensemble-docking calculation, a shell script is provided. This script assumes the presence of the *config.txt* file in the local directory and the receptor (rec_i.pdb, $i = 1, 2, \ldots, R$) and ligand (lig_j.pdb, $j = 1, 2, \ldots, L$) files in the *rec* and *lig* directories, respectively. The script runs the experiment iteratively.

```
N_replicas=6
MGL_DIR="$path/mgltools_x86_64Linux2_1.5.7/MGLToolsPckgs/
AutoDockTools/Utilities24" # Ensure this environment variable
is set
for i in $(ls rec/rec*.pdb)
do
filerec=$(basename ${i})
dir=$(echo $filerec | awk 'BEGIN{FS="."}{print $1}')
mkdir -p $dir; cd $dir
ln -sf../rec/$filerec rec.pdb
```

```
$MGL_DIR/prepare_receptor.py -r rec.pdb -o rec.pdbqt
for j in $(ls../lig/*.mol2)
do
filelig=$(basename ${j})
lig=$(echo $filelig | awk 'BEGIN{FS="."}{print $1}')
mkdir $lig
cd $lig
ln -s../../lig/$filelig lig.mol2
$MGL_DIR/prepare_ligand.py -Z -l lig.mol2 -o lig.pdbqt
for k in $(seq 1 $N_replicas)
do
mkdir -p $k || exit
cd $k || exit
ln -sf../lig.pdbqt.
ln -sf../../rec.pdbqt.
ln -sf../../../config.txt.
vina --config config.txt
#optional
grep -v -e BRANCH -e ROOT -e TORS docked.pdbqt > docked.pdb
cd..
done
cd..
done
cd..
done
```

Given the stochastic nature of the search algorithm implemented in the program, it is advisable to maximize the exhaustiveness parameter. This parameter, equivalent to the time allocated for the search, is adjusted heuristically based on factors such as the number of atoms and flexibility [275]. Additionally, running a single instance multiple times adjusted by the $N_replicas variable, can enhance the exploration.

With AutoDock Vina (using its default scoring function), there is no need to generate receptor grid maps separately. However, when using the AD4 scoring function, grid maps must be precomputed with *AutoGrid*, which is particularly advantageous for virtual screening of large compound libraries. In our experience, AutoDock Vina can be slow in such cases due to its serial execution and repeated grid calculations. To address this, we typically adopt a two-step approach: first, generating the grid maps using *AutoGrid*; then performing docking with AutoDock-GPU (https://github.com/ccsb-scripps/AutoDock-GPU), which docks exclusively into precomputed maps and achieves up to a ~ 50-fold speed increase.

The script below facilitates docking using Autodock-GPU for multiple ligand conformers against the same receptor:

```
# Activate conda environment
conda activate autodock_gpu

# Set up MGLTools environment
MGL_DIR=$path/mgltools_x86_64Linux2_1.5.7/MGLToolsPckgs/Auto-
DockTools/Utilities24

# Determine box dimensions based on the binding sites residue,
selection has to be done inside the.tcl script
vmd -dispdev none rec/rec.pdb -e bs_dim_center.tcl
boxscale=1
pts=$(head -n 1 bs_dimension.dat | awk 'BEGIN{FS=",";OFS=","}
{printf "%d,%d,%d\n", $1*'${boxscale}',$2*'${boxscale}',
$3*'${boxscale}'}')
cnt=$(tail -n 1 bs_dimension.dat)

# Prepare receptor
cd rec
python $MGL_DIR/prepare_receptor.py -r rec.pdb -o rec.pdbqt -U
nphs -A checkhydrogens

# Prepare grid parameter file
$ MGL_DIR/prepare_gpf.py -r rec.pdbqt -p npts="${pts}" -p
gridcenter="${cnt}" -p spacing=0.375 -p ligand_types="A C NA
OA N SA HD F" -o grid.gpf

# Run autogrid (check path)
$path/AUTODOCK/autogrid4 -p grid.gpf -l grid.glg
# Navigate to ligands directory
cd../ligands

# Loop through ligand files in the ligands directory
for ligand_file in *.pdbqt; do
# Create directory for each ligand file
ligand_dir="${ligand_file%.*}"
mkdir -p $ligand_dir
cd $ligand_dir
# Prepare docking parameter file
$MGL_DIR//prepare_dpf4.py -l../../ligands/$ligand_file -r../
../rec/rec.pdbqt -o dock.dpf
# Run autodock_gpu_8wi
autodock_gpu_8wi -ffile../../rec/rec.maps.fld -lfile../../li-
gands/$ligand_file -nrun 10
cd..
done
```

The script and files that have successfully executed in our lab are accessible at: https://doi.org/10.5281/zenodo.16720685. Check **Note 7** for the details of *bs_dim_center.tcl*.

4.4 MD Simulations

In this section, we provide sample protocols for studying proteins partially embedded in phospholipid bilayers, like RND transporters. Please note that the following examples should be customized to suit specific requirements, as certain protocol components and inputs heavily rely on the system being investigated.

4.4.1 Embedding the Protein in a Phospholipid Membrane Model

Selection of a Membrane Model: The phospholipid bilayer plays a critical role in the study of proteins that are fully or partially embedded within a membrane [188, 276, 277]. However, in certain circumstances and under specific assumptions, it may be reasonable to neglect the contribution of the membrane environment to the dynamics of the embedded protein domains [113, 115, 116, 122, 127]. The decision to use simplified models where the membrane or protein–membrane interactions are neglected should be made cautiously, considering different factors (e.g., system complexity, biological relevance, specific research questions, model validation). While using reduced models can offer computational advantages, researchers must weigh these benefits against the potential loss of biological relevance and accuracy. In many cases, particularly when studying membrane proteins or related phenomena, it is essential to include the phospholipid bilayer and consider the protein–membrane interactions to obtain meaningful and realistic insights. Therefore, researchers should carefully consider and justify the use of simplified models based on the specific context and objectives of their studies.

4.4.2 Reduced Model of Protein

As the recognition of substrates by the huge RND transporters is thought to occur predominantly in the large periplasmic loops of the protein [33, 36], it could be convenient to use a reduced model of the protein, generated by removing both the transmembrane domain of the protein and the lipid bilayer [115]. This approach simplifies the system and focuses on the critical periplasmic regions involved in substrate binding and transport. We used and validated this model in previous studies on AcrB [39, 113, 115, 116, 122, 127], significantly reducing the size of the system (from approximately 450,000 atoms to around 132,000 atoms), thus considerably saving computational time.

4.4.3 Insertion of the Protein into the Membrane

Different tools are available to prepare a full protein–membrane system for MD simulations, such as the CHARMM-GUI web server (http://www.charmm-gui.org) [278]. The server builds a complete system comprising the protein, water, ions, and a pre-equilibrated membrane, which can be downloaded by choosing different formats depending on the MD software (e.g., AMBER format files can be obtained). Alternatively, the Membrane builder plugin (http://sourceforge.net/p/membplugin/wiki/Home) implemented in VMD [200] can build a lipid bilayer as well. Different from the previous method, the plugin generates only the membrane structure, which must be equilibrated separately before protein insertion. A tutorial for the membrane equilibration protocol using the AMBER program can be found at: http://ambermd.org/tutorials/advanced/tutorial16. To parametrize lipids with the AMBER force field, the molecule and the corresponding atoms should have the correct nomenclature. An automatic and easy way is provided by the LAMBADA/InflateGro2 packages [199].

These methods, however, do not insert phospholipids inside proteins, as it occurs within the central cavity of RND transporters. Therefore, it is preferable to use the PACKMOL-Memgen tool implemented since AMBER22 [202]. The typical membrane composition for these types of simulations is a double layer of phosphatidylethanolamine (POPE) and palmitoyloleoylphosphatidylglycerol (POPG) with a ratio of 4:3, respectively [138, 279, 280]. We use the OPC water model [281] in combination with the ff19SB protein force field [165] (*see* **Note 8**). In the example below, we embedded a protonated RND transporter using PACKMOL-Memgen (*RND_transporter.pdb*):

```
$ packmol-memgen -p RND_transporter.pdb --salt --salt_c K+ --
salt_override --parametrize --ffwat opc --ffprot ff19SB --
gaff2 --lipids POPE:POPG --ratio 4:3 --dist 17.5 --dist_wat
25 --n_ter in --random --movebadrandom --lip_offset 1.1 --log
RND_transporter_pepgpmg.log
```

Using a different membrane model could lead in different choice of lipid offset value (*see* **Note 9**). The resulting PDB file of the embedded protein (*bilayer_RND_transporter.pdb*) needs to be fixed in the formatting, as PACKMOL-Memgen adds some “TER” entries within the lipid moieties. First, the PDB file should be divided into three different files: one containing the protein (*protein.pdb*), one containing the lipids (*lipids.pdb*), and one containing the solvent (*solvent.pdb*). Then, we remove the hydrogen atoms from the protein file:

```
$ pdb4amber --nohyd --in protein.pdb --out protein_noh.pdb
```

We edit the lipid file, removing the extra “TER”:

```
$ awk '{if ($0 ~ /C116 PA/) {c=1}; if (c==1) {print $0}}'
lipids.pdb | grep -v ^TER | awk '{if ($0 ~ /H18T OL/) {print
$0;print "TER"} else {print $0}}' > lipids_fix.pdb
```

In the solvent file, we align the “EPW” entry with the other OPC water atom names and remove the extra space after “WAT” (solvent_fix.pdb), and we merge all three files:

```
$ cat protein_noh.pdb lipids_fix.pdb solvent_fix.pdb > system.
pdb
```

1. Using this procedure, in the case of a transporter (like AcrB), a certain number of lipids will be inserted inside the central cavity lined by the transmembrane domains of the three monomers constituting the protein [28]. PACKMOL-Memgen only adds one ion type; therefore, we can use the formula reported in **Note 6** to compute how many ions to add to reach the ionic concentration of 0.15 M (#IONS). The addition of ions and the final generation of the topological and parameter file can be done using tleap:

```
source leaprc.protein.ff19SB
source leaprc.water.opc
source leaprc.lipid21
source leaprc.gaff2
loadamberprep MOL.prep
loadamberparams MOL.frcmod
com=loadpdb system.pdb
setbox com vdw 0
addionsrand com K+ #IONS
addionsrand com Cl- #IONS
savepdb com system_final.pdb
saveamberparm com system.parm7 system.rst7
quit
```

The MOL.* files refer to the ligand parameters. An output example of the aforementioned procedure is shown in Fig. 7.

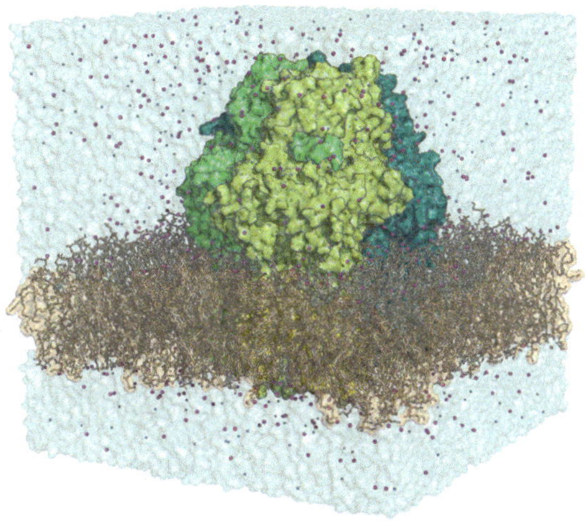

Fig. 7 Example of a solvated system, including water molecules, counter ions, and the lipid bilayer generated by PACKMOL-Memgen

4.4.4 Running MD Simulations

In the following, we will describe a protocol to perform unbiased MD simulations of RND transporters using AMBER [198], although the same procedure can be easily adapted to other programs, such as NAMD [282] or GROMACS [283].

Structural Optimization and System Equilibration: To perform the initial relaxation of a molecular system, particularly focusing on relieving steric clashes between the solvent and solute, soft restraints are applied to the solute's heavy atoms. Below is an example input file (relax1.in) for the AMBER MD engine (pmemd or sander) that performs this relaxation step. In detail, 10,000 cycles of structural optimization are performed (1.000 via the steepest descent algorithm, the remaining through the conjugate gradient one) while applying restraints with a force constant of 1 kcal mol^{-1} $Å^{-2}$:

```
# Relaxation waters and ions
&cntrl
imin=1, maxcyc=10000, ncyc=1000, cut=12.0, ntr=1, restraint_wt
= 1.0,
ntb=0, ntp=0, ntf=1, ntc=1, ntpr=1000, ntwr=1000,
restraintmask = "!(:WAT |:K+ |:Cl-) & !@H="/
```

This step is followed by two additional ones, where only the last line of the input file must be changed.

1. Restraints are applied on the protein backbone, on the ligand heavy atoms, and on the membrane (*relax2.in*):

```
restraintmask = "(:1-#NRESPROT@CA, C, O, N) | (:#NRESFIRSTLI-
PID-#NRESLIG & !@H=)
```

2. Same restraints, but for the protein, only the Cα atoms are considered (relax3.in):

```
restraintmask = "(:1-#NRESPROT@CA) | (:#NRESFIRSTLIPID-
#NRESLIG & !@H=)"
```

where #NRESPROT is the resid number of the last residue in the protein, and #NRESFIRSTLIPID and #NRESLIG are the resid numbers of the first phospholipid and of the ligand, respectively (*see* **Note 10**).

Finally, to optimize the specific interactions established between each ligand and its putative binding site before switching on the thermostat and the barostat in the heating and box equilibration phases, we perform a further optimization allowing only residues of the protein within a given cutoff from the ligand (usually 7 Å) to fully relax [115]. We can substitute the right line into the file relax1.in and get something like the following (*relax4.in*), assuming that the ligand residue is called MOL:

```
restraintmask = "!(:MOL<:7.0)"
```

Relaxation of the system can be finally performed by running the various steps described above (either with pmemd or sander):

```
$ pmemd -O -i relax1.in -o relax1.out -p system.parm7 -c
system.rst7 -ref system.rst7 -r relax1.rst7
$ pmemd -O -i relax2.in -o relax2.out -p system.parm7 -c
relax1.rst7 -ref relax1.rst7 -r relax2.rst7
$ pmemd -O -i relax3.in -o relax3.out -p system.parm7 -c
relax2.rst7 -ref relax2.rst7 -r relax3.rst7
$ pmemd -O -i relax4.in -o relax4.out -p system.parm7 -c
relax3.rst7 -ref relax3.rst7 -r relax4.rst7
```

Once the system has been relaxed, it can be heated to the desired temperature. In this example (file *heat.in*), we perform heating by linearly increasing the temperature of the system from 0 to 310 K in 2 ns of simulation time and using the NTP ensemble with anisotropic pressure coupling (*see* http://ambermd.org/tutorials/advanced/tutorial and **Note 11**), while keeping the Cα atoms of the protein and the P atoms of the membrane restrained (this is strongly suggested at this stage, since the wrong initial density of water—lower than 1 g/cm^3—usually leads to the appearance of artificial vacuum bubbles at the corners of the box):

```
#Heating to 310K in the NTP ensemble and with restraints on
protein and membrane
&cntrl
imin = 0, irest = 0, ntx = 1, ntb = 1, cut = 12.0, ig = -1, ntf
= 2, ntc = 2,
ntp = 2, taup = 5.0, ntt = 3, gamma_ln = 1.0, nstlim = 1000000,
ntpr = 10000, ntwx = 10000, ntwr = 10000, ntwe = 10000,
dt = 0.002, ioutfm=1, ntr = 1, nmropt=1
restraint_wt = 10.0, restraintmask = '@CA | @P31',
/
&wt type='TEMP0', istep1=0, istep2=1000000, value1=0.0, va-
lue2=310.0
/
```

While this step will bring the system to the right temperature, sometimes the structure of the solvent and of the membrane need to be further equilibrated before productive MD can be performed (this is particularly true for medium-/large-sized systems such as those considered here). Therefore, starting from the final conformation, we perform a few further steps of equilibration, by gradually relieving the restraints from both the protein and the membrane.

Namely, in file eq.uil1.in, a 1 ns long plain MD run is performed where the restraints are softened (note the change of the irest and ntx parameters, needed to read also the velocities from the output of the previous step):

```
irest = 1, ntx = 5,
restraint_wt = 1.0, restraintmask = '@CA | @P31'
```

Then, restraints are removed from the membrane (2 ns, file *eq.uil2.in*):

```
restraint_wt = 1.0, restraintmask = '@CA'
```

and finally, from every atom in the system (10 ns, file *eq.uil3.in*). Clearly, the user should visually inspect the output of all the steps above for convergence and structural integrity of all the macromolecules in the system. Eventually, if convergence is not reached in some stage, the user should change the timings of the corresponding run according to their need. Equilibration of the system can be finally performed by running:

```
$ pmemd -O -i heat.in -o heat.out -p system.parm7 -c relax3.
rst7 -ref relax3.rst7 -r heat.rst7
$ pmemd -O -i equil1.in -o equil1.out -p system.parm7 -c heat.
rst7 -ref heat.rst7 -r equil1.rst7
$ pmemd -O -i equil2.in -o equil2.out -p system.parm7 -c
equil1.rst7 -ref equil1.rst7 -r equil2.rst7
$ pmemd -O -i equil3.in -o equil3.out -p system.parm7 -c
equil2.rst7 -ref equil2.rst7 -r equil3.rst7
```

Productive MD: Once the system is reasonably equilibrated and structural parameters of the protein/ligand complex and of the membrane seem to be converged, the user can finally run productive dynamics (file *md.in*). The input is the same as in the previous step, with changes in the parameters taup from 5.0 to 1.0 and gamma_ln from 1.0 to 5.0. *See* **Note 12** about the use of a 4 fs timestep in AMBER22 and **Notes 13** and **14** about productive dynamics of protein/ligand complexes.

Notes on the MPI and GPU-Accelerated Codes: The AMBER package comes with MPI support for both pmemd and sander (pmemd.MPI and sander.MPI binaries), and GPU support for pmemd (pmemd.cuda and pmemd.cuda.MPI). If the parallel version has been installed, the user could consider using the mpirun command:

```
$ mpirun -np #PROCS pmemd.MPI (or sander.MPI, pmemd.cuda.
MPI)...
```

If the GPU/MPI version of pmemd is used, the user should specify the GPUs to be used:

```
$ export CUDA_VISIBLE_DEVICES=ID1, ID2, ID3,...
$ mpirun -np #GPUs pmemd.cuda.MPI...
```

Here, the first line sets the GPUs to be used, so that 0, 2, 3 will tell pmemd.cuda.MPI (*see* **Note 15**) to run on the GPUs number 0, 2, and 3 (*see* **Note 16**). If the user decides to use the GPU code, it is necessary to increase the skinnb value during the equilibration phase with periodic boundary conditions and to restart the MD simulation after a shorter time, typically 0.5 ns. This is because box dimensions are changing, but for performance reasons, the GPU code does not recalculate the non-bond list cells during a simulation, and thus the code could halt with an error related to the skinnb parameter. Thus, in the equilibration steps, one should add the following line at the end of each input file:

```
/&ewald skinnb=5, ! Increase skinnb to avoid skinnb errors/
```

and split all the single steps into shorter ones (0.5 ns each according to the Lipid14 tutorial). Once the system is well equilibrated, the box size fluctuations should be small, and so this should not be an issue during the production phase.

4.4.5 Post-Processing Analysis

In this section, we describe how to process the results of the simulations, focusing on the computation of the free energies of binding of ligands to RND transporters using the MM/GBSA method. More basic analyses can be easily performed with the plethora of tools available, e.g., with the *ptraj* or *cpptraj* tools of AMBER (see, e.g., the tutorial at this webpage: http://ambermd.org/tutorials/advanced/tutorial16)

4.4.6 Contact Analysis

The number of protein–ligand contacts is an index of the strength of the complex interaction [284]. We can assume that the higher the number of contacts, the more stable the ligand binding. Furthermore, contact analysis can indicate how (and where) a ligand binds the protein. For instance, since some RND transporters (e.g., AcrB and MexB) are known to have multiple sub-pockets in the binding site [243], contact analysis can identify if a compound is a cave/groove/interface-binder [116, 121, 129, 130]. This type of post-processing can be performed both on docking poses and on MD trajectory frames, or also on clusters. Various tools are available for conducting this analysis (e.g., ProLif [285], but it can also be done simply using VMD and the following Tcl script.

```
# set cutoff
set cutoff 3.5
set numFrames [molinfo top get numframes]
#set number of residue for ligands
set lignd [atomselect top "resname MOL"]
set l_resid [lsort -integer -uniq [$lignd get resid]]
set l_residue [lsort -integer -uniq [$lignd get residue]]
set l_resname [lsort -uniq [$lignd get resname]]
set n_ligs [llength $l_residue]
#set protein/nucleic chains
set pchains [lsort -uniq [[atomselect top protein] get chain]]
#############
set out [open "tracking.txt" w]
#foreach ligrid $l_residue {
 foreach ch $pchains {
 puts $out "###### $l_resname $ch ######"
 puts $out "#frame resids"
 for {set i 1} {$i <= $numFrames} {incr i} {
  set reg [atomselect top "protein and chain $ch and same
residue as within $cutoff of residue $l_residue"]
 set res_reg [lsort -uniq -integer [[atomselect top [$reg text]
frame $i] get resid]]
 $reg delete
 if {$res_reg != ""} {
 puts $out "$res_reg"
 }
 if {$res_reg == ""} {
 puts $out ""
 }
 unset res_reg
 }
 flush $out
 }
#}
close $out
############
quit
```

Here, we consider contacts within 3.5 Å from the ligand. This script generates a *tracking.txt* file containing protein residues that have been contacted by the ligand (residue name MOL in the script), considering all protein chains (in case of AcrB, three chains A, B, and C). To have separate files for the three chains, we can use the following script (*track_ch.sh*):

```
file=$1
# chain A
awk '{if ($3 == "A") {a=1}; if ($3 == "B") {a=2}; if (a==1)
{print}}' $file | tail -n 1 > chainA.txt
# chain B
awk '{if ($3 == "B") {a=1}; if ($3 == "C") {a=2}; if (a==1)
{print}}' $file | tail -n 1 > chainB.txt
# chain C
awk '{if ($3 == "C") {a=1}; if ($3 == "A") {a=2}; if (a==1)
{print}}' $file | tail -n 1 > chainC.txt
```

It is then easy to cycle these scripts on multiple conformations (e.g., from docking poses) and compute the overall contacts that a ligand establishes with protein residues.

4.4.7 MM/GBSA

The MM/GBSA method can be performed in various ways within the AMBER22 package. User is guided to refer to the Theory section and the AMBER manual for a more comprehensive description [286]. In our opinion, the simplest way is to use the *MMPBSA.py* Python script [287], for which we have described its implementation in the context of a single trajectory. This method requires three separate topology files for the entire complex, ligand, and receptor. If the MD simulation occurs in explicit water, an additional topology for the solvate complex is required.

It is important that these topologies must be consistent in having identical charges for the corresponding atoms, same force field parameters, and matching Bondi radii (PBRadii) [288–291]. Several AMBER tools, like Leap, ParmEd, or ante-MMPBSA.py, can be used to properly set these radii.

Below is an example of a one-line command demonstrating how to construct the topologies for the complex, ligand, and receptor using ante-MMPBSA.py, while setting the PBradii to mbondi3 [289, 291], which is essential for compatibility with the Generalized Born model by Mongan et al. [291], used when igb=8 is set in in the MMPBSA.py input, as detailed below:

```
$ante-MMPBSA.py -p solv_system.parm7 -c com_mbondi3.parm7
-r rec_mbondi3.parm7 -l lig_mbondi3.parm7
-s ":#FIRST-NON-SOLUTE-RESIDUE-10000000"
-m ":#FIRST-REC-RESIDUE-#LAST-REC-RESIDUE"
```

Ensuring these parameters match the chosen Generalized Born model is crucial for accurate free energy calculations. The provided scripts can be adjusted to accommodate Bondi radii for all available GB models in AMBER. Once these initial adjustments are made, the MM/GBSA calculation can proceed using the straightforward input example outlined below (mmpbsa.in):

```
#MMPBSA GB calculations
&general
startframe=1, interval=1, use_sander=1/
&gb
igb=8, saltcon=0.1/
&decomp
idecomp=2, dec_verbose=3, csv_format=0, print_res="X,Y-
Z,..."/
```

which is run through the command:

```
$ MMPBSA.py -O -i mmpbsa.in -o mmpbsa.out -cp com_mbondi3.
parm7 -rp rec_mbondi3.parm7 -lp lig_mbondi3.parm7 -y trajec-
tory
```

The explanation of the flags is:

- -O: Overwrite existing output files.
- -i mmpbsa.in: Input file specifying the calculation parameters.
- -o mmpbsa.out: Output file containing the results.
- -cp com_mbondi3.parm7: Topology file for the complex with mbondi3 radii.
- -rp rec_mbondi3.parm7: Topology file for the receptor with mbondi3 radii.
- -lp lig_mbondi3.parm7: Topology file for the ligand with mbondi3 radii.
- -y trajectory: Input trajectory file containing the solute (ligand, protein, and membrane).
- *saltcon* indicates the salt concentration of the system in *mol/l*.
- *&decomp* indicates that the free energy of binding will be decomposed on a per-residue basis.
- *print_res* = "X,Y-Z,...". Specifies the format for printing residues.

For a more customized calculation, consider exploring the advanced options in the manual.

5 Notes

1. A crucial factor for both MM-P(G)BSA methods is the solute dielectric constant, which heavily influences the calculated results. For this reason, careful selection based on the properties of the protein–ligand interface is essential [238].

2. Due to its reliance on spatial restraints derived from the template structure, MODELLER generally requires the target and template sequences to be of equal length. In cases where the target sequence is longer, the modeling of these additional residues may be less reliable.

3. The empty space "::::" in the alignment.ali file of MODELLER needs to be completed according to the reader needs by following a specific format where each line contains multiple fields separated by colons (:):

```
structureX:template_pdb_name:FIRST:A:LAST:A:::0:0
```

Here's a breakdown of the fields:

Field 1: 3D Structure Availability: This field indicates whether a 3D structure is available for the template. Field 2. PDB Code: This field contains the PDB identifier for the template structure. Fields 3–6. Residue and Chain Identifiers: These four fields specify the starting and ending residues of the template sequence used in the model. Fields 3 and 4 define the chain and residue number for the first residue, and Fields 5 and 6 define them for the last residue. Field 7. Protein Name (Optional): This field might optionally include the name of the protein the template structure represents. Field 8. Protein Source (Optional): This optional field can specify the source organism of the protein. Fields 9 and 10. Crystallographic Data (Optional): These two optional fields can hold the resolution (Field 9) and R-factor (Field 10) obtained from the crystallographic analysis of the template structure. For detailed information on the file format, refer to the MODELLER manual (https://salilab.org/modeller/8v2/manual/node176.html)

4. All the generated homology models can be visualized in VMD or PyMOL by typing the following command:

```
$ vmd -m template_name.pdb target_name.BXX{1..25}.pdb
$ pymol template_name.pdb target_name.BXX{1..25}.pdb
```

5. To perform minimization in vacuum, the sander program within AMBER22 can be utilized. Examples of minimization steps are provided in the Membrane section (Subheading 4.4.4).

6. To obtain the number of positive and negative ions to add to the system, the user can use this formula: NIons = 0.0187·[Molarity]·NWaterMol.

7. To define the docking gridbox, we use the VMD Tcl script *bs_dim_center.tcl*, which calculates both the geometric center and the number of grid points (in x, y, z) required to enclose

the binding site. The user needs to specify the binding site residues, and the script automatically determines an appropriately sized grid based on the spatial extent of these atoms, using a grid spacing of 0.37 Å or 0.25 Å. MiMB/vina/rec/bs_dim_center.tcl at main · athar91/MiMB (github.com)

8. The membrane size should be large enough so that the images of the protein, introduced with the boundary conditions, would not interact with the protein in the original box. A reasonably safe distance between the protein images is around 30 Å, and thus, the membrane size in the xy plane would be 30 Å plus the approximate diameter of the protein.
9. Using a lipid offset larger than 1 allows packing in pathological cases; sometimes 1.1 is enough, but in other cases, values of 1.05 or 1.15 were used. If ligands are present, ligand_param flag should be added.
10. We suppose here that the topology file has been generated so that the listing of residues proceeds sequentially from the protein to the phospholipids to the ligand.
11. In cases where the barostat is thought to induce instability in the system, the user can consider running this step in the NVT ensemble and then continue with the next step.
12. The AMBER22 version allows the users to use a 4 fs time step, for which it is also needed to reweight all the masses in the system, according to the AMBER22 manual. This last step can be performed through the HMassRepartition command within the tool parmed.py included in the AMBER22 suite.
13. Running MD simulations for more than the best docking pose—possibly up to the top 5 ones—is encouraged, as it is known that the scoring functions in docking are approximated [292].
14. Since there is no absolute method to assess the convergence of MD simulations, it is a good idea to perform, for each pose, several production runs using different initial velocities [293]. This is, of course, time consuming, and it should also be noted that convergence in MD simulations of even smaller proteins than those investigated here has proven to be very difficult [294]. Thus, it is a good practice to extend the simulation time of each trajectory as much as the user can afford.
15. It is more efficient to run the simulation on a single GPU https://ambermd.org/GPUPerformance.php
16. The tool NVIDIA system management interface (nvidia-smi) can be used to check the state of each NVIDIA GPU card.

Acknowledgments and Funding

This study was partially funded by the Italian Ministry of University and Research, PNRR, mission 4, component 2, investment 1.3, (Partenariati estesi alle università, ai centri di ricerca, alle aziende per il finanziamento di progetti di ricerca di base), title INF-ACT, project number PE00000007, CUP: F53C22000720007 (M. A., A.V.V., University of Cagliari). The authors also gratefully acknowledge the Health Extended ALliance for Innovative Therapies, Advanced Lab-research, and Integrated Approaches of Precision Medicine partnership (HEAL ITALIA), founded by the Italian Ministry of University and Research, PNRR, mission 4, component 2, investment 1.3, project number PE00000019 (University of Cagliari), and the Project e.INS Ecosystem of Innovation for Next Generation Sardinia—spoke 10-CUP F53C22000430001—MUR Code No. ECS00000038, funded by the Italian Ministry of University and Research (MUR) and by the European Union–NextGenerationEU.

Conflicts of Interest The authors have no conflicts of interest to declare that are relevant to the content of this chapter.

References

1. Agnew E, Dolecek C, Hasan R, Lahra M, Merk H, Perovic O, Sievert D, Smith R, Taylor A, Turner P (2021) Global antimicrobial resistance and use surveillance system (GLASS) report. World Health Organization
2. Fernandez G (2022) Turning the juggernaut, vol 6. Elsevier Sci Ltd The Boulevard, Langford Lane, Kidlington, Oxford
3. Blair JMA, Webber MA, Baylay AJ, Ogbolu DO, Piddock LJV (2015) Molecular mechanisms of antibiotic resistance. Nat Rev Microbiol 13(1):42–51
4. Salam MA, Al-Amin MY, Salam MT, Pawar JS, Akhter N, Rabaan AA, Alqumber MAA (2023) Antimicrobial resistance: a growing serious threat for global public health. MDPI, p 1946
5. Christaki E, Marcou M, Tofarides A (2020) Antimicrobial resistance in bacteria: mechanisms, evolution, and persistence. J Mol Evol 88(1):26–40
6. Butler MS, Henderson IR, Capon RJ, Blaskovich MAT (2023) Antibiotics in the clinical pipeline as of December 2022. J Antibiot 76(8):431–473
7. Theuretzbacher U, Outterson K, Engel A, Karlén A (2020) The global preclinical antibacterial pipeline. Nat Rev Microbiol 18(5):275–285
8. Poole K, Krebes K, Mcnally C, Neshat S (1993) Multiple antibiotic-resistance in pseudomonas-aeruginosa - evidence for involvement of an efflux operon. J Bacteriol 175(22):7363–7372. https://doi.org/10.1128/jb.175.22.7363-7372.1993
9. Piddock LJV (2006) Clinically relevant chromosomally encoded multidrug resistance efflux pumps in bacteria. Clin Microbiol Rev 19(2):382–402. https://doi.org/10.1128/cmr.19.2.382-402.2006
10. Nikaido H (2009) Multidrug resistance in bacteria. Annu Rev Biochem 78:119–146. https://doi.org/10.1146/Annurev.Biochem.78.082907.145923
11. Nikaido H, Pagès J-M (2012) Broad-specificity efflux pumps and their role in multidrug resistance of Gram-negative bacteria. FEMS Microbiol Rev 36(2):340–363. https://doi.org/10.1111/j.1574-6976.2011.00290.x
12. Ruggerone P, Murakami S, Pos KM, Vargiu AV (2013) RND efflux pumps: structural information translated into function and

inhibition mechanisms. Curr Top Med Chem 13(24):3079–3100
13. Blair JMA, Richmond GE, Piddock LJV (2014) Multidrug efflux pumps in Gram-negative bacteria and their role in antibiotic resistance. Future Microbiol 9(10): 1165–1177. https://doi.org/10.2217/fmb.14.66
14. Li X-Z, Plésiat P, Nikaido H (2015) The challenge of efflux-mediated antibiotic resistance in Gram-negative bacteria. Clin Microbiol Rev 28(2):337–418. https://doi.org/10.1128/cmr.00117-14
15. Du D, Wang-Kan X, Neuberger A, Van Veen HW, Pos KM, Piddock LJV, Luisi BF (2018) Multidrug efflux pumps: structure, function and regulation. Nat Rev Microbiol 16(9): 523–539
16. Huang L, Wu C, Gao H, Xu C, Dai M, Huang L, Hao H, Wang X, Cheng G (2022) Bacterial multidrug efflux pumps at the front-line of antimicrobial resistance: an overview. Antibiotics 11(4):520
17. Nikaido H (1996) Multidrug efflux pumps of gram-negative bacteria. J Bacteriol 178(20): 5853–5859. https://doi.org/10.1128/jb.178.20.5853-5859.1996
18. Tikhonova EB, Zgurskaya HI (2004) AcrA, AcrB, and TolC of Escherichia coli form a stable intermembrane multidrug efflux complex. J Biol Chem 279(31):32116–32124. https://doi.org/10.1074/jbc.M402230200
19. Pos KM (2009) Trinity revealed: stoichiometric complex assembly of a bacterial multidrug efflux pump. Proc Natl Acad Sci USA 106(17):6893–6894. https://doi.org/10.1073/pnas.0902837106
20. Du D, Wang Z, James NR, Voss JE, Klimont E, Ohene-Agyei T, Venter H, Chiu W, Luisi BF (2014) Structure of the AcrAB-TolC multidrug efflux pump. Nature 509(7501):512–515. https://doi.org/10.1038/nature13205
21. Zgurskaya HI, Weeks JW, Ntreh AT, Nickels LM, Wolloscheck D (2015) Mechanism of coupling drug transport reactions located in two different membranes. Front Microbiol 6: 100. https://doi.org/10.3389/fmicb.2015.00100
22. Du D, van Veen HW, Luisi BF (2015) Assembly and operation of bacterial tripartite multidrug efflux pumps. Trends Microbiol 23(5): 311–319. https://doi.org/10.1016/j.tim.2015.01.010
23. Nikaido H (2018) RND transporters in the living world. Res Microbiol 169(7–8): 363–371
24. Colclough AL, Alav I, Whittle EE, Pugh HL, Darby EM, Legood SW, McNeil HE, Blair JMA (2020) RND efflux pumps in Gram-negative bacteria; regulation, structure and role in antibiotic resistance. Future Microbiol 15(2):143–157
25. Alav I, Kobylka J, Kuth MS, Pos KM, Picard M, Blair JMA, Bavro VN (2021) Structure, assembly, and function of tripartite efflux and type 1 secretion systems in gram-negative bacteria. Chem Rev 121(9):5479–5596
26. Zgurskaya HI, Adamiak JW, Leus IV (2022) Making sense of drug-efflux transporters in the physiological environment. Curr Opin Microbiol 69:102179
27. Zgurskaya HI, Nikaido H (1999) Bypassing the periplasm: reconstitution of the AcrAB multidrug efflux pump of Escherichia coli. Proc Natl Acad Sci USA 96(13):7190–7195. https://doi.org/10.1073/Pnas.96.13.7190
28. Murakami S, Nakashima R, Yamashita E, Yamaguchi A (2002) Crystal structure of bacterial multidrug efflux transporter AcrB. Nature 419(6907):587–593. https://doi.org/10.1038/nature01050
29. Murakami S, Nakashima R, Yamashita E, Matsumoto T, Yamaguchi A (2006) Crystal structures of a multidrug transporter reveal a functionally rotating mechanism. Nature 443(7108):173–179. https://doi.org/10.1038/nature05076
30. Seeger MA, Schiefner A, Eicher T, Verrey F, Diederichs K, Pos KM (2006) Structural asymmetry of AcrB trimer suggests a peristaltic pump mechanism. Science 313(5791): 1295–1298. https://doi.org/10.1126/science.1131542
31. Sennhauser G, Bukowska MA, Briand C, Grutter MG (2009) Crystal structure of the multidrug exporter MexB from Pseudomonas aeruginosa. J Mol Biol 389(1):134–145. https://doi.org/10.1016/j.jmb.2009.04.001
32. Glavier M, Puvanendran D, Salvador D, Decossas M, Phan G, Garnier C, Frezza E, Cece Q, Schoehn G, Picard M (2020) Antibiotic export by MexB multidrug efflux transporter is allosterically controlled by a MexA-OprM chaperone-like complex. Nat Commun 11(1):4948
33. Kobylka J, Kuth MS, Müller RT, Geertsma ER, Pos KM (2020) AcrB: a mean, keen, drug efflux machine. Ann N Y Acad Sci 1459(1):38–68
34. Zgurskaya HI, Malloci G, Chandar B, Vargiu AV, Ruggerone P (2021) Bacterial efflux

transporters' polyspecificity–a gift and a curse? Curr Opin Microbiol 61:115–123

35. Mazzariol A, Cornaglia G, Nikaido H (2000) Contributions of the AmpC beta-lactamase and the AcrAB multidrug efflux system in intrinsic resistance of Escherichia coli K-12 to beta-lactams. Antimicrob Agents Chemother 44(5):1387–1390. https://doi.org/10.1128/Aac.44.5.1387-1390.2000
36. Elkins CA, Nikaido H (2002) Substrate specificity of the RND-type multidrug efflux pumps AcrB and AcrD of Escherichia coli is determined predominately by two large periplasmic loops. J Bacteriol 184(23): 6490–6498. https://doi.org/10.1128/jb.184.23.6490-6499.2002
37. Middlemiss JK, Poole K (2004) Differential impact of MexB mutations on substrate selectivity of the MexAB-OprM multidrug efflux pump of Pseudomonas aeruginosa. J Bacteriol 186(5):1258–1269. https://doi.org/10.1128/jb.186.5.1258-1269.2004
38. Kobayashi N, Tamura N, van Veen HW, Yamaguchi A, Murakami S (2014) β-Lactam selectivity of multidrug transporters AcrB and AcrD resides in the proximal binding pocket. J Biol Chem 289(15):10680–10690. https://doi.org/10.1074/jbc.M114.547794
39. Kinana AD, Vargiu AV, Nikaido H (2013) Some ligands enhance the efflux of other ligands by the Escherichia coli multidrug pump AcrB. Biochemistry 52(46): 8342–8351. https://doi.org/10.1021/bi401303v
40. Klenotic PA, Moseng MA, Morgan CE, Yu EW (2020) Structural and functional diversity of resistance–nodulation–cell division transporters. Chem Rev 121(9):5378–5416
41. Mehla J, Malloci G, Mansbach R, López CA, Tsivkovski R, Haynes K, Leus IV, Grindstaff SB, Cascella RH, D'Cunha N (2021) Predictive rules of efflux inhibition and avoidance in Pseudomonas aeruginosa. MBio 12(1): 10–1128. https://doi.org/10.1128/mBio.02785-20
42. Gervasoni S, Mehla J, Bergen CR, Leus IV, Margiotta E, Malloci G, Bosin A, Vargiu AV, Lomovskaya O, Rybenkov VV (2023) Molecular determinants of avoidance and inhibition of Pseudomonas aeruginosa MexB efflux pump. MBio 14(4):e01403–e01423
43. Aron Z, Opperman TJ (2018) The hydrophobic trap—the Achilles heel of RND efflux pumps. Res Microbiol 169(7–8):393–400
44. Athar M, Malloci G, Ruggerone P, Vargiu AV (2024) Molecular rationale for the specificities of Mex transporters of Pseudomonas aeruginosa. Biophys J 123(3):397a
45. Koronakis V, Sharff A, Koronakis E, Luisi B, Hughes C (2000) Crystal structure of the bacterial membrane protein TolC central to multidrug efflux and protein export. Nature 405(6789):914–919. https://doi.org/10.1038/35016007
46. Akama H, Kanemaki M, Yoshimura M, Tsukihara T, Kashiwagi T, Yoneyama H, Narita S, Nakagawa A, Nakae T (2004) Crystal structure of the drug discharge outer membrane protein, OprM, of Pseudomonas aeruginosa - dual modes of membrane anchoring and occluded cavity end. J Biol Chem 279(51):52816–52819. https://doi.org/10.1074/jbc.C400445200
47. Bavro VN, Pietras Z, Furnham N, Perez-Cano L, Fernandez-Recio J, Pei XY, Misra R, Luisi B (2008) Assembly and channel opening in a bacterial drug efflux machine. Mol Cell 30(1):114–121. https://doi.org/10.1016/j.molcel.2008.02.015
48. Daury L, Orange F, Taveau J-C, Verchère A, Monlezun L, Gounou C, Marreddy RKR, Picard M, Broutin I, Pos KM (2016) Tripartite assembly of RND multidrug efflux pumps. Nat Commun 7(1):10731
49. Cacciotto P, Basciu A, Oliva F, Malloci G, Zacharias M, Ruggerone P, Vargiu AV (2022) Molecular rationale for the impairment of the MexAB-OprM efflux pump by a single mutation in MexA. Comput Struct Biotechnol J 20:252–260
50. Poole K (2001) Multidrug resistance in gram-negative bacteria. Curr Opin Microbiol 4(5): 500–508. https://doi.org/10.1016/S1369-5274(00)00242-3
51. Higgins MK, Bokma E, Koronakis E, Hughes C, Koronakis V (2004) Structure of the periplasmic component of a bacterial drug efflux pump. Proc Natl Acad Sci USA 101(27):9994–9999. https://doi.org/10.1073/pnas.0400375101
52. Mikolosko J, Bobyk K, Zgurskaya HI, Ghosh P (2006) Conformational flexibility in the multidrug efflux system protein AcrA. Structure 14(3):577–587. https://doi.org/10.1016/j.str.2005.11.015
53. Xu Y, Lee M, Moeller A, Song S, Yoon B-y, Kim H-m, Jun S-y, Lee K, Ha N-c (2011) Funnel-like Hexameric assembly of the periplasmic adapter protein in the tripartite multidrug efflux pump in Gram-negative bacteria. J Biol Chem 286(20):17910–17920. https://doi.org/10.1074/jbc.M111.238535

54. Alvarez-Ortega C, Olivares J, Martínez JL (2013) RND multidrug efflux pumps: what are they good for? Front Microbiol 4:40475
55. Zgurskaya HI, Yamada Y, Tikhonova EB, Ge Q, Krishnamoorthy G (2009) Structural and functional diversity of bacterial membrane fusion proteins. Biochim Biophy Acta Proteins Proteomics 1794(5):794–807
56. Krishnamoorthy G, Tikhonova EB, Zgurskaya HI (2008) Fitting periplasmic membrane fusion proteins to inner membrane transporters: mutations that enable Escherichia coli AcrA to function with Pseudomonas aeruginosa MexB. J Bacteriol 190(2):691–698
57. Hobbs EC, Yin X, Paul BJ, Astarita JL, Storz G (2012) Conserved small protein associates with the multidrug efflux pump AcrB and differentially affects antibiotic resistance. Proc Natl Acad Sci 109(41):16696–16701. https://doi.org/10.1073/pnas.1210093109
58. Du D, Neuberger A, Orr MW, Newman CE, Hsu P-C, Samsudin F, Szewczak-Harris A, Ramos LM, Debela M, Khalid S (2020) Interactions of a bacterial RND transporter with a transmembrane small protein in a lipid environment. Structure 28(6):625–634
59. Wang Z, Fan G, Hryc CF, Blaza JN, Serysheva II, Schmid MF, Chiu W, Luisi BF, Du D (2017) An allosteric transport mechanism for the AcrAB-TolC multidrug efflux pump. elife 6. https://doi.org/10.7554/eLife.24905
60. Eicher T, Cha H-j, Seeger MA, Brandstätter L, El-Delik J, Bohnert JA, Kern WV, Verrey F, Grütter MG, Diederichs K (2012) Transport of drugs by the multidrug transporter AcrB involves an access and a deep binding pocket that are separated by a switch-loop. Proc Natl Acad Sci 109(15):5687–5692
61. Athar M, Gervasoni S, Catte A, Basciu A, Malloci G, Ruggerone P, Vargiu AV (2023) Tripartite efflux pumps of the RND superfamily: what did we learn from computational studies? Microbiology 169(3):001307
62. Symmons MF, Bokma E, Koronakis E, Hughes C, Koronakis V (2009) The assembled structure of a complete tripartite bacterial multidrug efflux pump. Proc Natl Acad Sci USA 106(17):7173–7178. https://doi.org/10.1073/pnas.0900693106
63. Jin-Sik K, Hyeongseop J, Saemee S, Hye-Yeon K, Kangseok L, Jaekyung H, Nam-Chul H (2015) Structure of the tripartite multidrug efflux pump AcrAB-TolC suggests an alternative assembly mode. Mol Cells 38(2):180–186
64. Sennhauser G, Amstutz P, Briand C, Storchenegger O, Grutter MG (2007) Drug export pathway of multidrug exporter AcrB revealed by DARPin inhibitors. PLoS Biol 5(1):106–113. https://doi.org/10.1371/journal.pbio.0050007
65. Phan G, Benabdelhak H, Lascombe MB, Benas P, Rety S, Picard M, Ducruix A, Etchebest C, Broutin I (2010) Structural and dynamical insights into the opening mechanism of P. aeruginosa OprM Channel. Structure 18(4):507–517. https://doi.org/10.1016/j.str.2010.01.018
66. Akama H, Matsuura T, Kashiwagi S, Yoneyama H, Narita SI, Tsukihara T, Nakagawa A, Nakae T (2004) Crystal structure of the membrane fusion protein, MexA, of the multidrug transporter in Pseudomonas aeruginosa. J Biol Chem 279(25):25939–25942. https://doi.org/10.1074/jbc.C400164200
67. Elkins CA, Nikaido H (2003) Chimeric analysis of AcrA function reveals the importance of its c-terminal domain in its interaction with the AcrB multidrug efflux pump. J Bacteriol 185(18):5349–5356. https://doi.org/10.1128/jb.185.18.5349-5356.2003
68. Trépout S, Taveau J-c, Benabdelhak H, Granier T, Ducruix A, Frangakis AS, Lambert O (2010) Structure of reconstituted bacterial membrane efflux pump by cryo-electron tomography. Biochim Biophys Acta Biomembr 1798(10):1953–1960. https://doi.org/10.1016/j.bbamem.2010.06.019
69. Nakashima R, Sakurai K, Yamasaki S, Nishino K, Yamaguchi A (2011) Structures of the multidrug exporter AcrB reveal a proximal multisite drug-binding pocket. Nature 480(7378):565–569. https://doi.org/10.1038/nature10641
70. Pei X-Y, Hinchliffe P, Symmons MF, Koronakis E, Benz R, Hughes C, Koronakis V (2011) Structures of sequential open states in a symmetrical opening transition of the TolC exit duct. Proc Natl Acad Sci USA 108(5):2112–2117. https://doi.org/10.1073/pnas.1012588108
71. Xu Y, Moeller A, Jun S-Y, Le M, Yoon B-Y, Kim J-S, Lee K, Ha N-C (2012) Assembly and channel opening of outer membrane protein in tripartite drug efflux pumps of gram-negative bacteria. J Biol Chem 287(15):11740–11750. https://doi.org/10.1074/jbc.M111.329375
72. Hinchliffe P, Symmons MF, Hughes C, Koronakis V (2013) Structure and operation of bacterial tripartite pumps. Ann Rev Microbiol

67:221–242. https://doi.org/10.1146/annurev-micro-092412-155718
73. Ornik-Cha A, Wilhelm J, Kobylka J, Sjuts H, Vargiu AV, Malloci G, Reitz J, Seybert A, Frangakis AS, Pos KM (2021) Structural and functional analysis of the promiscuous AcrB and AdeB efflux pumps suggests different drug binding mechanisms. Nat Commun 12(1):6919
74. Morgan CE, Glaza P, Leus IV, Trinh A, Su C-C, Cui M, Zgurskaya HI, Yu EW (2021) Cryoelectron microscopy structures of AdeB illuminate mechanisms of simultaneous binding and exporting of substrates. MBio 12(1): 10–1128
75. Su C-C, Morgan CE, Kambakam S, Rajavel M, Scott H, Huang W, Emerson CC, Taylor DJ, Stewart PL, Bonomo RA (2019) Cryo-electron microscopy structure of an Acinetobacter baumannii multidrug efflux pump. MBio 10(4):10–1128
76. Tam H-K, Malviya VN, Foong W-E, Herrmann A, Malloci G, Ruggerone P, Vargiu AV, Pos KM (2020) Binding and transport of carboxylated drugs by the multidrug transporter AcrB. J Mol Biol 432(4):861–877
77. Zwama M, Hayashi K, Sakurai K, Nakashima R, Kitagawa K, Nishino K, Yamaguchi A (2017) Hoisting-loop in bacterial multidrug exporter AcrB is a highly flexible hinge that enables the large motion of the subdomains. Front Microbiol 8:304424
78. Oswald C, Tam HK, Pos KM (2016) Transport of lipophilic carboxylates is mediated by transmembrane helix 2 in multidrug transporter AcrB. Nat Commun 7:13819. https://doi.org/10.1038/ncomms13819
79. Ababou A, Koronakis V (2016) Structures of gate loop variants of the AcrB drug efflux pump bound by erythromycin substrate. PLoS One 11(7):e0159154
80. Eicher T, Seeger MA, Anselmi C, Zhou W, Brandstätter L, Verrey F, Diederichs K, Faraldo-Gómez JD, Pos KM (2014) Coupling of remote alternating-access transport mechanisms for protons and substrates in the multidrug efflux pump AcrB. elife 3:e03145
81. Hung L-W, Kim H-B, Murakami S, Gupta G, Kim C-Y, Terwilliger TC (2013) Crystal structure of AcrB complexed with linezolid at 3.5 Å resolution. J Struct Funct Genom 14:71–75
82. Yamaguchi A, Nakashima R, Sakurai K (2015) Structural basis of RND-type multidrug exporters. Front Microbiol 6:327. https://doi.org/10.3389/fmicb.2015.00327
83. Monroe N, Sennhauser G, Seeger MA, Briand C, Grütter MG (2011) Designed ankyrin repeat protein binders for the crystallization of AcrB: plasticity of the dominant interface. J Struct Biol 174(2):269–281
84. Veesler D, Blangy S, Cambillau C, Sciara G (2008) There is a baby in the bath water: AcrB contamination is a major problem in membrane-protein crystallization. Acta Crystallogr Sect F 64(10):880–885
85. Das D, Xu QS, Lee JY, Ankoudinova I, Huang C, Lou Y, DeGiovanni A, Kim R, Kim S-H (2007) Crystal structure of the multidrug efflux transporter AcrB at 3.1 Å resolution reveals the N-terminal region with conserved amino acids. J Struct Biol 158(3): 494–502
86. Su C-C, Li M, Gu R, Takatsuka Y, McDermott G, Nikaido H, Yu EW (2006) Conformation of the AcrB multidrug efflux pump in mutants of the putative proton relay pathway. J Bacteriol 188(20):7290–7296
87. Tsutsumi K, Yonehara R, Ishizaka-Ikeda E, Miyazaki N, Maeda S, Iwasaki K, Nakagawa A, Yamashita E (2019) Structures of the wild-type MexAB–OprM tripartite pump reveal its complex formation and drug efflux mechanism. Nat Commun 10(1):1520
88. Vargiu AV, Ramaswamy VK, Malloci G, Malvacio I, Atzori A, Ruggerone P (2018) Computer simulations of the activity of RND efflux pumps. Res Microbiol 169(7–8): 384–392
89. Karplus M, Lavery R (2014) Significance of molecular dynamics simulations for life sciences. Israel J Chem 54(8–9):1042–1051. https://doi.org/10.1002/ijch.201400074
90. Mortier J, Rakers C, Bermudez M, Murgueitio MS, Riniker S, Wolber G (2015) The impact of molecular dynamics on drug design: applications for the characterization of ligand-macromolecule complexes. Drug Discov Today 20(6):686–702. https://doi.org/10.1016/j.drudis.2015.01.003
91. De Vivo M, Masetti M, Bottegoni G, Cavalli A (2016) Role of molecular dynamics and related methods in drug discovery. J Med Chem 59(9):4035–4061. https://doi.org/10.1021/acs.jmedchem.5b01684
92. Amaro RE, Mulholland AJ (2018) Multiscale methods in drug design bridge chemical and biological complexity in the search for cures. Nat Rev Chem 2(4):0148. https://doi.org/10.1038/s41570-018-0148
93. Do P-C, Lee EH, Le L (2018) Steered molecular dynamics simulation in rational drug design. J Chem Inf Model 58(8):1473–1482

94. Salo-Ahen OMH, Alanko I, Bhadane R, Bonvin AMJJ, Honorato RV, Hossain S, Juffer AH, Kabedev A, Lahtela-Kakkonen M, Larsen AS (2020) Molecular dynamics simulations in drug discovery and pharmaceutical development. PRO 9(1):71
95. Keller BG, Aleksic S, Donati L (2019) Markov state models in drug design. In: Biomolecular simulations in structure-based drug discovery, p 67
96. Sadybekov AV, Katritch V (2023) Computational approaches streamlining drug discovery. Nature 616(7958):673–685
97. King E, Aitchison E, Li H, Luo R (2021) Recent developments in free energy calculations for drug discovery. Front Mol Biosci 8: 712085
98. Carracedo-Reboredo P, Liñares-Blanco J, Rodríguez-Fernández N, Cedrón F, Novoa FJ, Carballal A, Maojo V, Pazos A, Fernandez-Lozano C (2021) A review on machine learning approaches and trends in drug discovery. Comput Struct Biotechnol J 19:4538–4558
99. Pinzi L, Rastelli G (2019) Molecular docking: shifting paradigms in drug discovery. Int J Mol Sci 20(18):4331
100. Frye L, Bhat S, Akinsanya K, Abel R (2021) From computer-aided drug discovery to computer-driven drug discovery. Drug Discov Today Technol 39:111–117
101. Ciccotti G, Dellago C, Ferrario M, Hernández ER, Tuckerman ME (2022) Molecular simulations: past, present, and future (a Topical Issue in EPJB). Eur Phys J B 95(1):3
102. Lanrezac A, Férey N, Baaden M (2022) Wielding the power of interactive molecular simulations. Wiley Interdiscipl Rev 12(4): e1594
103. Zhang J, Chen D, Xia Y, Huang Y-P, Lin X, Han X, Ni N, Wang Z, Yu F, Yang L (2023) Artificial intelligence enhanced molecular simulations. J Chem Theory Comput 19(14):4338–4350
104. Pandey M, Fernandez M, Gentile F, Isayev O, Tropsha A, Stern AC, Cherkasov A (2022) The transformational role of GPU computing and deep learning in drug discovery. Nat Mach Intell 4(3):211–221
105. Khalid S, Brandner AF, Juraschko N, Newman KE, Pedebos C, Prakaash D, Smith IPS, Waller C, Weerakoon D (2023) Computational microbiology of bacteria: advancements in molecular dynamics simulations. Structure 31:1320
106. Berglund NA, Piggot TJ, Jefferies D, Sessions RB, Bond PJ, Khalid S (2015) Interaction of the antimicrobial peptide polymyxin B1 with both membranes of E. coli: a molecular dynamics study. PLoS Comput Biol 11(4): e1004180
107. Khalid S, Piggot TJ, Samsudin F (2018) Atomistic and coarse grain simulations of the cell envelope of gram-negative bacteria: what have we learned? Accounts Chem Res 52(1): 180–188
108. Khalid S, Schroeder C, Bond PJ, Duncan AL (2022) What have molecular simulations contributed to understanding of Gram-negative bacterial cell envelopes? Microbiology 168(3):001165
109. Cacciotto P, Ramaswamy VK, Malloci G, Ruggerone P, Vargiu AV (2018) Molecular modeling of multidrug properties of resistance nodulation division (RND) transporters. In: Bacter Multidrug Export, vol 1700, pp 179–219. https://doi.org/10.1007/978-1-4939-7454-2_11
110. Ruggerone P, Vargiu AV, Collu F, Fischer N, Kandt C (2013) Molecular dynamics computer simulations of multidrug RND efflux pumps. Comput Struct Biotechnol J 5: e201302008. https://doi.org/10.5936/csbj.201302008
111. Schulz R, Vargiu AV, Collu F, Kleinekathofer U, Ruggerone P (2010) Functional rotation of the transporter AcrB: insights into drug extrusion from simulations. PLoS Comput Biol 6(6):e1000806. https://doi.org/10.1371/journal.pcbi.1000806
112. Schulz R, Vargiu AV, Ruggerone P, Kleinekathöfer U (2011) Role of water during the extrusion of substrates by the efflux transporter AcrB. J Phys Chem B 115(25): 8278–8287
113. Vargiu AV, Collu F, Schulz R, Pos KM, Zacharias M, Kleinekathöfer U, Ruggerone P (2011) Effect of the F610A mutation on substrate extrusion in the AcrB transporter: explanation and rationale by molecular dynamics simulations. J Am Chem Soc 133(28):10704–10707. https://doi.org/10.1021/ja202666x
114. Collu F, Vargiu AV, Dreier J, Cascella M, Ruggerone P (2012) Recognition of Imipenem and Meropenem by the RND-transporter MexB studied by computer simulations. J Am Chem Soc 134(46): 19146–19158. https://doi.org/10.1021/ja307803m
115. Vargiu AV, Nikaido H (2012) Multidrug binding properties of the AcrB efflux pump

characterized by molecular dynamics simulations. Proc Natl Acad Sci USA 109(50): 20637–20642. https://doi.org/10.1073/pnas.1218348109

116. Vargiu AV, Ruggerone P, Opperman TJ, Nguyen ST, Nikaido H (2014) Molecular mechanism of MBX2319 inhibition of Escherichia coli AcrB multidrug efflux pump and comparison with other inhibitors. Antimicrob Agents Chemother 58(10): 6224–6234. https://doi.org/10.1128/aac.03283-14

117. Blair JMA, Bavro VN, Ricci V, Modi N, Cacciotto P, Kleinekathoefer U, Ruggerone P, Vargiu AV, Baylay AJ, Smith HE, Brandon Y, Galloway D, Piddock LJV (2015) AcrB drug-binding pocket substitution confers clinically relevant resistance and altered substrate specificity. Proc Natl Acad Sci USA 112(11):3511–3516. https://doi.org/10.1073/pnas.1419939112

118. Schulz R, Vargiu AV, Ruggerone P, Kleinekathoefer U (2015) Computational study of correlated domain motions in the AcrB efflux transporter. Biomed Res Int 2015:487298

119. Kinana AD, Vargiu AV, May T, Nikaido H (2016) Aminoacyl β-naphthylamides as substrates and modulators of AcrB multidrug efflux pump. Proc Natl Acad Sci 113(5): 1405–1410. https://doi.org/10.1073/pnas.1525143113

120. Sjuts H, Vargiu AV, Kwasny SM, Nguyen ST, Kim H-S, Ding X, Ornik AR, Ruggerone P, Bowlin TL, Nikaido H, Pos KM, Opperman TJ (2016) Molecular basis for inhibition of AcrB multidrug efflux pump by novel and powerful pyranopyridine derivatives. Proc Natl Acad Sci 113(13):3509–3514. https://doi.org/10.1073/pnas.1602472113

121. Gervasoni S, Malloci G, Bosin A, Vargiu AV, Zgurskaya HI, Ruggerone P (2022) Recognition of quinolone antibiotics by the multidrug efflux transporter MexB of Pseudomonas aeruginosa. Phys Chem Chem Phys 24(27): 16566–16575

122. Vergalli J, Chauvet H, Oliva F, Pajović J, Malloci G, Vargiu AV, Réfrégiers M, Ruggerone P, Pagès J-M (2022) A framework for dissecting affinities of multidrug efflux transporter AcrB to fluoroquinolones. Commun Biol 5(1):1062. https://doi.org/10.1038/s42003-022-04024-1

123. Catte A, K. Ramaswamy V, Vargiu AV, Malloci G, Bosin A, Ruggerone P (2022) Common recognition topology of mex transporters of Pseudomonas aeruginosa revealed by molecular modelling. Front Pharmacol 13: 1021916

124. Vargiu AV, Phan G, Venter H, Broutin IJA (2022) The structural and functional study of efflux pumps belonging to the RND transporters family from Gram-Negative bacteria, vol 11. MDPI

125. Plé C, Tam H-K, Vieira Da Cruz A, Compagne N, Jiménez-Castellanos J-C, Müller RT, Pradel E, Foong WE, Malloci G, Ballée A (2022) Pyridylpiperazine-based allosteric inhibitors of RND-type multidrug efflux pumps. Nat Commun 13(1):115

126. Wang B, Weng J, Wang W (2015) Substrate binding accelerates the conformational transitions and substrate dissociation in multidrug efflux transporter AcrB. Front Microbiol 6: 302. https://doi.org/10.3389/fmicb.2015.00302

127. Atzori A, Malloci G, Cardamone F, Bosin A, Vargiu AV, Ruggerone P (2020) Molecular interactions of carbapenem antibiotics with the multidrug efflux transporter AcrB of Escherichia coli. Int J Mol Sci 21(3):860. https://doi.org/10.3390/ijms21030860

128. Atzori A, Malloci G, Prajapati JD, Basciu A, Bosin A, Kleinekathöfer U, Dreier J, Vargiu AV, Ruggerone P (2019) Molecular interactions of cephalosporins with the deep binding pocket of the RND transporter AcrB. J Phys Chem B 123(22):4625–4635

129. Malvacio I, Buonfiglio R, D'Atanasio N, Serra G, Bosin A, Di Giorgio FP, Ruggerone P, Ombrato R, Vargiu AV (2019) Molecular basis for the different interactions of congeneric substrates with the polyspecific transporter AcrB. Biochim Biophys Acta 1861(7):1397–1408

130. Atzori A, Malviya VN, Malloci G, Dreier J, Pos KM, Vargiu AV, Ruggerone P (2019) Identification and characterization of carbapenem binding sites within the RND-transporter AcrB. Biochim Biophys Acta 1861(1):62–74

131. Ramaswamy VK, Vargiu AV, Malloci G, Dreier J, Ruggerone P (2018) Molecular determinants of the promiscuity of MexB and MexY multidrug transporters of Pseudomonas aeruginosa. Front Microbiol 9: 364058

132. Schwede T, Kopp J, Guex N, Peitsch MC (2003) SWISS-MODEL: an automated protein homology-modeling server. Nucleic Acids Res 31(13):3381–3385. https://doi.org/10.1093/nar/gkg520

133. Bower MJ, Cohen FE, Dunbrack RL Jr (1997) Prediction of protein side-chain rotamers from a backbone-dependent rotamer library: a new homology modeling tool. J

Mol Biol 267(5):1268–1282. https://doi.org/10.1006/jmbi.1997.0926

134. Gabler F, Nam SZ, Till S, Mirdita M, Steinegger M, Söding J, Lupas AN, Alva V (2020) Protein sequence analysis using the MPI bioinformatics toolkit. Curr Protoc Bioinform 72(1):e108. https://doi.org/10.1002/cpbi.108
135. Abramson J, Adler J, Dunger J, Evans R, Green T, Pritzel A, Ronneberger O, Willmore L, Ballard AJ, Bambrick J (2024) Accurate structure prediction of biomolecular interactions with AlphaFold 3. Nature 630(8016):493–500
136. Qiao Z, Nie W, Vahdat A, Miller III TF, Anandkumar A (2024) State-specific protein–ligand complex structure prediction with a multiscale deep generative model. 1–14
137. Jumper J, Evans R, Pritzel A, Green T, Figurnov M, Ronneberger O, Tunyasuvunakool K, Bates R, Žídek A, Potapenko A (2021) Highly accurate protein structure prediction with AlphaFold. Nature 596(7873):583–589
138. Eswar N, Webb B, Marti-Renom MA, Madhusudhan MS, Eramian D, Shen M-y, Pieper U, Sali A (2006) Comparative protein structure modeling using modeller. Curr Protoc Bioinform 15(1):5.6.1–5.6.30. https://doi.org/10.1002/0471250953.bi0506s15
139. Šali A, Potterton L, Yuan F, van Vlijmen H, Karplus M (1995) Evaluation of comparative protein modeling by MODELLER. Proteins 23(3):318–326
140. Šali A, Blundell TL (1993) Comparative protein modelling by satisfaction of spatial restraints. J Mol Biol 234(3):779–815
141. ŠAli A, Overington JP (1994) Derivation of rules for comparative protein modeling from a database of protein structure alignments. Protein Sci 3(9):1582–1596
142. Du X, Li Y, Xia Y-L, Ai S-M, Liang J, Sang P, Ji X-L, Liu S-Q (2016) Insights into protein–ligand interactions: mechanisms, models, and methods. Int J Mol Sci 17(2):144
143. Jorgensen WL (2004) The many roles of computation in drug discovery. Science 303(5665):1813–1818. https://doi.org/10.1126/Science.1096361
144. Sousa SF, Fernandes PA, Ramos MJ (2006) Protein-ligand docking: current status and future challenges. Proteins 65(1):15–26. https://doi.org/10.1002/prot.21082
145. Rodrigues JP, Karaca E, Bonvin AM (2015) Information-driven structural modelling of protein-protein interactions. Methods Mol Biol 1215:399–424. https://doi.org/10.1007/978-1-4939-1465-4_18
146. Chen YC (2015) Beware of docking! Trends Pharmacol Sci 36(2):78–95. https://doi.org/10.1016/j.tips.2014.12.001
147. Trott O, Olson AJ (2010) Autodock Vina: improving the speed and accuracy of docking with a new scoring function, efficient optimization, and multithreading. J Comput Chem 31(2):455–461. https://doi.org/10.1002/jcc.21334
148. Amaro RE, Baron R, McCammon JA (2008) An improved relaxed complex scheme for receptor flexibility in computer-aided drug design. J Comput Aided Mol Des 22(9):693–705. https://doi.org/10.1007/s10822-007-9159-2
149. Huang S-Y, Zou X (2007) Ensemble docking of multiple protein structures: considering protein structural variations in molecular docking. Proteins 66(2):399–421. https://doi.org/10.1002/prot.21214
150. Basciu A, Athar M, Kurt H, Neville C, Malloci G, Muredda FC, Bosin A, Ruggerone P, Bonvin AMJJ, Vargiu AV (2025) Toward the prediction of binding events in very flexible, allosteric, multidomain proteins. J Chem Inform Model 65:2052
151. Basciu A, Malloci G, Pietrucci F, Bonvin AMJJ, Vargiu AV (2019) Holo-like and druggable protein conformations from enhanced sampling of binding pocket volume and shape. J Chem Inf Model 59(4):1515–1528
152. Wohlwend J, Corso G, Passaro S, Reveiz M, Leidal K, Swiderski W, Portnoi T, Chinn I, Silterra J, Jaakkola T (2024) Boltz-1 democratizing biomolecular interaction modeling. BioRxiv
153. Passaro S, Corso G, Wohlwend J, Reveiz M, Thaler S, Ram Somnath V, Getz N, Portnoi T, Roy J, Stark H (2025) Boltz-2: towards accurate and efficient binding affinity prediction. BioRxiv:2025–2006
154. Kuriata A, Gierut AM, Oleniecki T, Ciemny MP, Kolinski A, Kurcinski M, Kmiecik S (2018) CABS-flex 2.0: a web server for fast simulations of flexibility of protein structures. Nucleic Acids Res 46(W1):W338–W343. https://doi.org/10.1093/nar/gky356
155. Malloci G, Vargiu AV, Serra G, Bosin A, Ruggerone P, Ceccarelli M (2015) A database of force-field parameters, dynamics, and properties of antimicrobial compounds. Molecules 20(8):13997–14021. https://doi.org/10.3390/molecules200813997

156. Gervasoni S, Malloci G, Bosin A, Vargiu AV, Zgurskaya HI, Ruggerone P (2022) AB-DB: Force-Field parameters, MD trajectories, QM-based data, and descriptors of antimicrobials. Sci Data 9(1):148. https://doi.org/10.1038/s41597-022-01261-1
157. Riniker S, Landrum GA (2015) Better informed distance geometry: using what we know to improve conformation generation. J Chem Inf Model 55(12):2562–2574
158. Wang S, Witek J, Landrum GA, Riniker S (2020) Improving conformer generation for small rings and macrocycles based on distance geometry and experimental torsional-angle preferences. J Chem Inf Model 60(4): 2044–2058
159. Hawkins PCD, Skillman AG, Nicholls A (2007) Comparison of shape-matching and docking as virtual screening tools. J Med Chem 50(1):74–82
160. Takatsuka Y, Chen C, Nikaido H (2010) Mechanism of recognition of compounds of diverse structures by the multidrug efflux pump AcrB of Escherichia coli. Proc Natl Acad Sci USA 107(15):6559–6565. https://doi.org/10.1073/pnas.1001460107
161. Basciu A, Callea L, Motta S, Bonvin AMJJ, Bonati L, Vargiu AV (2022) No dance, no partner! A tale of receptor flexibility in docking and virtual screening. Annu Rep Med Chem 59:43–97
162. Unke OT, Chmiela S, Sauceda HE, Gastegger M, Poltavsky I, Schütt KT, Tkatchenko A, Müller K-RJCR (2021) Machine learning force fields. Chem Rev 121(16):10142–10186. https://doi.org/10.1021/acs.chemrev.0c01111
163. Dauber-Osguthorpe P, Hagler AT (2019) Biomolecular force fields: where have we been, where are we now, where do we need to go and how do we get there? J Comput Aided Mol Des 33(2):133–203
164. Van der Spoel D (2021) Systematic design of biomolecular force fields. Curr Opin Struct Biol 67:18–24. https://doi.org/10.1016/j.sbi.2020.08.006
165. Tian C, Kasavajhala K, Belfon KAA, Raguette L, Huang H, Migues AN, Bickel J, Wang Y, Pincay J, Wu Q (2019) ff19SB: amino-acid-specific protein backbone parameters trained against quantum mechanics energy surfaces in solution. J Chem Theory Comput 16(1):528–552. https://doi.org/10.1021/acs.jctc.9b00591
166. Huang J, Rauscher S, Nawrocki G, Ran T, Feig M, De Groot BL, Grubmüller H, MacKerell JAD (2017) CHARMM36m: an improved force field for folded and intrinsically disordered proteins. Nat Methods 14(1): 71–73. https://doi.org/10.1038/nmeth.4067
167. Jorgensen WL, Ghahremanpour MM, Saar A, Tirado-Rives J (2023) OPLS/2020 force field for unsaturated hydrocarbons, alcohols, and ethers. J Phys Chem B 128(1):250–262. https://doi.org/10.1021/acs.jpcb.3c06602
168. Jing Z, Liu C, Cheng SY, Qi R, Walker BD, Piquemal J-P, Ren P (2019) Polarizable force fields for biomolecular simulations: recent advances and applications. Annu Rev Biophys 48:371–394. https://doi.org/10.1146/annurev-biophys-070317-033349
169. Dickson CJ, Walker RC, Gould IR (2022) Lipid21: complex lipid membrane simulations with AMBER. J Chem Theory Comput 18(3):1726–1736. https://doi.org/10.1021/acs.jctc.1c01217
170. Cole DJ, Vilseck JZ, Tirado-Rives J, Payne MC, Jorgensen WL (2016) Biomolecular force field parameterization via atoms-in-molecule electron density partitioning. J Chem Theory Comput 12(5):2312–2323. https://doi.org/10.1021/acs.jctc.6b00027
171. Lin F-Y, MacKerell AD (2019) Force fields for small molecules. Biomol Simul 2022:21–54. https://doi.org/10.1007/978-1-4939-9608-7_2
172. Raniolo S, Limongelli V (2021) Improving small-molecule force field parameters in ligand binding studies. Front Mol Biosci 8: 760283. https://doi.org/10.3389/fmolb.2021.760283
173. Bugnon M, Goullieux M, Röhrig UF, Perez MAS, Daina A, Michielin O, Zoete V (2023) SwissParam 2023: a modern web-based tool for efficient small molecule parametrization. J Chem Inf Model 63(21):6469–6475. https://doi.org/10.1021/acs.jcim.3c01053
174. He X, Walker B, Man VH, Ren P, Wang J (2022) Recent progress in general force fields of small molecules. Curr Opin Struct Biol 72: 187–193. https://doi.org/10.1016/j.sbi.2021.11.011
175. Boothroyd S, Behara PK, Madin OC, Hahn DF, Jang H, Gapsys V, Wagner JR, Horton JT, Dotson DL, Thompson MW (2023) Development and benchmarking of open force field 2.0. 0: the Sage small molecule force field. J Chem Theory Comput 19(11): 3251–3275. https://doi.org/10.1021/acs.jctc.3c00039
176. Alessandri R, Barnoud J, Gertsen AS, Patmanidis I, de Vries AH, Souza PCT, Marrink SJ (2022) Martini 3 coarse-grained force

field: small molecules. Adv Theory Simul 5(1):2100391

177. Xue B, Yang Q, Zhang Q, Wan X, Fang D, Lin X, Sun G, Gobbo G, Cao F, Mathiowetz AM (2023) Development and comprehensive benchmark of a high-quality AMBER-consistent small molecule force field with broad chemical space coverage for molecular modeling and free energy calculation. J Chem Theory Comput 20(2):799–818. https://doi.org/10.1021/acs.jctc.3c00920
178. Wang J, Wolf RM, Caldwell JW, Kollman PA, Case DA (2004) Development and testing of a general amber force field. J Comput Chem 25(9):1157–1174. https://doi.org/10.1002/jcc.20035
179. Malloci G, Serra G, Bosin A, Vargiu AV (2016) Extracting conformational ensembles of small molecules from molecular dynamics simulations: ampicillin as a test case. Computation 4(1):5
180. Lawrence R, Athar M, Uddin MR, Adams C, Sousa JS, Durrant O, Lellman S, Sutton L, Keevil CW, Patel N (2025) Molecular basis for multidrug efflux by an anaerobic-associated RND transporter. Nature Communications 16(1):10601 https://doi.org/10.1038/s41467-025-65565-7
181. Lazarova M, Eicher T, Börnsen C, Zeng H, Athar M, Okada U, Yamashita E, Spannaus IM, Borgosch M, Cha H-j, Vargiu AV, Murakami S, Diederichs K, Frangakis AS, Pos KM (2025) Conformational plasticity across phylogenetic clusters of RND multidrug efflux pumps and its impact on substrate specificity. Nature Communications 16(1):11649. https://doi.org/10.1038/s41467-025-66751-3
182. Manrique PD, Leus IV, López CA, Mehla J, Malloci G, Gervasoni S, Vargiu AV, Kinthada RK, Herndon L, Hengartner NW (2024) Predicting permeation of compounds across the outer membrane of P. aeruginosa using molecular descriptors. Commun Chem 7(1):84. https://doi.org/10.1038/s42004-024-01161-y
183. Mustafa G, Nandekar PP, Mukherjee G, Bruce NJ, Wade RC (2020) The effect of force-field parameters on cytochrome P450-membrane interactions: structure and dynamics. Sci Rep 10(1):7284
184. Saeedimasine M, Montanino A, Kleiven S, Villa A (2019) Role of lipid composition on the structural and mechanical features of axonal membranes: a molecular simulation study. Sci Rep 9(1):8000
185. Harayama T, Riezman H (2018) Understanding the diversity of membrane lipid composition. Nat Rev Mol Cell Biol 19(5):281–296
186. Leonard AN, Wang E, Monje-Galvan V, Klauda JB (2019) Developing and testing of lipid force fields with applications to modeling cellular membranes. Chem Rev 119(9):6227–6269
187. Lyubartsev AP, Rabinovich AL (2016) Force field development for lipid membrane simulations. Biochim Biophys Acta 1858(10):2483–2497
188. Moqadam M, Tubiana T, Moutoussamy EE, Reuter N (2021) Membrane models for molecular simulations of peripheral membrane proteins. Adv Phys 6(1):1932589
189. Marrink SJ, Corradi V, Souza PCT, Ingolfsson HI, Tieleman DP, Sansom MSP (2019) Computational modeling of realistic cell membranes. Chem Rev 119(9):6184–6226
190. Chu H, Cao L, Peng X, Li G (2017) Polarizable force field development for lipids and their efficient applications in membrane proteins. Wiley Interdiscipl Rev 7(5):e1312
191. Lopes PEM, Guvench O, MacKerell AD (2015) Current status of protein force fields for molecular dynamics simulations. In: Molecular modeling of proteins, pp 47–71
192. Best RB (2019) Atomistic force fields for proteins. In: Biomolecular simulations: methods and protocols, pp 3–19
193. Cheatham TE 3rd, Case DA (2013) Twenty-five years of nucleic acid simulations. Biopolymers 99(12):969–977. https://doi.org/10.1002/bip.22331
194. Vargiu AV, Magistrato A (2014) Atomistic-level portrayal of drug-DNA interplay: a history of courtships and meetings revealed by molecular simulations. ChemMedChem 9(9):1966–1981. https://doi.org/10.1002/cmdc.201402203
195. Love O, Galindo-Murillo R, Zgarbová M, Šponer J, Jurečka P, Cheatham ITE (2023) Assessing the current state of Amber Force Field modifications for DNA— 2023 edition. J Chem Theory Comput 19(13):4299–4307
196. Poger D, Caron B, Mark AE (2016) Validating lipid force fields against experimental data: progress, challenges and perspectives. Biochim Biophys Acta 1858(7):1556–1565
197. Srinivasan S, Zoni V, Vanni S (2021) Estimating the accuracy of the MARTINI model towards the investigation of peripheral protein-membrane interactions. Biophys J 120(3):232a

198. Case DA, Aktulga HM, Belfon K, Ben-Shalom IY, Berryman JT, Brozell SR, Cerutti DS, Cheatham TE III, Cisneros GA, Cruzeiro VWD, Darden TA, Forouzesh N, Giambaşu G, Giese T, Gilson MK, Gohlke H, Goetz AW, Harris J, Izadi S, Izmailov SA, Kasavajhala K, Kaymak MC, King E, Kovalenko A, Kurtzman T, Lee TS, Li P, Lin C, Liu J, Luchko T, Luo R, Machado M, Man V, Manathunga M, Merz KM, Miao Y, Mikhailovskii O, Monard G, Nguyen H, O'Hearn KA, Onufriev A, Pan F, Pantano S, Qi R, Rahnamoun A, Roe DR, Roitberg A, Sagui C, Schott-Verdugo S, Shajan A, Shen J, Simmerling CL, Skrynnikov NR, Smith J, Swails J, Walker RC, Wang J, Wang J, Wei H, Wu X, Wu Y, Xiong Y, Xue Y, York DM, Zhao S, Zhu Q, Kollman PA (2023) Amber 2023. University of California, San Francisco

199. Schmidt TH, Kandt C (2012) LAMBADA and InflateGRO2: efficient membrane alignment and insertion of membrane proteins for molecular dynamics simulations. J Chem Inf Model 52(10):2657–2669. https://doi.org/10.1021/ci3000453

200. Guixa-Gonzalez R, Rodriguez-Espigares I, Ramirez-Anguita JM, Carrio-Gaspar P, Martinez-Seara H, Giorgino T, Selent J (2014) MEMBPLUGIN: studying membrane complexity in VMD. Bioinformatics 30(10):1478–1480. https://doi.org/10.1093/bioinformatics/btu037

201. Jo S, Kim T, Iyer VG, Im W (2008) CHARMM-GUI: a web-based graphical user interface for CHARMM. J Comput Chem 29(11):1859–1865. https://doi.org/10.1002/jcc.20945

202. Schott-Verdugo S, Gohlke H (2019) PACKMOL-Memgen: a simple-to-use, generalized workflow for membrane-protein–lipid-bilayer system building. J Chem Inf Model 59(6):2522–2528. https://doi.org/10.1021/acs.jcim.9b00269

203. Lomize MA, Irina D, Joo H, Mosberg HI, Lomize AL (2012) OPM database and PPM web server: resources for positioning of proteins in membranes. Nucleic Acids Res 40 (D1):D370–D376

204. Wu EL, Cheng X, Jo S, Rui H, Song KC, Davila-Contreras EM, Qi Y, Lee J, Monje-Galvan V, Venable RM, Klauda JB, Im W (2014) CHARMM-GUI Membrane Builder toward realistic biological membrane simulations. J Comput Chem 35(27):1997–2004. https://doi.org/10.1002/jcc.23702

205. Nagle JF, Tristram-Nagle S (2000) Lipid bilayer structure. Curr Opin Struct Biol 10(4):474–480

206. Leftin A, Brown MF (2011) An NMR database for simulations of membrane dynamics. Biochim Biophys Acta Biomembr 1808(3): 818–839. https://doi.org/10.1016/j.bbamem.2010.11.027

207. Benz RW, Castro-Román F, Tobias DJ, White SH (2005) Experimental validation of molecular dynamics simulations of lipid bilayers: a new approach. Biophys J 88(2):805–817. https://doi.org/10.1529/biophysj.104.046821

208. Patra M, Karttunen M, Hyvönen MT, Falck E, Lindqvist P, Vattulainen I (2003) Molecular dynamics simulations of lipid bilayers: major artifacts due to truncating electrostatic interactions. Biophys J 84(6): 3636–3645. https://doi.org/10.1016/S0006-3495(03)75094-2

209. Poger D, Caron B, Mark AE (2016) Validating lipid force fields against experimental data: progress, challenges and perspectives. Biochim Biophys Acta 1858(7):1556–1565. https://doi.org/10.1016/j.bbamem.2016.01.029

210. Poger D, Mark AE (2010) On the validation of molecular dynamics simulations of saturated and cis-monounsaturated phosphatidylcholine lipid bilayers: a comparison with experiment. J Chem Theory Comput 6(1): 325–336. https://doi.org/10.1021/ct900487a

211. Kučerka N, Katsaras J, Nagle JF (2010) Comparing membrane simulations to scattering experiments: introducing the SIMtoEXP software. J Membr Biol 235:43–50. https://doi.org/10.1007/s00232-010-9254-5

212. Shahane G, Ding W, Palaiokostas M, Orsi M (2019) Physical properties of model biological lipid bilayers: insights from all-atom molecular dynamics simulations. J Mol Model 25:76. https://doi.org/10.1007/s00894-019-3964-0

213. Moradi S, Nowroozib A, Shahlaei M (2019) Shedding light on the structural properties of lipid bilayers using molecular dynamics simulation: a review study. RSC Adv 9:4644–4658. https://doi.org/10.1039/C8RA08441F

214. Skjevik ÅA, Madej BD, Walker RC, Teigen K (2012) LIPID11: A modular framework for LIPID simulations using Amber. J Phys Chem B 116(36):11124–11136. https://doi.org/10.1021/jp3059992

215. Allen WJ, Lemkul JA, Bevan DR (2009) GridMAT-MD: a grid-based membrane analysis tool for use with molecular dynamics. J Comput Chem 30(12):1952–1958. https://doi.org/10.1002/jcc.21172
216. Buchoux S (2017) FATSLiM: a fast and robust software to analyze MD simulations of membranes. Bioinformatics 33(1). https://doi.org/10.1093/bioinformatics/btw563
217. Dickson CJ, Rosso L, Betz RM, Walkercd RC, Gould IR (2012) GAFFlipid: a General Amber Force Field for the accurate molecular dynamics simulation of phospholipid. Soft Matter 8(37):9617–9627. https://doi.org/10.1039/C2SM26007G
218. Mori T, Ogushi F, Sugita Y (2013) Analysis of lipid surface area in protein–membrane systems combining voronoi tessellation and monte carlo integration methods. J Comput Chem 33(3):286–293. https://doi.org/10.1002/jcc.21973
219. Gapsys V, de Groot BL, Briones R (2013) Computational analysis of local membrane properties. J Comput Aided Mol Des 27: 845–858. https://doi.org/10.1007/s10822-013-9684-0
220. Woo HJ, Roux B (2005) Calculation of absolute protein-ligand binding free energy from computer simulations. Proc Natl Acad Sci USA 102(19):6825–6830. https://doi.org/10.1073/pnas.0409005102
221. Jiao D, Golubkov PA, Darden TA, Ren P (2008) Calculation of protein-ligand binding free energy by using a polarizable potential. Proc Natl Acad Sci USA 105(17): 6290–6295. https://doi.org/10.1073/pnas.0711686105
222. Mitomo D, Fukunishi Y, Higo J, Nakamura H (2009) Calculation of protein-ligand binding free energy using smooth reaction path generation (SRPG) method: a comparison of the explicit water model, gb/sa model and docking score function. Genome Inform 23(1):85–97
223. Steinbrecher T, Labahn A (2010) Towards accurate free energy calculations in ligand protein-binding studies. Curr Med Chem 17(8):767–785
224. Rathore RS, Sumakanth M, Reddy MS, Reddanna P, Rao AA, Erion MD, Reddy MR (2013) Advances in binding free energies calculations: QM/MM-based free energy perturbation method for drug design. Curr Pharm Des 19(26):4674–4686
225. Sousa SF, Ribeiro AJM, Coimbra JTS, Neves RPP, Martins SA, Moorthy N, Fernandes PA, Ramos MJ (2013) Protein-ligand docking in the new millennium–a retrospective of 10 years in the field. Curr Med Chem 20(18):2296–2314
226. Gilson MK, Zhou HX (2007) Calculation of protein-ligand binding affinities. Annu Rev Biophys Biomol Struct 36:21–42. https://doi.org/10.1146/annurev.biophys.36.040306.132550
227. Hansen N, Van Gunsteren WF (2014) Practical aspects of free-energy calculations: a review. J Chem Theory Comput 10(7): 2632–2647. https://doi.org/10.1021/ct500161f
228. Gilson MK, Given JA, Bush BL, Mccammon JA (1997) The statistical-thermodynamic basis for computation of binding affinities: a critical review. Biophys J 72(3):1047–1069. https://doi.org/10.1016/S0006-3495(97)78756-3
229. Zwanzig RW (1954) High-temperature equation of state by a perturbation method. I Nonpolar gases. J Chem Phys 22(8): 1420–1426
230. Kirkwood JG (1935) Statistical mechanics of fluid mixtures. J Chem Phys 3(5):300–313
231. Darve E, Rodríguez-Gómez D, Pohorille A (2008) Adaptive biasing force method for scalar and vector free energy calculations. J Chem Phys 128(14):144120. https://doi.org/10.1063/1.2829861
232. Maragliano L, Vanden-Eijnden EJC (2006) A temperature accelerated method for sampling free energy and determining reaction pathways in rare events simulations. 426(1-3): 168–175
233. St-Pierre J-F, Karttunen M, Mousseau N, Rog T, Bunker A (2011) Use of umbrella sampling to calculate the entrance/exit pathway for Z-Pro-Prolinal inhibitor in prolyl oligopeptidase. J Chem Theory Comput 7(6): 1583–1594. https://doi.org/10.1021/ct1007058
234. Fajer M, Hamelberg D, McCammon JA (2008) Replica-exchange accelerated molecular dynamics (REXAMD) applied to thermodynamic integration. J Chem Theory Comput 4(10):1565–1569. https://doi.org/10.1021/ct800250m
235. Srinivasan J, Cheatham TE III, Cieplak P, Kollman PA, Case DA (1998) Continuum solvent studies of the stability of DNA, RNA, and phosphoramidate-DNA helices. J Am Chem Soc 120(37):9401–9409
236. Kollman PA, Massova I, Reyes C, Kuhn B, Huo SH, Chong L, Lee M, Lee T, Duan Y, Wang W, Donini O, Cieplak P, Srinivasan J,

Case DA, Cheatham TE (2000) Calculating structures and free energies of complex molecules: combining molecular mechanics and continuum models. Accounts Chem Res 33(12):889–897. https://doi.org/10.1021/ar000033j

237. Fogolari F, Brigo A, Molinari H (2003) Protocol for MM/PBSA molecular dynamics simulations of proteins. Biophys J 85(1): 159–166. https://doi.org/10.1016/S0006-3495(03)74462-2

238. Hou T, Wang J, Li Y, Wang W (2010) Assessing the performance of the MM/PBSA and MM/GBSA methods. 1. The accuracy of binding free energy calculations based on molecular dynamics simulations. J Chem Inf Model 51(1):69–82. https://doi.org/10.1021/ci100275a

239. Brandsdal BO, Österberg F, Almlöf M, Feierberg I, Luzhkov VB, Åqvist J (2003) Free energy calculations and ligand binding. Adv Protein Chem 66:123–158. https://doi.org/10.1016/s0065-3233(03)66004-3

240. Christ CD, Mark AE, Van Gunsteren WF (2010) Basic ingredients of free energy calculations: a review. J Comput Chem 31(8): 1569–1582. https://doi.org/10.1002/jcc.21450

241. Kongsted J, Ryde U (2009) An improved method to predict the entropy term with the MM/PBSA approach. J Comput Aided Mol Des 23(2):63–71. https://doi.org/10.1007/s10822-008-9238-z

242. Massova I, Kollman PA (1999) Computational alanine scanning to probe protein-protein interactions: a novel approach to evaluate binding free energies. J Am Chem Soc 121(36):8133–8143. https://doi.org/10.1021/Ja990935j

243. Ramaswamy VK, Vargiu AV, Malloci G, Dreier J, Ruggerone P (2017) Molecular rationale behind the differential substrate specificity of bacterial RND multi-drug transporters. Sci Rep 7(1):1–18

244. Sali A, Blundell TL (1993) Comparative protein modelling by satisfaction of spatial restraints. J Mol Biol 234(3):779–815

245. Marti-Renom MA, Stuart AC, Fiser A, Sanchez R, Melo F, Sali A (2000) Comparative protein structure modeling of genes and genomes. Annu Rev Biophys Biomol Struct 29:291–325. https://doi.org/10.1146/Annurev.Biophys.29.1.291

246. Fiser A, Do RKG, Sali A (2000) Modeling of loops in protein structures. Protein Sci 9(9): 1753–1773. https://doi.org/10.1110/ps.9.9.1753

247. Eberhardt J, Santos-Martins D, Tillack AF, Forli S (2021) AutoDock Vina 1.2. 0: new docking methods, expanded force field, and python bindings. J Chem Inf Model 61(8): 3891–3898

248. Santos-Martins D, Solis-Vasquez L, Tillack AF, Sanner MF, Koch A, Forli S (2021) Accelerating AutoDock4 with GPUs and gradient-based local search. J Chem Theory Comput 17(2):1060–1073

249. Humphrey W, Dalke A, Schulten K (1996) VMD: visual molecular dynamics. J Mol Graph 14(1):33–38. https://doi.org/10.1016/0263-7855(96)00018-5

250. Balabin IA (2010) Membrane Plug-in. 1.1 edn.

251. Kozakov D, Grove LE, Hall DR, Bohnuud T, Mottarella SE, Luo L, Xia B, Beglov D, Vajda S (2015) The FTMap family of web servers for determining and characterizing ligand-binding hot spots of proteins. Nat Protoc 10(5):733–755

252. Altschul SF, Gish W, Miller W, Myers EW, Lipman DJ (1990) Basic local alignment search tool. J Mol Biol 215(3):403–410. https://doi.org/10.1016/S0022-2836(05)80360-2

253. Gish W, States DJ (1993) Identification of protein coding regions by database similarity search. Nat Genet 3(3):266–272. https://doi.org/10.1038/ng0393-266

254. Madden TL, Tatusov RL, Zhang J (1996) Applications of network BLAST server. Methods Enzymol 266:131–141

255. Shen MY, Sali A (2006) Statistical potential for assessment and prediction of protein structures. Protein Sci 15(11):2507–2524

256. Kim S, Chen J, Cheng T, Gindulyte A, He J, He S, Li Q, Shoemaker BA, Thiessen PA, Yu B (2023) PubChem 2023 update. Nucleic Acids Res 51(D1):D1373–D1380

257. O'Boyle NM, Banck M, James CA, Morley C, Vandermeersch T, Hutchison GR (2011) Open babel: an open chemical toolbox. J Cheminform 3:33. https://doi.org/10.1186/1758-2946-3-33

258. Pedretti A, Villa L, Vistoli G (2004) VEGA–an open platform to develop chemo-bio-informatics applications, using plug-in architecture and script programming. J Comput Aided Mol Des 18:167–173

259. Frisch MJ, Trucks GW, Schlegel HB, Scuseria GE, Robb MA, Cheeseman JR et al (2016) Gaussian 16 Rev. A.03. Wallingford

260. Hohenberg P, Kohn W (1964) Inhomogeneous electron gas. Phys Rev 136:B864–B871

261. Kohn W, Sham L (1965) Self-consistent equations including exchange and correlation effects. Phys Rev 140:A1133–A1138
262. Tirado-Rives J, Jorgensen WL (2008) Performance of B3LYP density functional methods for a large set of organic molecules. J Chem Theory Comput 4(2):297–306
263. Becke AD (1992) Density-functional thermochemistry. I. The effect of the exchange-only gradient correction. J Chem Phys 96(3):2155–2160
264. Kim K, Jordan KD (1994) Comparison of density-functional and Mp2 calculations on the water monomer and dimer. J Phys Chem-Us 98(40):10089–10094. https://doi.org/10.1021/J100091a024
265. Pople JA (1999) Quantum chemical models (Nobel lecture). Angew Chem Int Ed 38(13–14):1894–1902
266. Tomasi J, Mennucci B, Cammi R (2005) Quantum mechanical continuum solvation models. Chem Rev 105(8):2999–3093. https://doi.org/10.1021/cr9904009
267. Singh UC, Kollman PA (1984) An approach to computing electrostatic charges for molecules. J Comput Chem 5:129–145
268. Laio A, Vandevondele J, Rothlisberger U (2002) D-RESP: dynamically generated electrostatic potential derived charges from quantum mechanics / molecular mechanics simulations. J Phys Chem B 106:7300–7307
269. Wang J, Wang W, Kollman PA, Case DA (2006) Automatic atom type and bond type perception in molecular mechanical calculations. J Mol Graph Model 25(2):247–260
270. Izadi S, Anandakrishnan R, Onufriev AV (2014) Building water models: a different approach. J Phys Chem Lett 5(21): 3863–3871
271. Liao C, Nicklaus MC (2009) Comparison of nine programs predicting p K a values of pharmaceutical substances. J Chem Inform Model 49(12):2801–2812
272. Marvin (2012) Marvin 14.8.25.0. http://www.chemaxon.com
273. Shelley JC, Cholleti A, Frye LL, Greenwood JR, Timlin MR, Uchimaya M (2007) Epik: a software program for pK a prediction and protonation state generation for drug-like molecules. J Comput Aided Mol Des 21: 681–691
274. Dolinsky TJ, Nielsen JE, McCammon JA, Baker NA (2004) PDB2PQR: an automated pipeline for the setup of Poisson-Boltzmann electrostatics calculations. Nucleic Acids Res 32(Web Server issue):W665–W667. https://doi.org/10.1093/nar/gkh381
275. Agarwal R, Smith JC (2023) Speed vs accuracy: effect on ligand pose accuracy of varying box size and exhaustiveness in AutoDock Vina. Mol Inform 42(2):2200188. https://doi.org/10.1002/minf.202200188
276. Goossens K, De Winter H (2018) Molecular dynamics simulations of membrane proteins: an overview. J Chem Inf Model 58(11): 2193–2202. https://doi.org/10.1021/acs.jcim.8b00639
277. Muller MP, Jiang T, Sun C, Lihan M, Pant S, Mahinthichaichan P, Trifan A, Tajkhorshid E (2019) Characterization of lipid–protein interactions and lipid-mediated modulation of membrane protein function through molecular simulation. Chem Rev 119(9): 6086–6161. https://doi.org/10.1021/acs.chemrev.8b00608
278. Li Y, Liu J, Gumbart JC (2021) Preparing membrane proteins for simulation using CHARMM-GUI. Methods Mol Biol 2302: 237–251. https://doi.org/10.1007/978-1-0716-1394-8_13
279. Murzyn K, Róg T, Pasenkiewicz-Gierula M (2005) Phosphatidylethanolamine-phosphatidylglycerol bilayer as a model of the inner bacterial membrane. Biophys J 88(2):1091–1103. https://doi.org/10.1529/biophysj.104.048835
280. Reading E, Ahdash Z, Fais C, Ricci V, Wang-Kan X, Grimsey E, Stone J, Malloci G, Lau AM, Findlay H, Konijnenberg A, Booth PJ, Ruggerone P, Vargiu AV, Piddock LJV, Politis A (2020) Perturbed structural dynamics underlie inhibition and altered efflux of the multidrug resistance pump AcrB. Nat Commun 11:5565. https://doi.org/10.1038/s41467-020-19397-2
281. Izadi S, Anandakrishnan R, Onufriev AV (2014) Building water models: a different approach. J Phys Chem Lett 5(21): 3836–3871. https://doi.org/10.1021/jz501780a
282. Phillips JC, Braun R, Wang W, Gumbart J, Tajkhorshid E, Villa E, Chipot C, Skeel RD, Kalé L, Schulten K (2005) Scalable molecular dynamics with NAMD. J Comput Chem 26(16):1781–1802. https://doi.org/10.1002/jcc.20289
283. Berendsen HJC, van der Spoel D, van Drunen R (1995) GROMACS: a message-passing parallel molecular dynamics implementation. Comput Phys Commun 91(1–3):43–56. https://doi.org/10.1016/0010-4655(95)00042-E
284. Wang DD, Chan M-T (2022) Protein-ligand binding affinity prediction based on profiles of

intermolecular contacts. Comput Struct Biotechnol J 20:1088–1096

285. Bouysset C, Fiorucci S (2021) ProLIF: a library to encode molecular interactions as fingerprints. J Chem 13(1):72
286. Case DA, Berryman JT, Betz RM, Cerutti DS, Cheatham TE III, Darden TA, Duke RE, Giese TJ, Gohlke H, Goetz AW, Homeyer N, Izadi S, Janowski P, Kaus J, Kovalenko A, Lee TS, LeGrand S, Li P, Luchko T, Luo R, Madej B, Merz KM, Monard G, Needham P, Nguyen H, Nguyen HT, Omelyan I, Onufriev A, Roe DR, Roitberg A, Salomon-Ferrer R, Simmerling CL, Smith W, Swails J, Walker RC, Wang J, Wolf RM, Wu X, York DM, Kollman PA (2015) AMBER 2015. University of California, San Francisco
287. Miller BR, McGee TD, Swails JM, Homeyer N, Gohlke H, Roitberg AE (2012) MMPBSA.py: an efficient program for end-state free energy calculations. J Chem Theory Comput 8(9):3314–3321. https://doi.org/10.1021/ct300418h
288. Tsui V, Case DA (2000) Molecular dynamics simulations of nucleic acids with a generalized born solvation model. J Am Chem Soc 122(11):2489–2498. https://doi.org/10.1021/ja9939385
289. Bondi A (1964) van der Waals Volumes and Radii. J Phys Chem 68(3):441–451. https://doi.org/10.1021/j100785a001
290. Onufriev A, Bashford D, Case DA (2004) Exploring protein native states and large-scale conformational changes with a modified generalized born model. Proteins-Struct Funct Bioinform 55(2):383–394. https://doi.org/10.1002/prot.20033
291. Mongan J, Simmerling C, McCammon JA, Case DA, Onufriev A (2007) Generalized Born model with a simple, robust molecular volume correction. J Chem Theory Comput 3(1):156–169. https://doi.org/10.1021/ct600085e
292. Morris GM, Huey R, Lindstrom W, Sanner MF, Belew RK, Goodsell DS, Olson AJ (2009) AutoDock4 and AutoDockTools4: automated docking with selective receptor flexibility. J Comput Chem 30(16):2785–2791. https://doi.org/10.1002/jcc.21256
293. Grossfield A, Zuckerman DM (2009) Quantifying uncertainty and sampling quality in biomolecular simulations. Annu Reports Comput Chem 5:23–48. https://doi.org/10.1016/S1574-1400(09)00502-7
294. Grossfield A, Feller SE, Pitman MC (2007) Convergence of molecular dynamics simulations of membrane proteins. Proteins 67(1):31–40. https://doi.org/10.1002/prot.21308

Part IV

Genetic Tools and Functional Studies of Multidrug Efflux Pumps

Chapter 11

Alanine Scanning Analysis of MexB in Multidrug-Resistant *Pseudomonas aeruginosa* Using the Native Type I–F CRISPR–Cas-Mediated Genome Editing

Wenrui Wan, Jiajun Ning, and Aixin Yan

Abstract

The rising multidrug resistance (MDR) of the Gram-negative opportunistic pathogen *Pseudomonas aeruginosa* poses a global public health threat. One key resistance mechanism employed by this pathogen is the overexpression of the resistance–nodulation–cell division (RND) multidrug efflux transporters. MexAB-OprM, a prototype RND efflux pump, confers resistance to most conventional antibiotics except aminoglycosides. Molecules capable of inhibiting efflux pumps, i.e., efflux pump inhibitors (EPIs), have shown promise as therapeutic agents capable of restoring antibiotic efficacy against these superbugs by targeting MexB. Nevertheless, the structure–activity relationship governing substrate transport of MexB in the native genetic background of clinical MDR *P. aeruginosa* isolates is poorly characterized due to the lack of efficient genetic tools, hindering targeted EPIs development.

In this chapter, we introduce a native Type I–F CRISPR-mediated precise genome editing technique that enables in situ alanine substitutions in MexB in clinical MDR *P. aeruginosa* genotypes. We outline detailed procedures to construct MexB alanine-substitution mutations and methods to examine the expression levels and efflux activities of the constructed MexB mutants. Methods to evaluate the effect of constructed alanine substitutions on antibiotic susceptibilities are also described. These methodologies are expected to provide advanced insights into the substrate recognition and selection mechanism of MexB in the native genetic background of clinical MDR *P. aeruginosa* isolates to facilitate the development of effective EPIs.

Key words *Pseudomonas aeruginosa*, Multidrug resistance, Multidrug efflux pump, Type I–F CRISPR–Cas, In situ genome editing, SDS–PAGE, Western blot

1 Introduction

The rising tolerance of Gram-negative bacteria to antibiotics across various classes represents one of the greatest global health challenges. The opportunistic pathogen *Pseudomonas aeruginosa* is particularly concerning due to its exceptional capacity to develop multidrug resistance (MDR) [1]. A major resistance mechanism of this notorious pathogen is the overexpression of efflux pumps belonging to the resistance–nodulation–cell division (RND) family.

Kunihiko Nishino and Mikio Tanabe (eds.), *Bacterial Multidrug Exporters: Methods and Protocols*, Methods in Molecular Biology, vol. 3044, https://doi.org/10.1007/978-1-0716-5316-6_11,
© The Author(s), under exclusive license to Springer Science+Business Media, LLC, part of Springer Nature 2026

These pumps are composed of an inner membrane transporter, a periplasmic adaptor protein, and an outer membrane channel protein, forming tripartite complexes to recognize and extrude a broad spectrum of antimicrobial drugs [2]. *P. aeruginosa* encodes 12 RND efflux pumps in its genome [3, 4]. Among them, MexB is the housekeeping RND multidrug efflux transporter and is widely considered as the archetype RND transporter for studying the efflux mechanisms in Gram-negative bacteria. Overexpression of this transporter can lead to resistance to almost all β-lactams, quinolones, tetracyclines, and macrolides in clinics [5].

The diminishing discovery of new antibiotics underscores the urgency to rejuvenate the efficacy of existing drugs that have lost their activity against MDR bacteria and safeguard the current antibiotic pipeline. Inhibition of drug efflux represents an important strategy to restore the efficacy of antimicrobial drugs, but the lack of insights into the substrate selection mechanism of RND transporters has hindered the development of efflux pump inhibitors (EPIs). Efforts of structural biologists over the last 20 years have revealed that MexB works by a functionally rotating mechanism, with each protomer transiting through three conformational states that open and collapse two binding pockets to shuttle substrates into the exit channel [6–10]. Nevertheless, a systematic understanding of the structure–activity relationship of MexB in the native genetic background of clinical MDR *P. aeruginosa* isolates is still lacking. Previous studies of RND transporters are mostly limited to *E. coli* AcrB or heterologous reconstitution of MexB in *E. coli*. However, the genetic background and environmental context of clinical MDR strains significantly influence the efflux actions of RND transporters and impact the efficacy of antibiotic therapies [11, 12]. In situ investigations in MDR *P. aeruginosa* isolates have been challenging due to the lack of genetic tools compatible with these strain backgrounds.

The advent of the CRISPR–Cas9 system has revolutionized genome editing in a plethora of organisms. However, its applications in bacterial genome editing are rather limited due to the low transformation efficiency and cytotoxicity issues caused by heterologous expression of the bulky Cas9 and the need for strain-specific optimization [13–16]. To date, successful Cas9-based genome editing in *P. aeruginosa* has been confined to the model strain PAO1 [17], with failed application in clinical MDR genotypes. Remarkably, *P. aeruginosa* genomes encode a large number of the Type I–F CRISPR–Cas systems, and the species is the model organism for studying the CRISPR subtype. Previously, our group has established a native Type I–F CRISPR-based genome editing technique in a clinical MDR *P. aeruginosa* strain PA154197, which was isolated from a bloodstream infection case in Hong Kong [5, 12]. Leveraging the high DNA-interference activity of the native Type I–F Cascade in the strain, this technique achieves genome editing by expressing a programmed mini-CRISPR targeting the desired genetic locus and simultaneously providing a repair donor containing the intended mutation in a single editing plasmid. Upon introduction of this single editing plasmid into *P. aeruginosa* cells, high-efficiency gene editing is achieved in one step [12, 18, 19]. Functional genomics investigations have revealed that the high resistance profile of PA154197 is mainly attributed to its overexpression of the housekeeping *mexB* [5], making this strain a paradigm to study the structure–activity relationship of MexB in a naturally evolved MDR genetic background.

In this chapter, we present alanine scanning analysis of PA154197 MexB employing the native Type I–F CRISPR–Cas-mediated precise point mutation strategy. We first introduce the strategies of a two-step In-Del method for precise point mutations in a batch on chromosomal MexB, followed by methods to validate the expression of the resulting MexB variants and their efflux activities. Lastly, methods to study the effect of alanine substitutions on antibiotic susceptibility changes of the resulting mutants are presented. As a highly conserved, chromosomally encoded resistant determinant in all *P. aeruginosa* strains, insights garnered from investigations of MexB in the paradigm clinical MDR genotype PA154197 are anticipated to have broad clinical relevance, informing rational design and modifications of EPIs for combating drug-resistant infections.

2 Materials

Prepare all solutions using milli-Q water and analytical grade reagents. Prepare and store all reagents at room temperature unless indicated otherwise.

2.1 Materials for the Construction of the Targeting Plasmid (pTargeting) and the Editing Plasmid (pEditing)

1. Plasmid pAY9868: platform vector with a BsaI endonuclease site flanked by two 28-bp repeat sequences (5′-GTTCACTG CCGTATAGGCAGCTAAGAAA-3′) specific to the Type I–F CRISPR–Cas system, a kanamycin-resistant gene, and a luminescence-encoding operon (*luxCDABE*) for the dual selection of transformants.
2. NanoVue Plus spectrophotometer (GE Healthcare, Little Chalfont, UK).
3. Restriction enzymes: BsaI and XhoI.
4. DNA extraction kit.
5. T4 polynucleotide kinase.
6. NaCl: 1 M solution in water.

7. Quick ligase and buffer: 66 mM Tris–HCl, 10 mM $MgCl_2$, 1 mM dithiothreitol, 1 mM ATP, and 7.5% polyethylene glycerol. Prepare at 25 °C and adjust to pH 7.6.
8. Competent cells of *E. coli* strain DH5α prepared by the $CaCl_2$ treatment method [20] and stored at −80 °C.
9. Taq DNA polymerase.
10. Agarose: molecular biology grade for 0.8% agarose gels.
11. Plasmid DNA miniprep kit.
12. High-fidelity DNA polymerase.
13. Gibson assembly kit.

2.2 Conjugation and Plasmid Curing

1. Competent cells of *E. coli* strain SM10 prepared using the $CaCl_2$ treatment method [20] and stored at −80 °C.
2. Vogel–Bonner minimal medium (10×): To prepare the stock solution, weigh 0.1925 g $MgSO_4$, 4.375 g citric acid monohydrate, 7 g $NaNH_4HPO{\cdot}4H_2O$, and 20 g K_2HPO_4 and dissolve them in 180 mL of water. Mix well with magnetic stirring until fully dissolved. Finalize to 200 mL with water and pass through a 0.2 μm membrane filter into a sterile Schott bottle.
3. Synergy HTX Plate Reader (BioTek, USA).

2.3 SDS–PAGE and Western Blotting

1. Ultrospec 2100 pro spectrophotometer (Amersham Biosciences, UK).
2. Complete bacterial protein extraction reagent: detergent in 20 mM Tris-HCl buffer, pH 7.5, containing lysozyme and universal nuclease.
3. Protein colorimetric assay kit: bovine serum albumin (BSA) as the standard.
4. SDS–PAGE: gels (10%), apparatus, accessories.
5. SDS sample buffer (4×): 250 mM Tris-HCl (pH 6.8), 40% glycerol, 8%SDS, 0.04% bromophenol blue. Prepare the stock solution by dissolving 3.03 g Tris base in 50 mL of water, adjusting the pH to 6.8 with HCl. Add 40 mL of glycerol, 8 g SDS, and 0.04 g bromophenol blue, mixing until completely dissolved. Bring up to 100 mL with water. Add 10% β-mercaptoethanol prior to use.
6. SDS–PAGE running buffer (10×): 25 mM Tris-HCl, 192 mM glycine, 0.1% SDS. To prepare the stock solution, weigh 288 g glycine and 60.4 g Tris base and dissolve them in 1.8 L of water. Add 20 g SDS and mix until completely dissolved. Finalize to 2 L with water.
7. TBST buffer: 50 mM Tris–HCl (pH 7.6), 150 mM NaCl, 0.05% Tween 20.

8. Primary antibody: Anti-FLAG polyclonal affinity antibody produced in rabbit (Sigma-Aldrich, USA) is used for MexB-3xFLAG. Anti-GroEL polyclonal antibody (Sigma-Aldrich, USA) is used for the internal control. Store at −20 °C.
9. Secondary antibody: Horseradish peroxidase-conjugated goat anti-rabbit antibody (Sigma-Aldrich, USA). Store at −20 °C.
10. ECL Western Blotting Detection Reagents (GE Healthcare, UK). Store at 4 °C.

2.4 Hoechst Efflux Assay

1. Black polystyrene 96-well cell culture plates.
2. D-glucose solution: 0.4% (w/v) D-glucose dissolved in PBS. Sterilize by passing through a 0.2 μm syringe filter.
3. Hoechst 33342 (Beyotime, China).
4. Synergy HTX plate reader (BioTek, USA).

2.5 MIC Measurement

1. Polystyrene 96-well cell culture plate with a low evaporation lid.
2. Synergy HTX plate reader (BioTek, USA).

3 Methods

3.1 Construction of the Targeting Plasmid (pTargeting) and the Editing Plasmid (pEditing)

Since genetic editing of point mutations has a stringent requirement for the distance between the PAM sequence and the intended mutation site, we developed a two-step In-Del approach to enable point mutation at any desired site and to facilitate batch mutations (Fig. 1). This approach includes two rounds of one-step editing. The first round is to insert a 32-bp tag sequence upstream of the intended editing residue, while the second round simultaneously removes the tag and substitutes the target residue with alanine (Fig. 1a). Perform the procedures described in this section twice to construct two editing plasmids with different mini-CRISPR and repair donor sequences for each mutant. In this chapter, we selected three regions in MexB to perform alanine scanning analysis, which are the deep binding pocket groove (Q176, V177, F178, G179, S180), the cave (Q273, D274A, S276, I277), and the switch loop motif (F615, F617, R620) (Fig. 2a). Here, we use MexB Q273A alanine substitution as an example to illustrate the detailed workflow.

3.1.1 Construction of the Targeting Plasmid (pTargeting) with the Desired Mini-CRISPR

1. Digest the pAY9868 platform plasmid using BsaI endonuclease and purify the product using a DNA extraction kit or ethanol precipitation. Examine the quality and quantity of the purified plasmid using a NanoVue Plus spectrophotometer.
2. Identify a 5′-CC-3′ protospacer adjacent motif (PAM) sequence within 50 bp upstream or downstream of the desired point mutation site in the *mexB* locus of *P. aeruginosa*

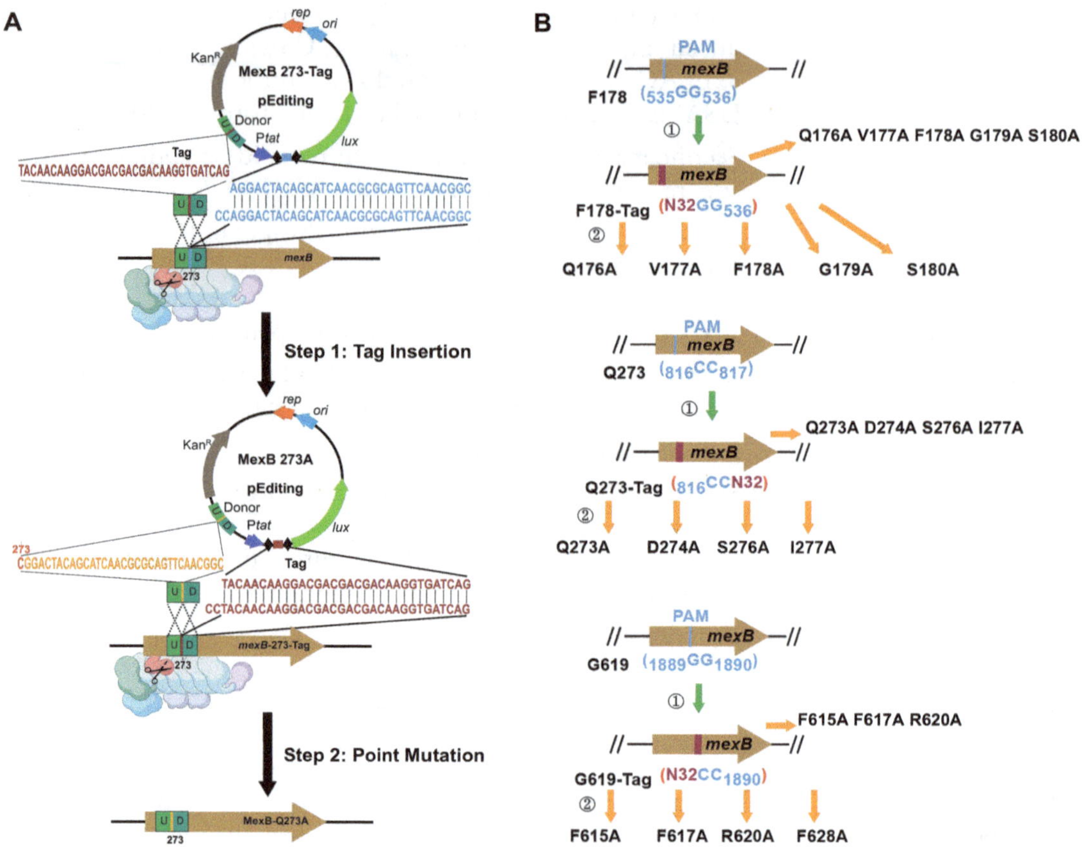

Fig. 1 Schematic diagram showing the native Type I–F CRISPR-mediated two-step In-Del strategy for constructing point mutations in the chromosomal *mexB* gene. (**a**) The construction workflow of the MexB Q273A variant as an example. A 32-bp tag sequence (red) is inserted first by targeting the 32-bp sequence (light blue) immediately downstream of a PAM (5′-$_{816}$CC-3′) that is within 50 bp upstream of Q273. This tag sequence (red) is incorporated in the repair donor of the editing plasmid (MexB 273-Tag pEditing) and is delivered to the wild-type PA154197 to generate the tag-inserted strain (*mexB*-273-Tag, middle panel). In the second step, a second editing plasmid (MexB 273A pEditing) that targets the tag sequence (red) and contains a repair donor carrying the desired Q273A point mutation (gray) is constructed. This plasmid is delivered to the tag-inserted strain (*mexB*-273-Tag), achieving simultaneous tag removal and point mutation (MexB-Q273A, right panel). Adjacent residues can share the same tag-inserted strain for the second round of editing. (**b**) Diagram showing the specific PAM locations, tag insertion sites, and the desired alanine substitutions that can be used to construct

PA154197 (*see* **Note 1**). Select 32 bp immediately downstream of the PAM sequence as the spacer sequence.

3. Append a four-nucleotide homologous arm to the 5′ end of the respective spacer sequence to create two 36-bp oligos: 5′-GAAA-top strand spacer-3′ and 5′-GAAC-bottom strand spacer-3′. Submit the designed two 36-bp oligos for synthesis commercially.
4. Upon receiving the synthesized two oligos, set up the following mixture: 1 μL of T4 polynucleotide kinase, 25 μL of quick

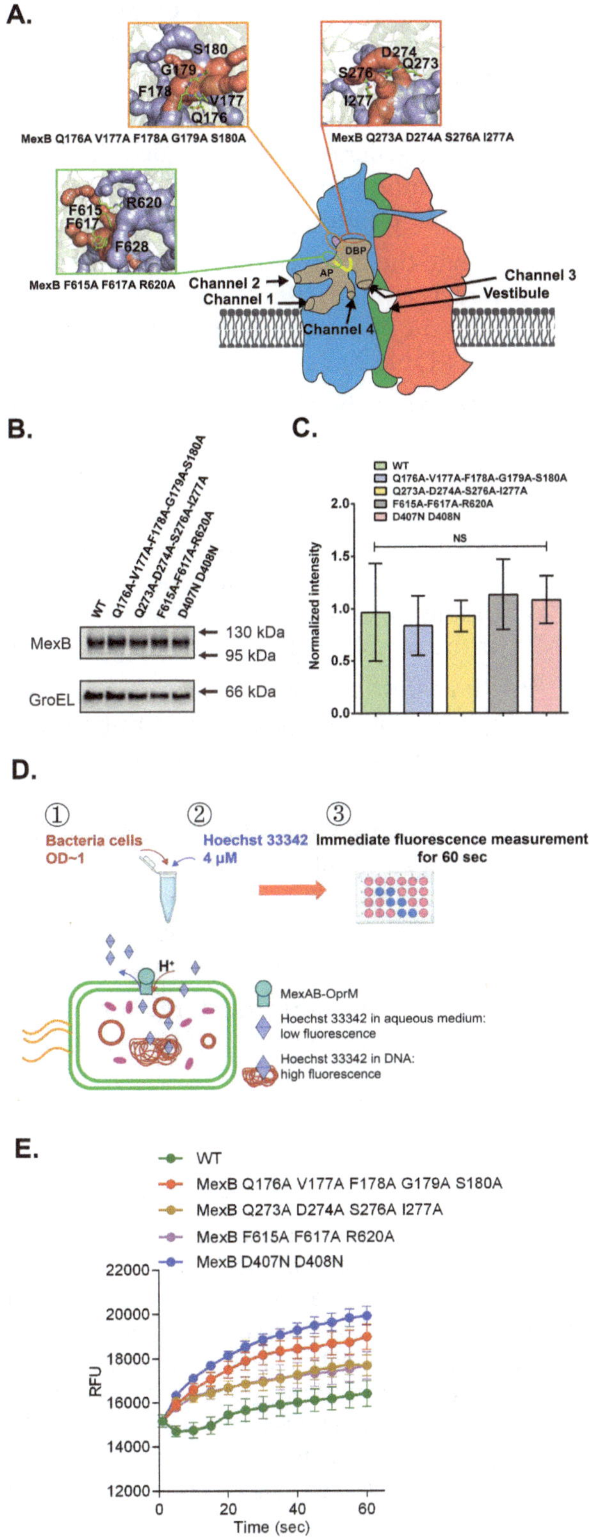

Fig. 2 Examination of the expression level and efflux activity of alanine-substitution mutants of MexB. (**a**) Schematic diagram showing the MexB mutant constructs. Highlighted residues were substituted with alanine.

ligase buffer, 20 μL of H_2O, and 2 μL of each of the synthesized oligos at 50 μM. Incubate at 37 °C for 1 h for phosphorylation.

5. Add 2.5 μL of 1 M NaCl and then heat the solution at 95 °C for 3 min, generating the annealed double-stranded spacer. Allow the solution to cool down to room temperature and dilute by 20-fold by adding 950 μL of H_2O.
6. Ligate the double-stranded spacer into the BsaI-digested plasmid vector pAY9868 using quick ligase, resulting a plasmid with the desired 88-bp mini-CRISPR array, termed as pTargeting (Table 2).
7. Transform the pTargeting into *E. coli* DH5α competent cells, plate transformants on LB plates containing 20 μg/mL kanamycin. Incubate plates at 37 °C overnight (15–20 h) to obtain colonies.
8. Perform colony PCR to validate the construction of pTargeting using Taq polymerase and primer pair PMS-F and Spacer-R (Table 1). A colony PCR using the platform plasmid pAY9868 as the template serves as the negative control.
9. Run a 0.8% agarose gel to examine the PCR product. The PCR product of the successfully constructed pTargeting should be 340 bp. PCR of the negative control pAY9868 should yield no products.
10. Inoculate two individual colonies validated above into 5 mL of LB broth supplemented with 20 μg/mL kanamycin to propagate the plasmids by growing the inoculum at 37 °C with 220 rpm agitation overnight. Extract pTargeting from the overnight culture with a miniprep kit.
11. Sequence the purified pTargeting with primer pairs PMS-F/R to confirm the desired incorporation of the spacer sequence (*see* **Note 2**).

Fig. 2 (continued) (**b**) Western blot analysis of the production of the MexB and various mutants using polyclonal anti-FLAG antibody. The expression level of GroEL is used as the internal control. (**c**) Relative production level of MexB-3xFLAG from various samples quantified using ImageJ software. Normalized intensities are deviated from the results of three biological replicates. (**d**) Workflow of the Hoechst efflux assay. Bacterial cells expressing wild-type and mutant MexB are harvested at OD_{600} ~ 1.0. Hoechst 33342 dye was added to the cell suspension to 4 μM, followed by immediate fluorescence reading for 60 s using a microplate reader. Higher fluorescence intensity indicates more dye molecules binding to the bacterial DNA. (**e**) Intracellular accumulation of the Hoechst 33342 in wild-type and mutant MexB. Fluorescence intensities were measured in real-time. The alanine-substitution mutants display impaired efflux capacity compared to the wild-type MexB. The MexB D407N D408N mutation served as a control for inactive MexB

3.1.2 Construction of the Editing Plasmid (pEditing)

1. Linearize the constructed pTargeting by XhoI digestion. Purify the linearized plasmid using a DNA extraction kit. Examine the quality and quantity of the purified plasmid using a NanoVue Plus spectrophotometer.
2. Select the 18-bp upstream sequence and the 18-bp downstream sequence of the desired editing site, respectively, as primers Donor-UR and Donor-DF.
3. Generate the complementary primer pairs (Donor-UF and Donor-DR) by inputting a 500–800 bp upstream sequence and a 500–800 bp downstream sequence that span the substitution site on the NCBI Primer-BLAST website (https://www.ncbi.nlm.nih.gov/tools/primer-blast/index.cgi).
4. Flank Donor-UF and Donor-DR by the upstream and downstream homologous arm, respectively. The homologous arm should overlap by 21 bp on the respective end of the XhoI site on the pTargeting, as indicated in Table 1.

Table 1
Oligonucleotides used for plasmid construction

PMS-F	5′-CCAGCTGGCAATTCCGA-3′
PMS-R	5′-AATCATCACTTTCGGGAAAG-3′
Spacer-F	5′-GAAANNNNNNNNNNNNNNNNNNNNNNNNNNNNNNNNNN-3′
Spacer-R	5′-GAACNNNNNNNNNNNNNNNNNNNNNNNNNNNNNNNNNN-3′
Donor-tag-UR	5′-CTGATCACCTTGTCGTCGTCGTCCTTGTTGTANNNNNNNNNNNNNN NNNN-3′
Donor-tag-DF	5′-TACAACAAGGACGACGACGACAAGGTGATCAGNNNNNNNNNNNNNN NNNN-3′
Donor-UF	5′-CCCTTTCGTCTTCACCTCGAGNNNNNNNNNNNNNNNNNNNN-3′
Donor-DR	5′-GATCAGGAACACCCCCTCGAGNNNNNNNNNNNNNNNNNNNN-3′
FLAG-spacer-F	5′-GAAAAAGCAGCAGGCGTCCGTCGAAAAGGGGCAATG-3′
FLAG-spacer-R	5′-GAACCATTGCCCCTTTTCGACGGACGCCTGCTGCTT-3′
FLAG-donor-tag-UR	5′-CTGATCACCTTGTCGTCGTCGTCCTTGTTGTAGGACGCCTCGTCCTT GAA-3′
FLAG-donor-tag-DF	5′-TACAACAAGGACGACGACGACAAGGTGATCAGAAGCAGCAGGCG TCCGTC-3′
FLAG-donor-UF	5′-CCCTTTCGTCTTCACCTCGAGTCAAGCAATTGCCGAAAGGC-3′
FLAG-donor-DR	5′-GATCAGGAACACCCCCTCGAGACTCTTCTGGTAGGTGCCCA-3′

5. Flank Donor-UR and Donor-DF by the 32-bp tag sequence at the 5′ end to generate Donor-Tag-UR and Donor-Tag-DF, as indicated in Table 1 (*see* **Note 3**).
6. Amplify the upstream and downstream donor fragments by PCR using a high-fidelity DNA polymerase with primer pairs (Donor-UF/Donor-Tag-UR for the upstream donor; Donor-Tag-DF/Donor-DR for the downstream donor), using the genomic DNA of PA154197 as the template (*see* **Note 4**).
7. Validate and purify the PCR products following the procedures described in Subheading 3.1.1. Examine the quality and quantity of the purified DNA fragments using a NanoVue Plus spectrophotometer.
8. Assemble the donor fragments into the linearized pTargeting by Gibson assembly, generating the editing plasmid (pEditing) (Table 2). The ratio of the DNA inserts vs. the vector is ~1:1:1 (*see* **Note 5**). An assembly reaction without the addition of donor fragments serves as the negative control.
9. Transform the reaction mixture into *E. coli* DH5a competent cells and plate the transformants on LB plates containing 20 μg/mL kanamycin. Incubate plates at 37 °C overnight (15–20 h) to obtain colonies. No colony should be recovered in the negative control plate (without the addition of donor fragments).

Table 2
Strains and plasmids

Strain or plasmid	Description
Strains	
DH5α	*E. coli* strain used for plasmid construction and propagation
SM10	*E. coli* strain used for plasmid delivery via conjugation
PA154197	An MDR *P. aeruginosa* clinical strain containing a native type I–F CRISPR–Cas system isolated from a bloodstream infection in Queen Mary Hospital, Hong Kong
Plasmids	
pAY9868	Platform vector with a BsaI endonuclease site flanked by two 28-bp repeats (5′-GTTCACTGCCGTATAGGCAGCTAAGAAA-3′) specific to the repeat sequence of the type I-F CRISPR–Cas system, a kanamycin-resistant gene, and a luciferase-encoding operon (*luxCDABE*) for the dual selection of transformants
pTargeting	pAY9868 containing the desired spacer cloned between the two 28-bp repeat sequences, expressing mini-CRISPR under the control of a strong promoter P*tat*
pEditing	pAY9868 containing the desired mini-CRISPR and the corresponding repair donors

10. Perform colony PCR to validate the construction of pEditing by using PMS-F and PMS-R. A colony PCR reaction using the undigested pTargeting as the template serves as the negative control.
11. Examine the PCR products with gel electrophoresis as described in Subheading 3.1.1. The PCR product of the successfully constructed pEditing is 1468 bp if the upstream and downstream donors are 500 bp each. The negative control reaction should yield a 468-bp product.
12. Inoculate two individual colonies validated above and send the extracted pEditing (MexB 273-Tag in the case of MexB Q273A mutation construction) from the overnight culture for sequencing to confirm the desired incorporation of the donor fragments, as described in Subheading 3.1.1.

3.2 Conjugation and Plasmid Curing

1. Transform the validated pEditing into *E. coli* SM10 competent cells and plate the transformants on LB plates containing 500 μg/mL kanamycin. Incubate plates at 37 °C overnight (15–20 h) to obtain colonies.
2. Perform colony PCR to validate the successful transformation of pEditing using PMS-F and PMS-R, as described in Subheading 3.2.1. Inoculate a validated colony into 3 mL of LB broth supplemented with 100 μg/mL kanamycin and incubate overnight at 37 °C with agitation.
3. Inoculate the recipient strain (wild-type PA154197) in LB broth and incubate overnight in a shaking incubator at 42 °C (*see* **Note 6**).
4. Mix 120 μL of the recipient strain and 1200 μL of SM10 cells harboring the constructed pEditing in a sterile centrifuge tube.
5. Centrifuge at 13,000 rpm for 1 min. Remove most of the supernatant, leaving around 30 μL to resuspend the pellets.
6. Pipette onto an LB plate in one compact droplet. Wait until the droplet dries, then invert the plate and incubate at 37 °C for 24 h (*see* **Note 7**).
7. After the incubation is completed, collect the cell mixtures using a sterile pipette tip and resuspend in a sterile centrifuge tube containing 300 μL of PBS buffer. Vortex to mix thoroughly.
8. Perform serial dilution to 10^{-2} and 10^{-3} in two separate sterile tubes. Plate the diluted mixtures on two different VBMM (Vogel–Bonner minimal medium) agar plates supplemented with 500 μg/mL of kanamycin and incubated for 24 h.
9. Isolate five recovered single colonies and inoculate them into 100 μL of LB broth supplemented with 100 μg/mL of kanamycin in a 96-well plate. Incubate at 37 °C with 220-rpm agitation for 2 h.

10. Measure the luminescence intensity using the Synergy HTX Plate Reader (BioTek, USA). Perform PCR using cell cultures with high luminescence values (>500) as the template and send the PCR products for DNA sequencing to validate the genome editing results (*see* **Note 8**).
11. Streak the successfully constructed strain carrying the desired point mutation on a fresh LB agar plate without kanamycin supplementation. Incubate the plate at 37 °C overnight. Plasmid curing is confirmed by inoculating the colony into LB broth supplemented with 100 μg/mL kanamycin. No growth after overnight culture indicates the loss of the pEditing (*see* **Note 9**).
12. A single colony with the desired mutation can now be used for the second-round editing or other assays.
13. To complete the construction of the alanine-substitution mutation, conduct second-round editing following **steps 1–11**. In the case of MexB Q273A, use the pEditing MexB 273A as the editing plasmid and PA154197 MexB 273-Tag as the recipient strain. Employing this strategy, a large number of alanine substitution mutants were efficiently constructed (Fig. 1b).

3.3 SDS–PAGE and Western Blotting

Construct the PA154197 MexB-3xFLAG strain by inserting the three tandem FLAG epitope tag immediately upstream of the stop codon of the *mexB* gene first. Use the same CRISPR-mediated genome editing method by following the same procedures as described in Subheadings 3.1 and 3.2. Primers used are indicated in Table 1.

3.3.1 Sample Preparation

1. Subculture the overnight cell culture of a desired construct with 1/100 dilution into fresh LB broth, grow at 37 °C with agitation (220 rpm) for around 6 h. Measure OD_{600} of the cell culture using a Ultrospec 2100 pro spectrophotometer (Amersham Biosciences, UK). Normalize to $OD_{600} = 1.0$ with fresh LB broth.
2. Centrifuge at 13,000 rpm for 1 min to remove supernatant. Freeze and store the cell pellets at −80 °C overnight.
3. Thaw the cell pellets and resuspend in 200 μL of complete bacterial protein extraction reagent. Vortex briefly and stand 15 min at room temperature.
4. Collect supernatant by centrifuging at 13,000 rpm for 5 min.
5. Measure the concentration of the total protein using a colorimetric assay kit. Use bovine serum albumin (BSA) as the protein standard.

6. In a clean tube, add 25 μL 2× sample buffer and 20 μg of protein sample. Bring up to 50 μL with PBS buffer.
7. Heat at 98 °C for 10 min. Centrifuge 13,000 rpm for 2 min to remove the debris.

3.3.2 SDS–PAGE, Electrophoretic Transfer, and Antibody Probing

SDS–PAGE, electrophoretic transfer, and antibody probing are conducted following the standard procedures. Briefly, separate the loaded proteins by running a 10% Tris–glycine polyacrylamide gel and transfer the blots to a PVDF membrane by electroblotting. Block the membrane using 5% non-fat milk in TBST buffer, followed by immunoblotting using polyclonal anti-FLAG antibody (Sigma-Aldrich, USA) to detect MexB-3xFLAG and horseradish peroxidase-conjugated goat anti-rabbit antibody (Sigma-Aldrich, USA) as the secondary antibody. Detect GroEL as the internal control with polyclonal anti-GroEL antibody (Sigma-Aldrich, USA). Develop the film using ECL Western Blotting Detection Reagents (GE Healthcare, UK) to obtain the readout. The procedures that need special attention are presented here:

1. Freshly prepared polyacrylamide gel is recommended. If the gel is not used immediately, wrap the whole setup with a wet paper towel and seal it in a plastic bag. Store at 4 °C for up to 3 days.
2. Primary antibody solution is prepared in 1% non-fat milk in TBST. For detection of MexB-3xFLAG, add 5 μL of polyclonal anti-FLAG antibody into 25 mL of solution (1:5000 dilution). For detection of GroEL control, add 0.5 μL of polyclonal anti-GroEL antibody into 75 mL of solution (1:150,000 dilution) (*see* **Note 10**).
3. Prepare 25 mL of secondary antibody solution in 1% non-fat milk in TBST by adding 5 μL of goat anti-rabbit antibody (1: 5000 dilution) (*see* **Note 10**).
4. The relative expression level of MexB variants (Fig. 2a) is quantified using software ImageJ (Fig. 2b, c) and expressed as the mean of at least two replicates. Figure 2b shows the results of three mutations (Q176A V177A F178A G179A S180S, Q273A D274A S276A I277A, and F615A F617A R620A) with the greatest MIC changes (Fig. 3). All three mutations exhibit expression levels comparable to the wild type.

3.4 Hoechst Efflux Assay

Hoechst 33342 is a lipophilic fluorescent dye that emits blue fluorescence when bound to dsDNA but remains non-fluorescent in an extracellular aqueous environment. It is actively transported by MexB; therefore, the intracellular accumulation of Hoechst 33342 across a period of time can be used to reflect the efflux activity of MexB (Fig. 2d, e).

Decreased MIC
No change in MIC
Increased MIC

MIC $\log_2$ Fold Changes (WT/Mutant)	β-lactams							Fluoroquinolones						Other antibiotics			
	Carbapenem		Mono bactam	Penicillins		Cephalosporins											
	MEM	IPM	ATM	PIP	CAR	CAZ	CRO	LVX	CIP	NOR	MXF	ENR	TMP	ERY	TET	CHL
Q176A	0	0	0	1	0	-1	0	0	0	0	1	0	0	0	1	0
V177A	0	0	1	0	0	1	1	0	0	0	0	0	0	0	0	0
F178A	2	0	3	3	2	1	2	1	1	1	2	2	2	1	2	2
G179A	2	0	3	1	2	1	3	0	0	0	1	0	1	0	1	0
S180A	0	0	0	0	0	-1	0	0	0	0	0	0	0	0	0	0
Q176A V177A F178A G179A S180A	5	1	7	5	9	5	6	1	1	1	2	2	1	1	3	0
Q273A	1	0	0	0	0	0	1	0	1	0	0	0	0	0	0	0
D274A	-1	0	0	0	0	2	0	0	1	0	0	0	0	0	0	0
S276A	0	0	0	0	0	-1	0	0	1	0	0	0	0	0	0	0
I277A	1	0	0	0	0	-1	0	0	1	0	0	0	0	0	0	0
Q273A D274A S276A I277A	1	0	0	1	0	3	1	0	1	0	1	1		0	0	0
F615A	0	0	0	3	0	0	0	1	0	1	0	1	1	0	1	1
F617A	1	0	1	1	0	2	1	1	1	0	0	0	1	0	1	0
R620A	1	0	2	1	2	1	1	1	2	1	1	0	1	0	1	0
F628A	2	0	2	2	0	2	1	1	2	1	1	1	1	0	1	0
F615A F617A R620A	2	1	0	4	3	1	2	3	3	2	4	4	2	1	2	1
D407N D408N	5	1	6	5	9	4	5	1	1	1	1	1	2	1	2	1

Fig. 3 Profile of antimicrobial susceptibility changes of the constructed MexB alanine-substitution mutants relative to the wild-type MexB shown as $\log_2$ fold changes of MIC. A positive $\log_2$ fold change (red shadings) indicates decreased resistance to the drug compared to the wild-type MexB, and a negative $\log_2$ fold change (green shadings) indicates increased resistance. MEM meropenem, IPM imipenem, ATM aztreonam, PIP piperacillin, CAR carbenicillin, CAZ ceftazidime, CRO ceftriaxone, LVX levofloxacin, CIP ciprofloxacin, NOR norfloxacin, MXF moxifloxacin, ENR enrofloxacin, TMP trimethoprim, ERY erythromycin, TET tetracycline, CHL chloramphenicol

1. Grow the constructed strains carrying the desired alanine substitutions (Fig. 2a) overnight in LB broth.
2. Subculture the overnight cultures with a 1/100 dilution ratio into LB broth, grow at 37 °C with agitation (220 rpm) until OD_{600} ~ 1.5.
3. Spin down the cells at 13,000 rpm and wash the pellet twice with sterilized PBS buffer.
4. Resuspend the pellet with sterile PBS buffer with 0.4% D-glucose to OD_{600} ~ 1.0.
5. Add Hoechst 33342 (Beyotime, China) to the final concentration of 4 μM.
6. Transfer 200 μL of the cell suspension into each well of a black 96-well microplate.
7. Measure the fluorescence emission at 460 nm (excitation at 355 nm) immediately, every 5 s for 60 s using the Synergy HTX Plate Reader (BioTek, USA). As shown in Fig. 2e, all three mutations, Q176A V177A F178A G179A S180S, Q273A D274A S276A I277A, and F615A F617A R620A, exhibited reduced efflux capacity compared to the wild type.

3.5 Antibiotic Sensitivity Testing

The minimum inhibitory concentration (MIC) is measured in polystyrene 96-well cell culture plates following the standard protocol described by Wiegand and colleagues with slight modifications [21]. Only the procedures that need special attention are presented here:

1. Prepare antibiotics solution in MH broth at concentrations of 1137.8, 568.9, 284.4, 142.2, 71.1, 35.5, 17.8, 8.9, 4.4, 2.2, 1.1, 0.55, and 0.28 μg/mL, respectively. This will result in final antibiotic concentrations after adding cell culture of 1024, 512, 256, 128, 64, 32, 16, 8, 4, 2, 1, 0.5, and 0.25 μg/mL, respectively.
2. Perform the assay using the wild-type *P. aeruginosa* PA154197 as a MexB-active control and the proton transport relay mutation variant MexB D407N D408N as a MexB-inactive control (Fig. 3).
3. When reading the microplate to determine MIC values, check the positive control well for cell growth. Turbidity should be present, indicating adequate growth in the MIC panel without antibiotics supplementation.
4. Check the negative control well. It should be clear, indicating free of contamination as bacterial cells are not present. As shown in Fig. 3, multiple alanine substitutions in the deep binding pocket groove (Q176A, V177A, F178A, G179A, S180A) significantly reduced MIC values against beta-lactam antibiotics, suggesting their role in recognizing these drugs. In contrast, multiple alanine substitutions in the switch loop motif (F615A, F617A, R620A) reduced MIC values against all antibiotics tested, indicating an essential role in effluxing diverse antibiotic substrates.

In summary, in this chapter, we detailed the alanine scanning approach to investigate the structure–activity relationship of chromosomal MexB in the native genetic context of a clinical MDR *P. aeruginosa* isolate. Alanine substitutions can be extended to all amino acids of MexB. Results obtained from these comprehensive substitution-based analyses could facilitate our understanding of the substrate recognition and selection mechanism of MexB. Insights gained hold promise to be extrapolated to other homologous RND transporters, providing valuable information to guide the rational design and modifications of EPIs.

4 Notes

1. The approaches and methodologies presented can be applied to all *P. aeruginosa* clinical isolates encoding an active, native Type I–F CRISPR–Cas system in the genome.

2. Use the same pTargeting for the second round of editing as the protospacer is the same (the 32-bp short tag) for every mutant at this step. There is no need to construct a different pTargeting.
3. This step is for the first round of editing to insert the 32-bp short tag. For the second round of editing to achieve the desired amino acid substitution, flank Donor-UR and Donor-DF with an 18-bp homologous arm encompassing the desired point mutation at the 5′ end. For the second round of editing to achieve 3xFLAG insertion, flank Donor-UR and Donor-DF with the 3xFLAG sequence at the 5′ end, as indicated in Table 1.
4. Adjust the annealing temperature to 5 °C below the Tm of the primer with the lower Tm. Set up a temperature gradient to increase the value by 5–10 °C if necessary for amplifying the repair donors.
5. The optimal mass of vector and each insert is [0.02 × number of base pairs] ng (~0.03 pmol).
6. For the second round of editing to achieve simultaneous amino acid substitution and tag removal, use the tag-inserted strain generated in the first round as the recipient strain.
7. It is optional to reduce the drying time of the plate using a fume hood. Avoid droplet spreading over the plate, as this will greatly reduce the efficiency of the conjugation process.
8. The primer pair used for amplification should specifically anneal to the genomic DNA of PA154197, avoiding binding to the pEditing. Failure to do so may lead to the unintended amplification of donor sequences from the plasmid, yielding false positive sequencing results.
9. In some cases, 2–3 rounds of curing are required for complete loss of the plasmid.
10. The antibody solution can be stored at 20 °C and reused twice.

Acknowledgments

This work is supported by the Research Grants Council of Hong Kong (CRF C7033-20G and GRF 17115223 to A.Y.).

Conflicts of Interest The authors have no conflicts of interest to declare that are relevant to the content of this chapter.

References

1. Zahedi Bialvaei A et al (2021) Expression of RND efflux pumps mediated antibiotic resistance in Pseudomonas aeruginosa clinical strains. Microb Pathog 153:104789
2. Kobylka J et al (2020) AcrB: a mean, keen, drug efflux machine. Ann N Y Acad Sci 1459(1):38–68
3. Lister PD, Wolter DJ, Hanson ND (2009) Antibacterial-resistant Pseudomonas aeruginosa: clinical impact and complex regulation of chromosomally encoded resistance mechanisms. Clin Microbiol Rev 22(4):582–610
4. Morita Y, Tomida J, Kawamura Y (2012) MexXY multidrug efflux system of Pseudomonas aeruginosa. Front Microbiol 3
5. Cao H et al (2019) Uncoupled quorum sensing modulates the interplay of virulence and resistance in a multidrug-resistant clinical Pseudomonas aeruginosa isolate belonging to the MLST550 clonal complex. Antimicrob Agents Chemother 63(4):e01944-18
6. Nakashima R et al (2013) Structural basis for the inhibition of bacterial multidrug exporters. Nature 500(7460):102–106
7. Sennhauser G et al (2009) Crystal structure of the multidrug exporter MexB from Pseudomonas aeruginosa. J Mol Biol 389(1):134–145
8. Sakurai K et al (2019) Crystal structures of multidrug efflux pump MexB bound with high-molecular-mass compounds. Sci Rep 9(1):4359
9. Tsutsumi K et al (2019) Structures of the wild-type MexAB-OprM tripartite pump reveal its complex formation and drug efflux mechanism. Nat Commun 10(1):1520
10. Glavier M et al (2020) Antibiotic export by MexB multidrug efflux transporter is allosterically controlled by a MexA-OprM chaperone-like complex. Nat Commun 11(1):4948
11. Vogwill T et al (2014) Testing the role of genetic background in parallel evolution using the comparative experimental evolution of antibiotic resistance. Mol Biol Evol 31(12): 3314–3323
12. Xu Z et al (2019) Native CRISPR-Cas-mediated genome editing enables dissecting and sensitizing clinical multidrug-resistant P. aeruginosa. Cell Rep 29(6):1707–1717.e3
13. Ji W et al (2014) Specific gene repression by CRISPRi system transferred through bacterial conjugation. ACS Synth Biol 3(12):929–931
14. Vento JM, Crook N, Beisel CL (2019) Barriers to genome editing with CRISPR in bacteria. J Ind Microbiol Biotechnol 46(9–10): 1327–1341
15. Jiang Y et al (2017) CRISPR-Cpf1 assisted genome editing of Corynebacterium glutamicum. Nat Commun 8(1):15179
16. Cho S et al (2018) High-level dCas9 expression induces abnormal cell morphology in Escherichia coli. ACS Synth Biol 7(4): 1085–1094
17. Chen W et al (2018) CRISPR/Cas9-based genome editing in Pseudomonas aeruginosa and cytidine deaminase-mediated base editing in Pseudomonas species. iScience 6:222–231
18. Xu Z, Li Y, Yan A (2020) Repurposing the native type I-F CRISPR-Cas system in Pseudomonas aeruginosa for genome editing. STAR Protoc 1(1):100039
19. Xu Z et al (2021) A transferrable and integrative type I-F Cascade for heterologous genome editing and transcription modulation. Nucleic Acids Res 49(16):e94–e94
20. Reid GA (1991) Molecular cloning: a laboratory manual, 2nd edn: by J. Sambrook, E. F. Fritsch and T. Maniatis, Cold Spring Harbor Laboratory Press, 1989. Elsevier Ltd, pp 213–214
21. Wiegand I, Hilpert K, Hancock RE (2008) Agar and broth dilution methods to determine the minimal inhibitory concentration (MIC) of antimicrobial substances. Nat Protoc 3(2): 163–175

Chapter 12

Genetic Tools and Approaches to Study Bacterial Efflux Pumps

Megan R. Caswell, Mallory Wright, and Georgina Cox

Abstract

The substrate specificities and functions of bacterial efflux pumps are difficult to define due to the large number encoded in bacterial genomes, variations in gene expression, and functional redundancies. In this chapter, we describe the use of efflux-deficient *Escherichia coli* mutants as simplified genetic backgrounds for studying efflux pumps. We present detailed methodologies for introducing individual efflux pump-encoding genes under constitutive promoters, and we discuss how these strains can be used to examine the physicochemical substrate profiles of drug efflux pumps using susceptibility testing and chemical properties analysis.

Key words *Escherichia coli*, Efflux Pump, Genomic Integration, Susceptibility Testing, Substrate Profiles

1 Introduction

Efflux pumps are present across all three domains of life, with bacteria possessing extensive and largely uncharacterized efflux networks [1]. The high number of functionally redundant and poly-specific efflux pumps encoded in bacterial genomes, combined with differential gene expression levels, hinders functional studies. In *Escherichia coli* K-12 strains, sequence annotation has identified 35 known or putative drug efflux pumps belonging to five major protein families: the ATP-binding cassette (ABC) superfamily, the resistance–nodulation–cell division (RND) superfamily, the major facilitator superfamily (MFS), the small multidrug resistance (SMR) family, and the multi-antimicrobial extrusion (MATE) family [1, 2].

Most efflux pumps are conserved across the *E. coli* species [2] and can be grouped into two structural categories: tripartite efflux assemblies and single-component inner membrane pumps [3]. In *E. coli*, at least eight efflux pumps form tripartite complexes with

Kunihiko Nishino and Mikio Tanabe (eds.), *Bacterial Multidrug Exporters: Methods and Protocols*, Methods in Molecular Biology, vol. 3044, https://doi.org/10.1007/978-1-0716-5316-6_12,
© The Author(s), under exclusive license to Springer Science+Business Media, LLC, part of Springer Nature 2026

the outer membrane channel TolC and periplasmic adaptor proteins [4–6]; *tolC*-deficient mutants are commonly used as efflux-deficient models since the deletion of this outer membrane channel disrupts multiple tripartite efflux systems and increases susceptibility to various compounds [7]. However, *tolC*-inactivation also leads to pleiotropic effects, including fitness defects, membrane stress, and metabolic imbalances under nutrient-limited conditions, partly due to the periplasmic accumulation of enterobactin caused by iron limitation [6]. We have also demonstrated that Δ*tolC* mutants exhibit altered outer membrane integrity and mechanical properties, increasing their sensitivity to compounds independently of efflux inactivation [5].

We generated extensively efflux pump-deficient mutants to overcome these limitations, creating simplified genetic backgrounds for reintroducing and functionally analyzing individual efflux pumps [4, 5]. Our initial work introduced *E. coli* Efflux KnockOut-35 (EKO-35), an extensively efflux-deficient mutant lacking 35 inner membrane efflux pumps [4]. This strain was the foundation for developing the Efflux Platform, which was designed to reintroduce individual efflux pumps. We have since constructed the next generation of this mutant, EKO-$35_{2.0}$, to strengthen the platform by removing nonsynonymous genomic mutations that may impact metabolism-dependent antibiotics and limit the identification of physiological efflux substrates [8]. This mutant was generated using CRISPR–Cas9-mediated counterselection [9], and three tandem stop codons were introduced at the beginning of each efflux gene to inactivate. We also constructed the tripartite efflux (TE) mutant lacking the eight efflux pump genes that form complexes with TolC while retaining single-component efflux pumps [5], a valuable comparator to a Δ*tolC* strain. Together, these efflux-deficient strains provide simplified genetic backgrounds for studying efflux pumps. This chapter describes how these mutants are used to study individual efflux pumps and investigate their physicochemical substrate profiles.

2 Materials

2.1 *Escherichia coli* Strains

We use *E. coli* K-12 substr. BW25113, the parent strain of the Keio knockout collection with the following genotype: F^- λ^- *Δ(araD-araB)567* Δ*lacZ4787(::rrnB-3) rph-1* Δ*(rhaD-rhaB)568 hsdR514*. This strain can be obtained from various sources (e.g., Horizon Discovery). The domestication of *E. coli* K-12 has led to the accumulation of mutations that alter cell envelope composition. One such mutation is the disruption of *wbbL*, a key LPS biosynthetic gene, by an insertion element, producing truncated LPS [10]. Additionally, *E. coli* K-12 strain BW25113 carries a mutation in the fatty acid regulator FabR (G42V), leading to an increased proportion of unsaturated fatty acids in the inner membrane [11]. We have repaired the *wbbL* locus using duplication-insertion recombineering and corrected the *fabR* missense mutation, generating *E. coli* strain BW25113$_{WF}$, which we now use as our wild-type *E. coli* strain for genetic manipulation and efflux studies [8].

We have used two methods for inactivating efflux pump-encoding genes in *E. coli*: λ-Red recombineering and CRISPR–Cas9–mediated counterselection by introducing three tandem stop codons at the beginning of each gene [4]. Our preferred method is CRISPR–Cas9–mediated counterselection [9] since λ-Red recombineering can result in unintended genomic deletions and mobilization due to the introduction of adjacent Flp recognition target (FRT) sites when multiple genes are removed. The EKO-35$_{2.0}$ strain [8] was generated solely using CRISPR–Cas9–mediated counterselection (*E. coli* BW25113$_{WF}$ efflux-deficient derivative with three tandem stop codons inserted into: *acrB; acrD; acrF; mdtF; macB; emrB; mdtL; mdtK; bcr; ydeA; mdtM; yddA; yebQ; emrE; mdtD; sugE; ynfM; emrD; ydeF; mdtJ; ydiM; mdtB; mdlA; emrY; mdfA; fsr; mdtG; mdtH; yieO; mdlB, mdtO, yojI, yajR, ydhC; cusA*). *E. coli* K-12 BW25113$_{WF}$, EKO-35$_{2.0}$ [8], and the TE mutant [5] strains can be obtained from Dr. Cox. While this methodology is designed for *E. coli*, it could, in principle, be adapted for other bacterial species and efflux-deficient mutants.

Finally, cloning is required to introduce efflux pumps into these backgrounds, for which we use chemically competent *E. coli* DH5α and *E. coli* TOP10 (Thermo Fisher Scientific).

2.2 Plasmids

The pINT2 and pGDP2 plasmids [12] express efflux pump-encoding genes from the chromosome and plasmids, respectively (Fig. 1). These plasmids harbor the same multiple cloning site (MCS) and were modeled on a common pET vector. Both use the constitutive P_{LacI} promoter for gene expression. pGDP2 allows for plasmid-mediated expression, increasing gene copy number, and kanamycin is used for selection. The pKD46 is a temperature-sensitive plasmid used for λ-Red recombineering and integrating the efflux pump-encoding genes into the bacterial genome. pCP20 is a temperature-sensitive plasmid encoding FLP recombinase that facilitates the excision of FRT-flanked antibiotic resistance cassettes from bacterial genomes [13].

2.3 Primers

Primers are designed to amplify the genes of interest while incorporating restriction sites for cloning into the MCS of the pINT2 and pGDP2 plasmids (Table 1). pINT-rev and pLac-check primers confirm the successful ligation of efflux-encoding genes into the pINT2 and pGDP2 plasmids. For genomic integration of efflux pump-encoding genes, the genes and the adjacent kanamycin resistance cassette are amplified from pINT2 using the F1C-Fwd/F4C-Rev primers. Genomic integrations are verified using PolB-Fwd and pINT-Rev primers.

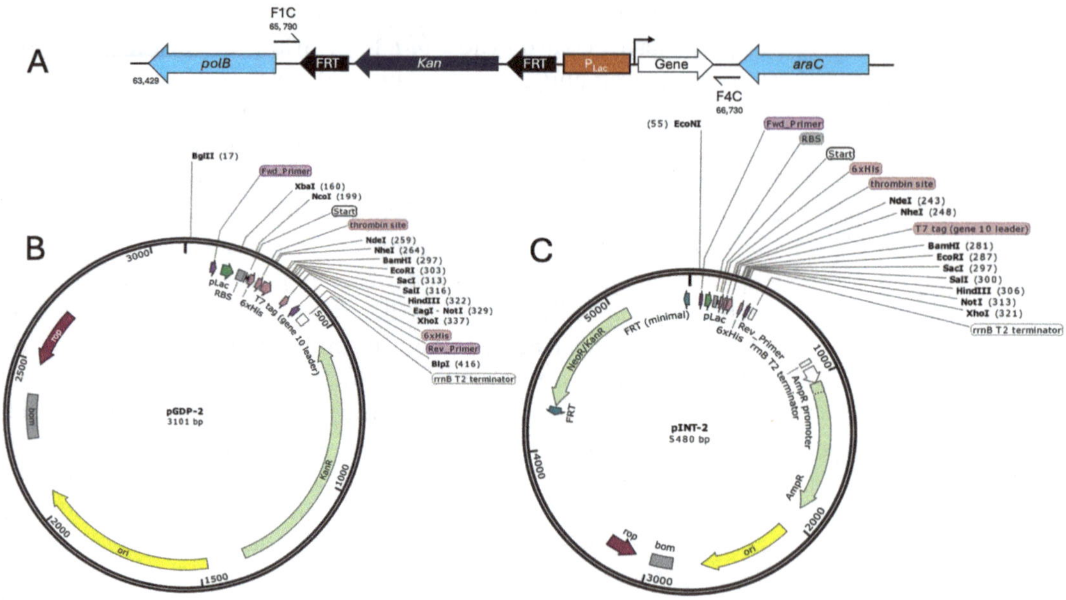

Fig. 1 Genetic tools for the expression of efflux pump-encoding genes. (**a**) Efflux genes are ligated into the pINT2 plasmid, which is then used as a template for amplification using the F1C and F4C primers, generating an amplicon for integration into the *E. coli* genome using homologous recombination. The *polB* and *araC* flanking regions (blue), FRT sites for removal of the Kan (black) cassette, P_{LacI} promoter (brown), inserted efflux gene (white), and annealing sites of the F1C and F4C primers used for amplification are shown. Plasmid maps of (**b**) pGDP2 and (**c**) pINT2

Table 1
Primers for the introduction of efflux pump-encoding genes into *E. coli*

Primer	Sequence (5′- 3′)	Source
F1C-Fwd	CCATCAAAAAACCAGGCTTGAG	4
F4C-Rev	GCATTCTGTAACAAAGCGGGAC	4
PolB-Fwd	CAGTTCGAAATCAAGCGAGGAG	4
pINT-Rev	CCAACTCAGCTTCCTTTCGG	4
pLac-check	GCTCAACGGCCTCAACCTAC	4

2.4 Media

Lysogeny broth (LB): 10 g/L tryptone, 5 g/L yeast extract, 10 g/L NaCl, and agar are used for routine culturing; cation-adjusted Mueller–Hinton II broth (MHB II): 17.5 g/L acid hydrolysate of casein, 2 g/L beef extract, and 1.5 g/L starch are used for susceptibility testing.

2.5 Commercial Kits for DNA Purification

GeneJET PCR Purification Kit, GeneJET Gel Extraction Kit, and GeneJET Plasmid Miniprep Kit.

2.6 Gel Electrophoresis

TAE running buffer: 40 mM Tris–HCl, 20 mM acetic acid, 1 mM Na_2EDTA (pH 8.0); RedSafe nucleic acid staining solution; 1 kb plus DNA ladder; and 6× loading dye (0.25% bromophenol blue, 0.25% xylene cyanol FF, 30% glycerol in water).

2.7 Antibiotics for Selection

Ampicillin and kanamycin.

2.8 Enzymes for Cloning and Genomic Integrations

*Nde*I, *Xho*I, *Dnp*I, *Bam*HI, *Hind*III, T4 ligase, and 2× GB-AMP high-fidelity PaCeR polymerase Master Mix (GeneBio Systems).

2.9 Chemicals and Solutions

1 M arabinose, filter sterilized; 10% (*v/v*) glycerol sterilized by autoclaving; agarose; phosphate-buffered saline (PBS): 137 mM NaCl, 2.7 mM KCl, 10 mM Na_2PO_4, 1.8 mM K_2PO_4, pH 7.4; dimethylsulfoxide (DMSO); and desired compounds for susceptibility testing. We routinely screen the following antimicrobials when assessing the antimicrobial substrate profiles of efflux pumps: rifampicin, vancomycin, fosfomycin, ampicillin, oxacillin, meropenem, chloramphenicol, puromycin, azithromycin, erythromycin, telithromycin, tetracycline, spectinomycin, kanamycin, hygromycin B, streptomycin, linezolid, minocycline, fusidic acid, ciprofloxacin, norfloxacin, nalidixic acid, novobiocin, trimethoprim, doxorubicin, daunorubicin, ethidium bromide, acriflavine, benzalkonium chloride, deoxycholate, sodium taurodeoxycholate, sodium cholate, glycocholate, chenodeoxycholate, chlorhexidine, sulfathiazole, and spermine [4].

2.10 Equipment and Software

1. Water bath suitable for incubation at 43 °C.
2. Electroporater (e.g., Bio-Rad MicroPulser), 1 mm electroporation cuvettes (Fisher).
3. Swinging bucket centrifuge.
4. NanoDrop 2000 spectrophotometer (Thermo Fisher Scientific).
5. Thermal cycler (e.g., T100 Bio-Rad).
6. Gel electrophoresis apparatus.
7. Transilluminator to visualize electrophoresis gels before DNA extraction (e.g., the BluPAD LED Transilluminator, FroggaBio).
8. Bio-Rad ChemiDoc XRS+ imaging system for imaging electrophoresis gels.

9. 384-well flat-bottom microtiter plate, sterile, polystyrene (Corning).
10. Shaking incubator suitable for microtiter plates (e.g., INFORS Multitron).
11. Sterile Erlenmeyer flasks.
12. A liquid handler such as the OT-2 Robot (Opentrons).
13. Centrifuge compatible with microtiter plates.
14. Microcentrifuge.
15. Microplate reader for OD_{600nm} measurements (e.g., BioTek Synergy Neo2).
16. GraphPad Prism.
17. DataWarrior V6.1.0.

3 Methods

3.1 Rational

In this section, we describe the necessary methodology required for introducing efflux pump-encoding genes into efflux-deficient strains under the control of constitutive promoters and then describe approaches to study their substrate specificity. These methodologies encompass a range of molecular biology techniques essential for cloning and genomic manipulation. They include preparing chemically and electrocompetent cells for efficient DNA uptake, colony PCR for screening transformants, restriction enzyme DNA digestion, ligation and cloning for constructing recombinant plasmids, and gel electrophoresis for analyzing DNA.

Subheading 3.2.1 first describes single-copy integration into the genomes of the $EKO\text{-}35_{2.0}$ and TE mutant strains. Profiling both mutants provides preliminary insight into whether inner membrane pumps are involved in functional interplay with tripartite efflux systems [14–16], as the TE mutant retains these proteins [5]. Genomic integration offers the advantage of stable gene maintenance, with the possibility of eliminating selective markers. Genes are integrated using pINT2, a recombineering vector designed for chromosomal integration into the arabinose operon of *E. coli* (Fig. 1) [12]. Following the ligation of the target gene into pINT2, the construct is then used as a template to generate an amplicon for homologous recombination into *araC*. The antibiotic resistance cassette used for selection can be eliminated following genomic integration. This approach enables stable and constitutive expression of integrated genes from the P_{LacI} promoter.

As an alternative to chromosomal integration, genes can be ligated into the pGPD2 plasmid (Fig. 1) for expression from a low-copy-number plasmid, increasing expression levels (*see* Subheading 3.2.2); however, kanamycin is required to maintain the plasmid. Both approaches result in gene expression under the control of the constitutive P_{LacI} promoter, eliminating the need for an inducer and enabling comparisons between pumps. When studying multicomponent efflux pump systems (e.g., tripartite complexes), we amplify and ligate the inner membrane efflux pump and the periplasmic adaptor. Several efflux systems form complexes with TolC, which the mutant strains retain.

Following strain generation, susceptibility profiling enables examining efflux pump substrate profiles. The physicochemical properties of each compound are determined using DataWarrior [17]. The methods used in this process are outlined in Subheading 3.3.

3.2 Expression of Efflux Pump-Encoding Genes in Efflux-Deficient Mutants

3.2.1 Genomic Integration

Obtain the Efflux Gene Sequence of Interest Using Ecocyc [18]

1. Design a forward primer to anneal to the 5′ end of your gene of interest. The primer should include the 5′ restriction enzyme site (*see* **Note 1**), the start codon, and 15–25 nucleotides of the gene sequence to ensure proper annealing. The reverse primer should anneal to the 3′ end of your gene of interest. It should include the 3′ restriction enzyme site for ligation, the stop codon, and 18–25 nucleotides of the gene sequence. In addition, 2–4 extra nucleotides (often referred to as "overhangs") upstream of the restriction enzyme site should be included to ensure efficient cutting by the restriction enzymes. Check for primer specificity, avoiding secondary structures or primer-dimer formation. Primers should typically be 20–25 bases long, with a 40–60% GC content and a melting temperature (T_m) of 55–65 °C and within 5 °C of each other.

Amplify Target Genes of Interest from *E. coli* Genomic DNA

1. Extract crude template DNA from *E. coli*.
   (a) Add 50 μL of mH_2O into a microcentrifuge tube.
   (b) Using a sterile pipette tip, pick a well-isolated colony from an agar plate, resuspend by pipetting up and down until the water becomes slightly cloudy.
   (c) Boil the cells for 2 min.
   (d) Centrifuge the tubes to remove the cell debris, leaving the crude extract DNA in the supernatant.
2. Amplify the target gene from the crude template DNA using PaCeR polymerase (*see* **Note 2**) and gene-specific primers, following the manufacturer's recommended protocol.
3. Confirm the size of the amplicons by electrophoresis (load 2–5 μL of the PCR reaction) and visualize the gel using an LED transilluminator or Bio-Rad ChemiDoc XRS+ imaging system.

Ligate Efflux Pump-Encoding Genes into pINT2

1. Purify the amplicons generated in subheading "Amplify Target Genes of Interest from *E. coli* Genomic DNA" using the GeneJET PCR Purification Kit according to the manufacturer's suggested guidelines.
   (a) Quantify DNA following purification using a NanoDrop spectrophotometer.
2. Digest 1000 ng of purified amplicons with restriction enzymes according to the manufacturer's suggested guidelines (*see* **Note 3**).
3. Purify the digested amplicons using the GeneJET PCR Purification Kit according to the manufacturer's suggested guidelines.
4. Use a Miniprep Kit to isolate the pINT2 plasmid according to the manufacturer's suggested guidelines from a 3 mL LB (100 μg/mL ampicillin) overnight culture of an *E. coli* cloning host harboring the plasmid.
5. Digest the pINT2 plasmid with restriction enzymes according to the manufacturer's guidelines (*see* **Note 4**).
6. Ligate the restriction enzyme-digested efflux pump-encoding genes into the digested pINT2 plasmid using T4 ligase according to the manufacturer's suggested guidelines (*see* **Note 4**).
7. Transform the ligation mixture into chemically competent *E. coli* DH5α or TOP10 cells (*see* **Note 5**):
   (a) Thaw 50 μL aliquots of the cells on slushy ice.
   (b) Add 10 μL of the pINT2 ligation reaction to the cells. Mix gently.
   (c) Incubate the tube on ice for 30 min.
   (d) Heat shock cells for 45–60 s in a 43 °C water bath.
   (e) Place tubes on ice for 2 min.
   (f) Add 1 mL of pre-warmed LB.
   (g) Incubate at 37 °C for 1 h with aeration (220 rpm).
   (h) To collect the cells, centrifuge the transformation reaction at 6000 × *g* for 5 min at room temperature. Remove 750 μL of the supernatant and resuspend the pellet in the remaining supernatant.
   (i) Spread the cells onto the surface of a pre-warmed LB agar plate supplemented with 100 μg/mL ampicillin. Allow the plate to dry for 30 min.
   (j) Incubate overnight at 37 °C.
8. Verify successful clones using colony PCR and the pLac-check/pINT-Rev primers (*see* **Note 6**) or a gene-specific forward primer in combination with the pINT-Rev primer. Confirm with Sanger sequencing.

Prepare Electrocompetent Cells of the Efflux-Deficient Mutant Strains (EKO-$35_{2.0}$ or the TE Mutant)

1. Streak plate the efflux-deficient mutants from a cryogenic stock to obtain isolated colonies.
2. Inoculate a single colony into 3 mL of LB and incubate overnight at 37 °C with aeration (220 rpm).
3. Add 500 μL of the overnight culture to 50 mL of LB in a 250 mL Erlenmeyer flask.
4. Incubate at 37 °C with aeration (220 rpm) for 2–3 h until an OD_{600nm} of 0.55–0.6 is achieved.
5. Transfer the culture into a 50 mL sterile conical tube and incubate on ice for 5–10 min.
6. Harvest the cells by centrifugation at 4 °C at 6000 × *g* for 10 min.
7. Discard the supernatant and gently resuspend the cells in 40 mL of sterile 10% *v/v* glycerol, pre-equilibrated on ice.
8. Harvest the cells by centrifugation at 4 °C at 6000 × *g* for 10 min.
9. Repeat the wash step two more times.
10. Once the final supernatant has been discarded, resuspend the cells in 400 μL of 10% (*v/v*) glycerol, pre-equilibrated on ice.
11. Aliquot 50 μL of the cell suspension into microcentrifuge tubes and place them on slushy ice.
12. For storage, flash freeze in liquid nitrogen and store at −80 °C.

Isolate the pKD46 Plasmid

1. Use a plasmid miniprep kit to isolate the pKD46 plasmid according to the manufacturer's suggested guidelines from a 3 mL LB (100 μg/mL ampicillin) overnight culture of *E. coli* DH5α or TOP10 cells harboring the plasmid.

Transform Electrocompetent Cells of the Efflux-Deficient Mutants with pKD46

1. Incubate 10–50 ng of pKD46 (*see* **Note 7**) with 50 μL of electrocompetent cells on ice for 5–10 min. Simultaneously, chill a 1-mm electroporation cuvette on ice.
2. Transfer the bacterial cell–plasmid mixture into the chilled electroporation cuvette and tap the cuvette gently on the benchtop.
3. Set the electroporator to "Ec1." Press the "measurements" button to display the "time-ms" value.
4. Pulse and record the time constant displayed. If the reading says "Arc," repeat this step.
5. Remove the cuvette from the holder, and immediately add 1 mL of LB.
6. Transfer this suspension to a fresh microcentrifuge tube and incubate for 1 h at 30 °C with 220 rpm to recover the cells.

7. Use a sterile hockey stick to evenly spread 200 μL of cells onto the surface of LB agar supplemented with 100 μg/mL ampicillin. Allow the agar plate to dry, and incubate overnight at 30 °C.
8. Resulting colonies harbor the pKD46 plasmid. Cryostocks of the efflux-deficient mutants harboring pKD46 can be generated and stored.

Generate the Amplicon for Genomic Integration

1. Use the constructs generated in subheading "Ligate Efflux Pump-Encoding Genes into pINT2" as the template.
2. Using PCR and the F1C-Fwd/F4C-Rev primers with PaCeR polymerase, amplify the *polB* and *araC* flanking regions, the gene of interest, and the kanamycin resistance cassette from the pINT2 constructs (50–100 ng) (Fig. 1).
3. Confirm the size of the amplicons by gel electrophoresis (*see* **Notes 8–10**).
4. Use the GeneJET PCR Purification Kit to purify the amplicon following the manufacturer's suggested guidelines.
5. Digest the amplicon overnight with the *Dpn*I restriction enzyme to remove any contaminating template DNA, according to the manufacturer's suggested guidelines (*see* **Notes 3** and **12**).
6. Purify using the GeneJET PCR Purification Kit, according to the manufacturer's suggested guidelines, and elute with mH_2O.

Integrate the Amplicon into the Genome of the Efflux-Deficient Mutants

1. In a 250 mL Erlenmeyer flask, inoculate 50 mL of LB with 500 μL of overnight cultures of the efflux-deficient mutants harboring pKD46 (prepared in subheading "Transform Electrocompetent Cells of the Efflux-Deficient Mutants with pKD46"). Add 50 μL of ampicillin (100 μg/mL) and 100 μL of 1 M arabinose to induce the recombinase.
2. Incubate the cultures at 30 °C with aeration (220 rpm) until an OD_{600nm} of 0.5 is achieved (~3–4 h).
3. Follow the electrocompetent cell preparation protocol as described in subheading "Transform Electrocompetent Cells of the Efflux-Deficient Mutants with pKD46".
4. Place an aliquot of the cells on ice and chill the electroporation cuvettes on ice.
5. Add 350–500 ng of the *Dpn*I-digested amplicon, prepared in subheading "Generate the Amplicon for Genomic Integration", to the thawed electrocompetent cells and incubate on ice for 5–10 min.

6. Follow the electroporation protocol described in subheading "Transform Electrocompetent Cells of the Efflux-Deficient Mutants with pKD46", **steps 1–5** (*see* **Note** 7).
7. Transfer the cell suspension to a microcentrifuge tube and recover the cells for 1 h at 37 °C, 220 rpm.
8. Collect the cells by centrifugation at 6000 × *g* for 5 min at room temperature.
9. Remove 500 μL of the supernatant and resuspend the cell pellet in the remaining supernatant.
10. Spread 250 μL of the cell suspension onto LB agar supplemented with 50 μg/mL kanamycin and allow the remaining 250 μL to incubate overnight at room temperature before plating onto the surface of LB agar supplemented with 50 μg/mL kanamycin. Incubate both plates overnight at 37 °C.

Patch Resulting Colonies onto Two Agar Plates

1. LB agar with 50 μg/mL kanamycin. Incubate overnight at 37 °C.
   (a) Growth is expected, indicating that the kanamycin resistance cassette and gene were successfully integrated.
2. LB agar supplemented with 100 μg/mL ampicillin. Incubate overnight at 30 °C.
   (a) Growth is not expected, indicating that the pINT2 template is absent and that pKD46 was cured.

Verify Successful Genomic Integrations

1. Perform PCR using crude extract template DNA generated as described in subheading "Amplify Target Genes of Interest from *E. coli* Genomic DNA", the PolB-Fwd and F4C-Rev primers, and PaCeR polymerase according to the manufacturer's suggested guidelines.

Remove the Kanamycin Resistance Cassette from the Genomic Integrations Using pCP20

1. Use the ThermoScientific GeneJET Plasmid Miniprep Kit to isolate the pCP20 plasmid according to the manufacturer's suggested guidelines from a 3 mL LB (100 μg/mL ampicillin) overnight culture of *E. coli*DH5α or TOP10 cells harboring the plasmid.
2. Inoculate 3 mL of LB, supplemented with 50 μg/mL kanamycin, with one colony of the successfully integrated strain. Incubate overnight at 37 °C.
3. Generate electrocompetent cells of the integrated strains following the protocol outlined in subheading "Prepare Electrocompetent Cells of the Efflux-Deficient Mutant Strains (EKO-$35_{2.0}$ or the TE Mutant)", with 50 μg/mL kanamycin instead of 100 μg/mL ampicillin.

4. Transform the cells with 50 ng of pCP20 using the protocol outlined in subheading "Transform Electrocompetent Cells of the Efflux-Deficient Mutants with pKD46" and ampicillin for selection, incubate at 30 °C.

Remove the Kanamycin Resistance Cassette

1. Inoculate 3 mL of LB, supplemented with 100 μg/mL ampicillin, with one colony from subheading "Remove the Kanamycin Resistance Cassette from the Genomic Integrations Using pCP20". Incubate at 30 °C overnight to enable the flippase to remove the cassette.
2. Isolate colonies from the overnight culture on LB agar and incubate at 43 °C overnight to cure the plasmid.
3. Patch colonies onto:
   (a) LB agar. Incubate overnight at 37 °C. Growth is expected.
   (b) LB agar supplemented with 50 μg/mL kanamycin. Incubate overnight at 37 °C. Growth is not expected, indicating that the kanamycin resistance cassette was successfully removed.
   (c) LB agar supplemented with 100 μg/mL ampicillin. Incubate overnight at 30 °C. Growth is not expected, indicating the pCP20 plasmid was successfully cured.
4. After removing the kanamycin resistance cassette, amplify the integrated gene and the P_{LacI} promoter region using PCR and the PolB-Fwd/F4C-Rev primers. Verify using Sanger sequencing or long-read sequencing technology from Oxford Nanopore Technologies.

3.2.2 Plasmid-Mediated Expression Using pGDP2

1. Amplify the efflux pump-encoding gene, digest it, and ligate it into pGDP2, as described in Subheading 3.2.1 for pINT2 (*see* **Note 11**).
2. Verify successful transformants by PCR using pLac-check/pINT-Rev primers, followed by Sanger sequencing or long-read sequencing technology from Oxford Nanopore Technologies.
3. As described in Subheading 3.2.1, transform into electrocompetent efflux-deficient mutant cells using kanamycin for selection. Generate cryogenic stocks.

3.3 High-Throughput Susceptibility Profiling

1. Streak strains (*see* **Note 13**) onto LB agar and incubate at 37 °C for 18 h. When profiling pGDP2-harboring strains, kanamycin is used for selection.
2. Inoculate LB medium with isolated colonies and incubate overnight at 37 °C for 18 h at 220 rpm.
3. Collect 1 mL of bacterial cells by centrifugation at 6000 × *g* for 5 min.

4. Resuspend the cells to an OD_{600nm} of 1.0 in PBS.
5. Dilute the cell suspension 1:400 in Mueller–Hinton II broth in a sterile Erlenmeyer flask.
6. Stock solutions for each antimicrobial compound should be 1.25 times the highest desired concentration. Dissolve compounds in an appropriate solvent (e.g., water, ethanol, or DMSO, *see* **Note 14**). Dilute compound stocks in MHB II to achieve a final concentration of 2% (*v/v*).
7. Dispense MHB II-diluted compounds in duplicate two-fold serial dilutions in a 384-well microtiter plate using a liquid handler such as the Opentrons OT-2 Robot. Each well should contain 40 μL of compound diluted in MHB II.
   (a) Include solvent controls in wells that contain no compound.
8. Use the liquid handler to dispense 10 μL of the standardized cell suspensions to each well, with a final volume of 50 μL.
9. Centrifuge plates at 500 × *g* for 1 min at room temperature to collect any droplets.
10. Measure the OD_{600nm} using a microplate reader, representing the time-zero reading subtracted from the final measurements.
11. Seal microtiter plates with tape and incubate at 37 °C with aeration (900 rpm) for 18 h.
12. Following incubation, measure the OD_{600nm} using a microplate reader set to 37 °C and subtract the time-zero OD_{600nm} values for background correction.
13. Analyze data using Microsoft Excel and GraphPad Prism to determine each compound's minimum inhibitory concentration (MIC). Calculate fold change compared to the parental strains; a > 4-fold increase in resistance upon expressing an efflux pump-encoding gene is considered significant.
14. Summarize the physicochemical properties of substrates identified for each efflux pump locus.
   (a) Retrieve the SMILES chemical notation for each compound from PubChem and compile in Microsoft Excel.
   (b) Input the resulting spreadsheet into DataWarrior [17] to calculate chemical properties, including molecular weight, polar surface area, hydrophobicity, and aqueous solubility.
   (c) Visualize the distribution of each property for each pump using box plots generated in GraphPad Prism.

4 Notes

1. For ligation into the vectors pINT2 and pGPD2, we preferentially use the restriction enzymes *Nde*I (5′-CA/TATG-3′) and *Xho*I (5′-C/TCGAG-3′). However, if *Nde*I and *Xho*I have recognition sites within genes of interest, we substitute them with *Hind*III (cut site 5′-A/AGCTT-3′) and *Bam*HI (cut site 5′-G/GATCC-3′).
2. We use PaCeR polymerase because it amplifies longer amplicons, which is important when amplifying the efflux pumps with an adjacent periplasmic adaptor protein gene.
3. We typically incubate restriction digestions for 18 h at 37 °C. Do not over-incubate, as this may lead to unexpected cleavage patterns.
4. We suggest digesting 500 ng–1 μg of plasmid. For ligation of the amplicon into the plasmid, we recommend using a 1:3 molar ratio of vector:insert DNA with an incubation time of 2 h at 22 °C in the thermal cycler on the "incubate" setting.
5. To generate chemically competent cells, refer to the Inoue method for high-efficiency transformation of *Escherichia coli* with plasmids [19]. Aliquot 50 μL of the cell mixture into microcentrifuge tubes and flash freeze in liquid nitrogen. Store at −80 °C until use.
6. The PCR product amplified from pINT2 using pLac-check/pINT-Rev primers should be 284 bp + gene size.
7. Do not add more than 7 μL of DNA to avoid arcing.
8. When primers F1C-Fwd/F4C-Rev are used, the expected amplicon size is 1937 bp + gene size.
9. If the band is unclear, verify that the plasmid miniprep was performed correctly and repeat the process if needed.
10. If non-specific products are present, optimize the primer annealing temperature.
11. The digested amplicons ligated into pINT2 may also be ligated into pGPD2.
12. *Dpn*I is a restriction enzyme that cleaves methylated sequences, making it practical to remove template DNA (e.g., the pINT2 constructs used to generate amplicons for genomic integrations). Templates must be derived from *dam* + *E. coli* strains.
13. We include the following parental strains as controls: the wild-type parental strain (*E. coli* BW25113$_{WF}$), efflux-deficient mutants lacking efflux genes (TE mutant and EKO-35$_{2.0}$), and an *E. coli* BW25113$_{WF}$ Δ*tolC* mutant.
14. The concentration of DMSO should not exceed 1% (*v/v*) of the final well volume.

Conflicts of Interest

The authors have no conflicts of interest to declare that are relevant to the content of this chapter.

References

1. Paulsen IT, Sliwinski MK, Saier MH Jr (1998) Microbial genome analyses: global comparisons of transport capabilities based on phylogenies, bioenergetics and substrate specificities. J Mol Biol 277:573–592
2. Teelucksingh T, Thompson LK, Cox G (2020) The evolutionary conservation of *Escherichia coli* drug efflux pumps supports physiological functions. J Bacteriol 202. https://doi.org/10.1128/JB.00367-20.12
3. Van Bambeke F, Balzi E, Tulkens APM (2000) Antibiotic Efflux Pumps. Biochem Pharmacol 60:457–470
4. Teelucksingh T, Thompson LK, Zhu S et al (2022) A genetic platform to investigate the functions of bacterial drug efflux pumps. Nat Chem Biol 18:1399–1409
5. Zhu S, Alexander MK, Paiva TO et al The inactivation of *tolC* sensitizes *Escherichia coli* to perturbations in lipopolysaccharide transport. iScience 27:109592
6. Wright M, Kaur M, Thompson LK et al (2025) A historical perspective on the multifunctional outer membrane channel protein TolC in *Escherichia coli*. NPJ Antimicrob Resist 3:6
7. Krishnamoorthy G, Wolloscheck D, Weeks JW et al (2016) Breaking the permeability barrier of *Escherichia coli* by controlled hyperporination of the outer membrane. Antimicrob Agents Chemother 60:7372–7381
8. Thompson LK, Cox G (2024) The next generation of the Efflux Platform: a genetic toolkit to study compound transport across the *Escherichia coli* cell envelope. NIH BioProject: Accession: PRJNA1101995 ID: 1101995. National Center for Biotechnology Information. https://www.ncbi.nlm.nih.gov/bioproject/?term=PRJNA1101995
9. Jiang Y, Chen B, Duan C et al (2015) Multigene editing in the *Escherichia coli* genome via the CRISPR-Cas9 system. Appl Environ Microbiol 81:2506–2514
10. Browning DF, Wells TJ, França FLS et al (2013) Laboratory adapted *Escherichia coli* K-12 becomes a pathogen of *Caenorhabditis elegans* upon restoration of O antigen biosynthesis. Mol Microbiol 87:939–950
11. Zhu K, Zhang Y-M, Rock CO (2009) Transcriptional regulation of membrane lipid homeostasis in *Escherichia coli*. J Biol Chem 284:34880–34888
12. Cox G, Sieron A, King AM et al (2017) A common platform for antibiotic dereplication and adjuvant discovery. Cell Chem Biol 24:98–109
13. Datsenko KA, Wanner BL (2000) One-step inactivation of chromosomal genes in *Escherichia coli* K-12 using PCR products. Proc Natl Acad Sci USA 97:6640–6645
14. Goetz JA, Kuehfuss NM, Botschner AJ et al (2022) Exploring functional interplay amongst *Escherichia coli* efflux pumps. Microbiology 168:001261
15. Lee A, Mao W, Warren MS et al (2000) Interplay between efflux pumps may provide either additive or multiplicative effects on drug resistance. J Bacteriol 182:3142–3150
16. Tal N, Schuldiner S (2009) A coordinated network of transporters with overlapping specificities provides a robust survival strategy. Proc Natl Acad Sci USA 106:9051–9056
17. Sander T, Freyss J, von Korff M et al (2015) DataWarrior: an open-source program for chemistry aware data visualization and analysis. J Chem Inf Model 55:460–473
18. Karp PD, Paley S, Caspi R et al (2023) The EcoCyc Database (2023). EcoSal Plus 11: eesp00022023
19. Inoue H, Nojima H, Okayama H (1990) High efficiency transformation of *Escherichia coli* with plasmids. Gene 96:23–28

Chapter 13

Mapping Genes Controlling Bacterial Cell Permeability Using TraDISort

Liping Li, Evan Gibbs, Francesca Short, Claire Maher, Amy K. Cain, Ian T. Paulsen, and Karl A. Hassan

Abstract

Bacterial cell envelopes have evolved to control the passage of small molecules into and out of the cell. Understanding these barriers and the pathways that small molecules can take as they cross the barrier could help to improve outcomes in broad areas of applied microbiology, e.g., to improve antibiotic uptake for infectious disease treatment, to control pollutant uptake in bioremediation, and to enhance bioproduct export in industrial fermentation. We developed an innovative high-throughput screening method that can be used to discover novel bacterial genes that affect intracellular accumulation of small molecules, such as efflux or uptake transporter genes, or genes controlling cell membrane composition. This method is a combination of Transposon Directed Insertion site Sequencing and fluorescence-activated cell sorting through flow cytometry (TraDISort). TraDISort has broad applications beyond studies of cell permeability, but in this chapter, we provide a basic protocol to study the penetration of fluorescent compounds.

Key words Transposon mutagenesis, TraDIS, Cell permeability, Antibiotic, Fluorescence-activated cell sorting (FACS), Metabolite accumulation, Efflux pumps, Antimicrobial resistance

1 Introduction

There are many instances where it may be desirable to control the cellular accumulation of small molecules in bacteria, such as enhancing the uptake of pollutants for bioremediation or antibiotics for bacterial killing [1]. The accumulation pathways taken by different small molecules differ depending on their chemical properties and the properties of the cell envelope in the target organism [2]. Two primary cell envelope types, Gram-positive and Gram-negative, and other less common arrangements exist in bacteria. Gram-positive bacterial cell envelopes contain a thick layer of peptidoglycan, on the order of 80 nm, and a phospholipid cytoplasmic membrane bilayer. Gram-negative bacteria contain a relatively thin peptidoglycan layer of around 8 nm, a cytoplasmic membrane of

Kunihiko Nishino and Mikio Tanabe (eds.), *Bacterial Multidrug Exporters: Methods and Protocols*, Methods in Molecular Biology, vol. 3044, https://doi.org/10.1007/978-1-0716-5316-6_13,
© The Author(s), under exclusive license to Springer Science+Business Media, LLC, part of Springer Nature 2026

similar composition to that in Gram-positives, and an outer membrane surrounding the peptidoglycan layer. The Gram-negative outer membrane is an asymmetric lipid bilayer with an outer leaflet composed of lipopolysaccharides (LPSs) or lipooligosaccharides (LOSs) and an inner leaflet composed of phospholipids. The outer LPS/LOS leaflet of Gram-negative outer membranes acts as a strong permeable barrier, preventing the passage of most small molecules, especially hydrophilic solutes, including many antimicrobials and chemical toxins [3].

Proteins residing in the outer membrane, e.g., general porins and other uptake systems, have been found to facilitate the passage of small molecules that are sufficiently hydrophilic, including various antibiotics [4]. Large or hydrophobic molecules are either excluded or may diffuse across the outer membrane, albeit relatively less frequently. Similarly, to pass into the cytoplasm, small molecules must either be carried by an uptake transporter or diffuse across the cytoplasmic membrane [2]. Small molecules that do enter the cell may be subject to efflux by one or more efflux pumps. Different pumps are involved in the transport of compounds across cytoplasmic membranes in Gram-positive and Gram-negative bacteria, and across Gram-negative outer membranes [5].

Due to the importance of transport proteins in controlling the movement of small molecules across bacterial cell envelopes, the regulatory factors controlling their expression can have a major influence on accumulation efficiency. The expression of efflux pumps and outer membrane channels is sensitive to the presence of antimicrobials or membrane stress in many bacteria, such that efflux pumps may be transiently overexpressed, and channels, especially general porins, may be repressed [5]. These changes in expression facilitate adaptive resistance via active antimicrobial export or reduced uptake, respectively. Prolonged antimicrobial/toxic compound exposure may select for mutations that lead to longer-term gene expression changes promoting either efflux pump overexpression or porin repression. Such mutations may be in the promoter region of the transporter gene, in a gene encoding a regulator of the transporter, or in the transporter gene itself. The mutations may be minor, consisting of a single nucleotide polymorphism or a single nucleotide insertion/deletion. However, an awareness of these mutations is essential, as they will have a significant impact on the identification of the major transport proteins and regulators in studies using functional genomic approaches, such as TraDISort (*see* **Note 1**) or transcriptomics.

Measuring the intracellular concentrations of small molecules poses experimental challenges, especially for studies aiming to screen individual cells rather than bulk populations of cells, e.g., from a bacterial culture. A useful approach applied extensively, especially in studies of efflux pumps, has been the use of dyes that are differentially fluorescent inside and outside cells, such as stains used in fluorescence microscopy [6]. For example, ethidium bromide will fluoresce more highly when intercalated into nucleic acids inside the cell than it will outside, and Hoeschst H33342 fluoresces more highly when present in the hydrophobic membrane interior than in solution. The fluorescence intensity of these dyes can thus be used as a proxy for their cellular concentrations.

We developed a high-throughput approach to screen thousands of individual bacterial mutants for compound accumulation that combines transposon sequencing technology and fluorescence-activated cell sorting (FACS) [7]. In this approach, mutant populations are generated in a strain of interest by transposon mutagenesis, and the mutant pool is exposed to a fluorescent compound and subjected to FACS to collect both high- and low-fluorescent mutant pools. The genomic locations of the transposon insertions within the enriched mutant pools are determined by Illumina sequencing using the TraDIS (Transposon Directed Insertion site Sequencing) approach [7–9].

TraDISort can detect genes encoding known or novel transport proteins that recognize the small molecule of interest, as well as their regulatory genes. For example, efflux pump mutants, positive regulators of efflux pumps and negative regulators of uptake transporters are expected to be enriched in the high fluorescent pools because these mutants will accumulate more of the compound. Conversely, mutants of uptake transporters, including porins and active permeases, and repressors of efflux pumps are expected to be enriched in the low-fluorescent pools because they will accumulate less than average amounts of compound. TraDISort has further potential for the discovery of novel genes affecting cell membrane permeability, such as those that control the composition of the cell membranes.

An advantage of TraDISort lies in its high-throughput capacity to screen hundreds of thousands of mutants simultaneously, its ability to detect subtle phenotypic changes through shifts in fluorescence intensity within individual live cells, and its strong statistical power derived from a saturated transposon mutant library. This approach provides direct, high-confidence correlations between phenotype and genotype.

The protocol begins with exposing the mutant library to a fluorescent dye at a non-inhibitory concentration. After incubation to allow the dye to reach equilibrium, the individual mutant cells are then subjected to FACS based on intracellular fluorescence intensity, then sorted and sequenced to determine the individual genes that affect the fluorescence intensity/accumulation of the compound. Here, we present the detailed TraDISort procedure, using the accumulation of ethidium bromide within *A. baumannii* cells as a representative example.

2 Materials

1. Saturated transposon insertion mutant library glycerol stock (*see* **Note 2**).
2. TraDIS primers (*see* **Note 3**).
3. Fluorescent dyes (*see* **Note 4**).
4. Cell culture media: Muller–Hinton broth (Sigma-Aldrich).
5. Flow cytometer reagents (sheath fluid): phosphate-buffered saline (PBS), flow cytometry buffer (1% NaCl) (*see* **Note 5**).
6. Genomic DNA extraction kit: e.g., QIAGEN DNeasy Blood and Tissue Kit (use an extraction protocol effective for the organism of choice).
7. Quant-iT™ PicoGreen® dsDNA Assay Kit.
8. TraDIS toolkit materials (*see* **Note 3**).
9. 1.7 mL tubes.
10. Barrier filter tips, DNase- and RNase-free.
11. Pipettes.
12. Flow cytometry tubes.
13. 50 mL cell culture flasks.
14. Spectrophotometer cuvettes (*see* **Note 6**).

2.1 Equipment

1. Cell incubator (e.g., Infors Multitron Standard).
2. Benchtop centrifuge.
3. Spectrophotometer or Nanodrop spectrophotometer (*see* **Note 6**).
4. ThermoFisher Scientific Qubit Fluorometer (e.g., Qubit 4).
5. FACS instrument that can detect the organism and fluorophore of choice, e.g., BD Influx cell sorter (Becton Dickinson) equipped with 488 nm laser for studies of *A. baumannii* loaded with ethidium bromide (*see* **Note 7**).
6. Illumina sequencing platform or TraDIS service provider (*see* **Note 3**).

2.2 Software

1. BD FACS Software Sorter Software.
2. TraDIS toolkit (*see* **Note 3**).
3. ChromoCorrect.

2.3 Bio-TraDIS Pipeline

1. Raw FASTQ files from TraDIS sequencing.
2. Bio-TraDIS software suite.
3. FastQC (for quality control).
4. Reference genome in .fasta format.

5. Reference genome in .embl format.
6. conda environment with Bio-TraDIS and associated bioinformatics tools (e.g., smalt for mapping, R for analysis) or access to Galaxy.org server.
7. Artemis genome browser.

3 Methods

1. Acquire a random transposon mutant library with good genome coverage on insertions or generate a library using a protocol relevant to the organism of choice, e.g., a mini-Tn5 transposome [10] (*see* **Note 2**).
2. For each of three replicates, grow an overnight culture of 10^9 cells in 10 mL of MH broth, at 37 °C with shaking.
3. In triplicate, dilute each overnight culture 1:100 in 15 mL fresh MH broth, and grow to mid-log OD_{600} 0.5–0.6 (*see* **Note 8**; takes 1.5–2 h for *A. baumannii*).
4. Two samples (positive control and FACS sample) are incubated with the fluorescent compound at room temperature for 20 min. If the compound has antimicrobial activity, the concentration used should not threaten the viability of mutants (we recommend to not exceed 1/16th of the inhibitory concentration for wild-type cells). The third culture (negative control) is left on the bench for 20 min. Grow the positive control and the negative control for the same amount of time as the cells collected in the FACS for recovery (**step 9**).
5. Dilute FACS sample 1/40 in MH with the same concentration of dye, then filter it by 10 μm membrane before FACS (*see* **Note 8**).
6. Run the filtrate through the FACS instrument. Gate on the single population (called singles in Fig. 1a) to discriminate against aggregated and dead cells using the forward scatter versus forward scatter pulse width parameters (Fig. 1a). To obtain uniform-sized cells, another gate (called cells in Fig. 1b) is applied on the gated single population using forward scatter (FSC) and side scatter parameters (SSCs) (Fig. 1b) (*see* **Note 9**).
7. Within the gated population of uniformly sized cells, based on forward and side scatter, the 2–5% of cells with the highest fluorescence and the 2–5% with the lowest fluorescence, along with cells showing intermediate fluorescence intensity, should be sorted into three separate population pools (Fig. 2) using the highest purity mode (one-drop single).

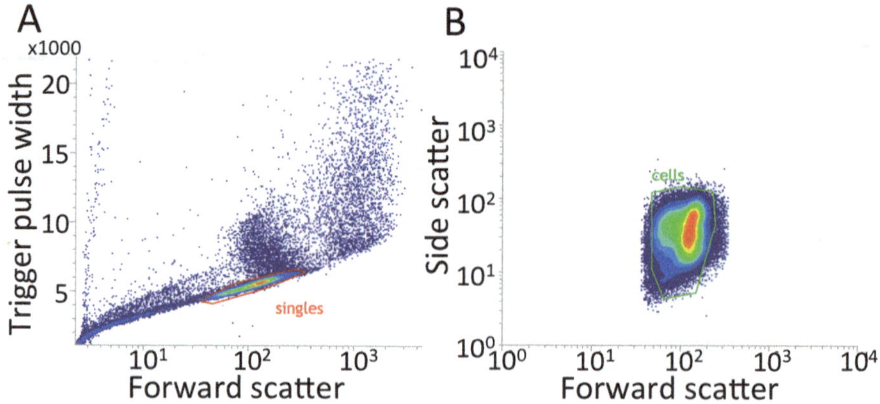

Fig. 1 Example of targeting uniform cell size for further cell sorting using a two-step gating method. (**a**) The trigger pulse width represents the signal width of pulses above the threshold. Gate "singles" captures the population of single cells, excluding aggregates and dead cells. (**b**) The forward scatter (FSC) parameter is indicative of cell size, while the side scatter (SSC) parameter correlates with cell granularity or internal complexity. Gate "cells" select uniform-size cells within the "singles" gate, indicating the targeted cells for analysis. (This figure has been reproduced from the supplemental materials in [7] under a Creative Commons license)

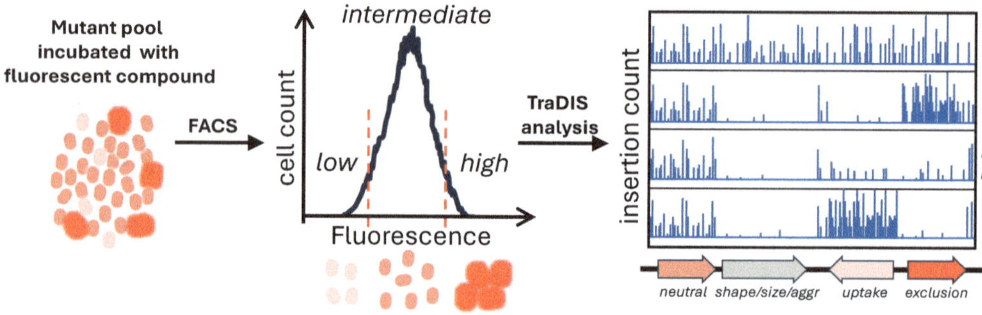

Fig. 2 TraDISort workflow for analysis of fluorescent compound accumulation. A mutant cell population is loaded with the compound of interest, e.g., ethidium bromide (left panel). The population is sorted by FACS, initially gated to obtain only single uniformly sized cells (Fig. 1), then into populations based on fluorescence intensity: low (bottom 5%), high (top 5%), and intermediate (central 90%; middle panel). The cell populations are subsequently subjected to genomic DNA extraction, TraDIS, and data analysis to map the locations and frequencies of transposon insertions (blue dashes, right panel). Genes that play no role in controlling compound accumulation or cell shape, size, or aggregation (neutral) are expected to display a stable insertion frequency in all cell populations. Genes that function in cell size, morphology, or aggregation control (shape/size/aggr) will be depleted in all sorted populations compared to the base library (input) due to exclusion by filtration or pre-sorting gates (Fig. 1). Mutations in genes whose function is in the uptake of the compound (uptake), such as those encoding uptake permeases, porins, or repressors of efflux pumps, should be enriched in the low-fluorescent pool, depleted in the high fluorescent pool, and moderately depleted in the intermediate pool. Mutations in genes whose function is in the efflux or exclusion of compounds (exclusion), such as those encoding efflux pumps, activators of efflux pumps, and systems involved in the production of cell permeability barriers, should be enriched in the high-fluorescent pool, depleted in the low-fluorescent pool, and moderately depleted in the intermediate pool

8. Fluorescence is detected using an appropriate filter according to the emission and excitation spectra of the fluorescent dye used (580/30 bandpass filter in BD Influx for ethidium bromide detection; *see* **Note 7**). A minimum of 0.5 million cells should be collected for each of the three cell fractions. Cells with medium fluorescence intensity may be collected at the end, as the collection time is much shorter.
9. Grow the cells, including diluted controls, for 3–12 h in permissive growth conditions, e.g., the media used in initial culturing (*see* **Note 10**) to obtain sufficient DNA for TraDIS.
10. Isolate genomic DNA using an appropriate extraction procedure, e.g., the QIAGEN DNeasy Blood and Tissue Kit. The concentration and purity of the DNA are measured by the Nanodrop spectrophotometer and Quant-iT™ PicoGreen® dsDNA Assay Kit. The remaining culture is frozen in 20% glycerol at −80 °C as a backup for later use.
11. All five sets of gDNA samples per biological replicate are used for TraDIS to recover transposon insertion reads. The reads are mapped to the host genome and analyzed according to the instructions described in the TraDIS toolbox or from a DNA sequencing center (Fig. 2; *see* **Note 3**) [11].

3.1 Data Analysis

1. TraDIS sequencing data analysis (*see* **Note 3**).

The raw sequencing data in unaligned fastq format are analyzed by the Bio-TraDIS analysis pipeline. Please refer to the TraDIS toolkit for the detailed procedure [11]. Galaxy.org also has the Bio-TraDIS pipeline installed. This process will map the transposon-tagged reads to the target genome and generate plot files that contain insertion counts per nucleotide, which can be visualized through the Artemis genome browser. The transposon insertion read counts are further analyzed through tradis_comparisons.R (part of the Bio-TraDIS pipeline) to identify the significance of read differences for each gene between the input mutant pool and the FACS-sorted output mutant pool.

3.2 Bio-TraDIS Pipeline Protocol

Procedure

Step 1: Concatenate Raw FASTQ Files (if needed).

1. If reads are split across multiple files, concatenate them into a single file.

    cat Sample_L001.fastq Sample_L002.fastq Sample_L003.fastq Sample_L004.fastq > Sample_combined.fastq

Step 2: Quality Control (QC) Using FastQC.

1. Run FastQC to assess the quality of the concatenated FASTQ files.

fastqc Sample_combined.fastq

2. Inspect the FastQC output .html file for quality metrics such as adapter contamination, quality score distribution, and GC content.
3. Optionally, trim or clean the reads if quality issues are detected (e.g., using a sequence read trimming tool).

Step 3: Map Reads to the Reference Genome.

1. Index the reference genome if not already indexed.

   smalt index ref_index reference_genome.fasta

2. Map the reads using Bio-TraDIS's bacteria_tradis tool, specifying the reference genome and transposon tag sequence (commonly "TAAGAGACAG" or another insertion tag).

   bacteria_tradis -v -f Sample_combined.fastq -t TAA-GAGACAG -s -r reference_genome.fasta

   -v: Verbose output
   -f: Input file
   -t: Transposon tag sequence
   -s: Use smalt for mapping
   -r: Reference genome

3. Verify that .bam, .stats, and .insert_site_plot.gz files are generated for further analysis.

Step 4: Generate Insertion Site Plots.

1. Run tradis_gene_insert_sites to create gene insertion plots for downstream analyses.

   tradis_gene_insert_sites -trim3 0.1 reference_genome.embl Sample_combined.insert_site_plot.gz

   -trim3 0.1: Optional trimming of transposon ends for better specificity

Step 5: Compare Conditions (e.g., Control vs. Treated).

1. Prepare lists of .csv files for control and treated conditions:

   ls *control*.csv > control_list.txt
   ls *treated*.csv > treated_list.txt

2. Run tradis_comparison.R to compare insertion patterns between conditions.

   tradis_comparison.R -f -t 10 -o comparison_output.csv -p comparison_output.pdf —controls control_list.txt —conditions treated_list.txt

   -f: Optional, specifies filtering settings
   -t: 10: Sets threshold for significance
   -o: Output .csv file
   -p: Output .pdf file for plots

Step 6: View Results.

1. Use Artemis (or another genomic viewer) to visualize mapped reads, essential genes, and insertion sites. Load the .embl reference genome and .gz insert site plot files.
   1. Review output files for:

      Comparative results: Shows differentially essential genes under treated vs. control conditions.
   2. (Optional) Visualize your TraDIS comparison data as a function of chromosomal location to detect and correct for any chromosomal location bias using the ChromoCorrect tool [12]. This can be done by downloading the R Package/R Shiny app (https://github.com/BarquistLab/ChromoCorrect) or online (https://thecainlab.shinyapps.io/ChromoCorrect/), to visualize and normalize your tradis_gene_insert_sites.csv files or any read count tables.
   3. TraDISort result validation.

      Genes that are identified by TraDISort as determinants affecting the fluorescent dye uptake can be validated through the targeted deletion or insertion mutants. BD Influx cell sorter should be able to detect that the target gene mutant population has a different fluorescent intensity compared to the wild-type strain. These mutants may exist in an accessible collection, such as the *E. coli* Keio collection [13] or may need to be generated in the strain of interest.

4 Notes

1. Strain selection will have an influence on the outcome of TraDISort accumulation experiments. For example, if fixed mutations are present in an efflux pump repressor gene that lead to constitutive expression of the efflux pump, transposon inactivation of the efflux pump gene should allow increased accumulation of the small molecule and thus selection of the mutant, but it is unlikely that insertions in the inactive repressor gene will result in changes to accumulation. In contrast, if the repressor gene is active, insertions in this gene will result in reductions in accumulation, due to increased expression of the pump, but insertions in the pump may have a less profound impact on accumulation because its expression is under active repression.
2. A saturated random transposon insertion mutant library comprising even insertion distribution across the genome and high

mutant density is important for TraDISort [11, 14]. Typically, a library with more than 100,000 unique insertion mutants is recommended, but depending on the number of cells that are sorted, some bottlenecking may occur, and lower numbers of mutants may be sufficient.

3. The TraDIS toolkit contains the TraDIS sequencing protocol and software to process raw fastq data, map and count transposon insertion reads, and statistically analyze transposon insertion read differences between experimental conditions. A DNA sequencing center that provides TraDIS sequencing as a service, such as the Ramaciotti Centre for Genomics, UNSW, Sydney, can be used to generate TraDIS sequence reads for downstream processing.
4. The dyes could be naturally fluorescent or fluorescently labeled. Ideally, these dyes will differ in fluorescent properties inside and outside cells or be accumulated in cells so that fluorescence is above background. Alternatively, a biosensor that detects a non-fluorescent molecule of interest and provides a fluorescent readout (e.g., from fluorescent protein expression) could be incorporated into the library.
5. Both PBS and NaCl are feasible for this purpose. If aiming for sorting 150,000 cells, you will collect around 150 μL PBS buffer (or 1% NaCl), which will not affect downstream experiments.
6. For measuring cell density (OD_{600}), the Eppendorf BioPhotometer can measure 50–2000 μL of cell culture, and the Nanodrop spectrophotometer can measure as low as 2 μL. Larger volumes could read OD_{600} with higher accuracy, and thus a standard photometer is preferred, such as the Eppendorf BioPhotometer. However, with a limited amount of dye, the Nanodrop spectrophotometer is preferable.
7. For most commercial fluorochromes, the excitation (Ex) and emission (Em) spectra can be obtained from the manufacturers, but the properties of some dyes may not be clear. To determine the appropriate laser used for FACS, the maximal Ex and Em wavelength can be measured by a benchtop fluorimeter.
8. It is important that the cell cultures (at OD_{600} 0.6) are diluted (1:100 in the case of *A. baumannii* BAL062) further to remove cell aggregates that will block the flow cytometer nozzle and to make sure the culture contains free-living single cells at the exponential phase. The dilution factor could be adjusted according to the cell aggregation phenotype of different bacterial organisms or strains. The dye concentration in the cell culture should be consistent throughout the whole sorting experiment.

9. The cell culture contains some dead cells and cells with abnormal morphology, which will also affect intracellular fluorescence intensity. The morphological differences may or may not be a result of transposon insertion, but these cells need to be excluded from TraDIS analysis because the fluorescence intensity of these cells/aggregates is not directly comparable to fluorescence in typical cells. A control experiment can be performed to identify mutations that lead to aberrant cell morphology or aggregation, or data can be analyzed to identify mutants that are highly depleted in both high- and low-fluorescent pools (Fig. 2). The BD Influx cell sorter and other sorters can detect the input cell population distribution in terms of cell morphology. As shown in Fig. 1a, the majority of the cells (colored red, green, and light blue) have similar morphology and are different from the cells presented as dark blue dots, which could be either dead cells, aggregates, or other abnormal cells. Through gating cell size and the side scatter parameters, the FACS instrument can selectively sort free-living cells with uniform size. The filter membrane used in Procedure 5 is to remove big cell aggregates.
10. The reason for further culturing cells after sorting is to get enough cells for total DNA extraction.

Acknowledgments

TraDISort work in our laboratories has been supported by Project Grants from the National Health and Medical Research Council of Australia to ITP and KAH (GNT1060895 and GNT1120298), an Australian Research Council Future Fellowship to KAH (FT180100123), an ARC Discovery Project to KAH (DP210100553), a US AFOSR grant to KAH, AKC, ITP, and EG (FA2386-24-1-4057), and the Australian Research Council Centre of Excellence in Synthetic Biology (CE200100029).

Conflict of Interest The authors have no conflicts of interest to declare that are relevant to the content of this chapter.

References

1. Kohanski MA, Dwyer DJ, Collins JJ (2010) How antibiotics kill bacteria: from targets to networks. Nat Rev Microbiol 8(6):423–435
2. Maher C, Hassan KA (2023) The gram-negative permeability barrier: tipping the balance of the in and the out. MBio 14(6): e0120523
3. Zgurskaya HI, Lopez CA, Gnanakaran S (2015) Permeability barrier of gram-negative cell envelopes and approaches to bypass it. ACS Infect Dis 1(11):512–522
4. Pages JM, James CE, Winterhalter M (2008) The porin and the permeating antibiotic: a

selective diffusion barrier in gram-negative bacteria. Nat Rev Microbiol 6(12):893–903

5. Du D, Wang-Kan X, Neuberger A, van Veen HW, Pos KM, Piddock LJV, Luisi BF (2018) Multidrug efflux pumps: structure, function and regulation. Nat Rev Microbiol 16(9): 523–539
6. Blair JM, Piddock LJ (2016) How to measure export via bacterial multidrug resistance efflux pumps. MBio 7(4)
7. Hassan KA, Cain AK, Huang TT, Liu Q, Elbourne LDH, Boinett CJ, Brzoska AJ, Li LP, Ostrowski M, Nhu NTK et al (2016) Fluorescence-based flow sorting in parallel with transposon insertion site sequencing identifies multidrug efflux systems in Acinetobacter baumannii. MBio 7(5)
8. Paulsen IT, Cain AK, Hassan KA (2017) Physical enrichment of transposon mutants from saturation mutant libraries using the TraDI-Sort approach. Mob Genet Elements. https://doi.org/10.1080/2159256X.2017.1313805
9. Cain AK, Barquist L, Goodman AL, Paulsen IT, Parkhill J, van Opijnen T (2020) A decade of advances in transposon-insertion sequencing. Nat Rev Genet 21(9):526–540
10. Warner IA, Kok WJ, Martinelli N, Yang Z, Goodall ECA, Henderson I (2023) Microbial primer: transposon directed insertion site sequencing (TraDIS): a high throughput method for linking genotype to phenotype. Microbiology (Reading) 169(11)
11. Barquist L, Mayho M, Cummins C, Cain AK, Boinett CJ, Page AJ, Langridge GC, Quail MA, Keane JA, Parkhill J (2016) The TraDIS toolkit: sequencing and analysis for dense transposon mutant libraries. Bioinformatics 32(7):1109–1111
12. Sullivan GJ, Barquist L, Cain AK (2024) A method to correct for local alterations in DNA copy number that bias functional genomics assays applied to antibiotic-treated bacteria. mSystems 9(4):e0066523
13. Baba T, Ara T, Hasegawa M, Takai Y, Okumura Y, Baba M, Datsenko KA, Tomita M, Wanner BL, Mori H (2006) Construction of Escherichia coli K-12 in-frame, single-gene knockout mutants: the Keio collection. Mol Syst Biol 2006(2):0008
14. Langridge GC, Phan MD, Turner DJ, Perkins TT, Parts L, Haase J, Charles I, Maskell DJ, Peters SE, Dougan G et al (2009) Simultaneous assay of every Salmonella Typhi gene using one million transposon mutants. Genome Res 19(12):2308–2316

Chapter 14

Evaluating the Efficacy of Drug Efflux Pump Inhibitors Through Growth Curve Analysis

Seiji Yamasaki, Martijn Zwama, and Kunihiko Nishino

Abstract

Drug efflux pumps are ubiquitous in bacterial genomes, and it has been shown that each bacterium harbors multiple of them. These pumps play crucial roles not only in drug resistance but also in pathogenicity, virulence, biofilm formation, and resistance to metabolic products. Therefore, they represent promising targets for the development of new drugs and efflux pump inhibitors aimed at combating bacterial infections by suppressing multiple key functions. Despite ongoing research efforts, clinically effective inhibitors remain elusive, which has prompted research and development involving large-scale compound screening in various research groups. However, to evaluate the inhibitory activity of candidate compounds, Minimum Inhibitory Concentration (MIC) assays are often unsuitable, as they require substantial quantities of compounds that may be rare or expensive. To address this, this chapter describes a method for conveniently evaluating the pump-inhibitory activity of multiple inhibitor substances at different concentrations using growth curve measurements, which require only small amounts of compounds.

Key words RND family, Multidrug exporter, Multidrug efflux pump, Inhibitor, Growth curve

1 Introduction

Drug efflux pumps are present in the genomes of all bacteria, and it has been shown that each bacterium harbors a number of drug efflux pumps [1, 2]. These efflux pumps are a major factor in bacterial inherent drug resistance (intrinsic resistance) and can lead to multidrug resistance (MDR) when overexpressed [3] (acquired resistance) due to their ability to recognize and efflux various antibiotics [4, 5]. In Gram-negative bacteria, efflux pumps belonging to the resistance–nodulation–division (RND) superfamily export drugs across the inner and outer membrane [6, 7] and are known to be one of the main causes of multidrug resistance due to their broad substrate recognition and strong efflux capacity [1]. Moreover, these RND pumps are involved not only in drug resistance but also in pathogenicity, virulence, biofilm formation, and resistance to metabolic products, among others [1, 2, 8–

Kunihiko Nishino and Mikio Tanabe (eds.), *Bacterial Multidrug Exporters: Methods and Protocols*, Methods in Molecular Biology, vol. 3044, https://doi.org/10.1007/978-1-0716-5316-6_14,
© The Author(s), under exclusive license to Springer Science+Business Media, LLC, part of Springer Nature 2026

13]. Therefore, these pumps are attractive drug targets for developing new drugs and efflux pump inhibitors (EPIs) that can suppress multiple important functions to combat bacterial infections. In this chapter, we use the RND pump MexB from *Pseudomonas aeruginosa*, which forms the tripartite complex MexAB–OprM as a probe.

Once an inhibitor is developed, it could serve as a broad-spectrum drug effective against a diverse array of resistant bacteria. This includes strains employing resistance mechanisms other than high expression of drug efflux pumps, such as reduced drug affinity due to mutations or drug inactivation facilitated by modifying or degrading enzymes.

Given the current absence of clinically effective inhibitors [14, 15], numerous research groups are actively engaged in research and development and large-scale compound screenings for potential candidates. However, the next step of evaluation (by Minimum Inhibitory Concentration (MIC) assays) poses challenges as it usually requires large quantities of compounds. Because of their scarcity and high cost, these compounds are often unsuitable for evaluating the pump-inhibitory activity of each candidate. Here, we describe a method that demands only small amounts of compounds to conveniently assess the pump-inhibitory activity of multiple individual compounds by using growth curve measurements [16]. Furthermore, this method allows for the confirmation of inhibitor effectiveness at multiple concentrations to check the dose-dependent responses.

2 Materials

2.1 Bacterial Strains

1. The strain *Escherichia coli* MG1655Δ*acrB*Δ*tolC*/pMMB67HE is used as a control.
2. The strain *E. coli* MG1655Δ*acrB*Δ*tolC*/pMMB67HE-*mexA-mexB-oprM* is used as the strain expressing the efflux pump system (MexAB–OprM) targeted by the inhibitor.

2.2 Pre-culture

1. Lauria–Bertani (LB) broth, Miller: 10.0 g/L Tryptone, 5.0 g/L Yeast Extract, 10.0 g/L Sodium Chloride. Prepared with distilled water and used after autoclaving.
2. 25 mg/mL ampicillin solution. Prepared with distilled water.
3. Mini Recipro Shaker NA-201 N (Nissinrika) is used as a test tube shaker.

2.3 Growth Curve Analysis

1. Lauria–Bertani (LB) broth, Miller: 10.0 g/L Tryptone, 5.0 g/L Yeast Extract, 10.0 g/L Sodium Chloride. Prepared with distilled water and used after autoclaving.

2. Thermo Scientific Nunc Edge 2.0 96-well Non-Treated Flat-Bottom Plate (Thermo Fisher Scientific) is used for 96-well plates.
3. 2.5 μg/mL aztreonam solution. Prepared with distilled water. Aztreonam is the antibiotic used in conjunction with the efflux pump inhibitor.
4. 0, 20, 80, 320 μg/mL [[2-({[((3*R*)-1-{8-{[(4-*tert*-butyl-1,3-thiazol-2-yl)amino] carbonyl}-4-oxo-3-[(*E*)-2-(1*H*-tetrazol-5-yl)vinyl]-4*H*-pyrido[1,2-*a*]pyrimidin-2-yl}piperidin-3-yl) oxy]carbonyl}amino)ethyl](dimethyl)ammonio]acetate (ABI-PP, D13-9001) [17] in DMSO solution. ABI-PP is used as the efflux pump inhibitor.
5. Infinite M200 Pro (Tecan) plate reader is used to measure the OD_{600} (growth curves).

3 Methods

1. **Steps 1–12** are all under sterile conditions. Add 1992 μL of LB broth to two small glass test tubes.
2. Add 8 μL of 25 mg/mL ampicillin to each of the two test tubes (2 mL) to achieve a final concentration of 100 μg/mL (*see* **Notes 1** and **2**).
3. Inoculate MG1655Δ*acrB*Δ*tolC*/pMMB67HE and MG1655Δ*acrB*Δ*tolC*/pMMB67HE-*mexA*-*mexB*-*oprM*, respectively, in each of the tubes.
4. Cover the tubes with a cap.
5. Incubate the tubes overnight in an incubator at 37 °C while shaking (180 rpm linear reciprocal motion) (*see* **Note 3**).
6. Add 10 mL of LB broth to two larger glass test tubes.
7. Add the pre-cultures of the MG1655Δ*acrB*Δ*tolC*/pMMB67HE strain and MG1655Δ*acrB*Δ*tolC*/pMMB67HE-*mexA*-*mexB*-*oprM* strain to the fresh LB broth to achieve an OD_{600} of 0.05 (*see* **Notes 4** and **5**).
8. Add 10 μL of 2.5 μg/mL aztreonam solution (final concentration 0.125 μg/mL) directly to the bottom of each well of the 96-well plate (*see* **Note 6**).
9. Add 5 μL of 0, 20, 80, 320 μg/mL ABI-PP in DMSO solution (final concentration 0, 0.5, 2, 8 μg/mL) along the left side of each well of the 96-well plate (*see* **Note** 7).
10. Add 185 μL of LB broth containing MG1655Δ*acrB*Δ*tolC*/pMMB67HE or MG1655Δ*acrB*Δ*tolC*/pMMB67HE-*mexA*-*mexB*-*oprM* (OD_{600} of 0.05) to each well (*see* **Note 8**).

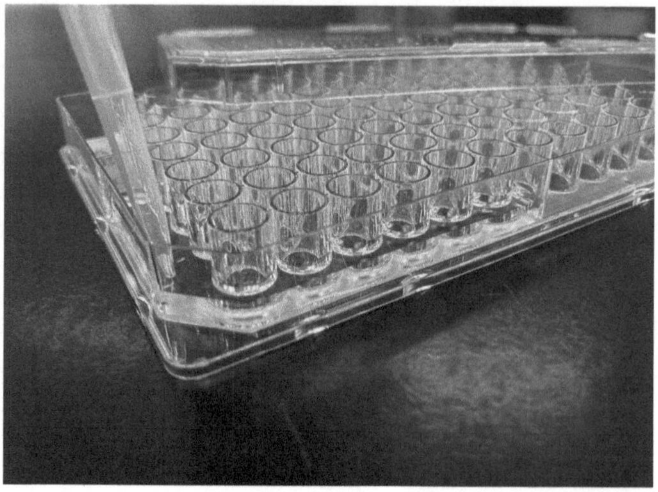

Fig. 1 Adding 2 mL of sterile water to each of the four corner grooves of the 96-well plate. Thermo Scientific Nunc Edge 2.0 96-well Plate (Thermo Fisher Scientific) can reduce evaporation of culture medium from the wells during measurement by adding sterile water in each of the four corner grooves

11. Add 2 mL of sterile water to each of the four corner grooves of the 96-well plate (*see* **Note 9**) (Fig. 1).
12. Cover the 96-well plate and then place it in the plate reader.
13. In the plate reader, shake the plate (400 rpm orbital motion) at 37 °C and measure the OD_{600} every 15 min (*see* **Note 10**) (Fig. 2).

4 Notes

1. Choose the antibiotic accordingly, depending on the resistance marker on the plasmid and the resistance profile of the bacteria used.
2. Regardless of the resistance marker on the plasmid, the efflux pump-deficient strain may be more sensitive to the drug than wild-type strains. Therefore, if it is difficult to culture in the usual antibiotic concentration, use a lower concentration of the drug that still prevents the growth of the non-resistant parent strain, for example, 50 instead of 100 μg/mL ampicillin.
3. To prevent the formation of a pellet by bacterial growth only at the bottom of the tube, fix the tubes at an angle during shaking incubation, depending on the characteristics of the strain used.
4. If the measurement is performed for a long time and the effect of evaporation of the culture medium is significant, adjust the initial value of OD_{600} according to the characteristics of the strain, antibiotic, and inhibitor used.

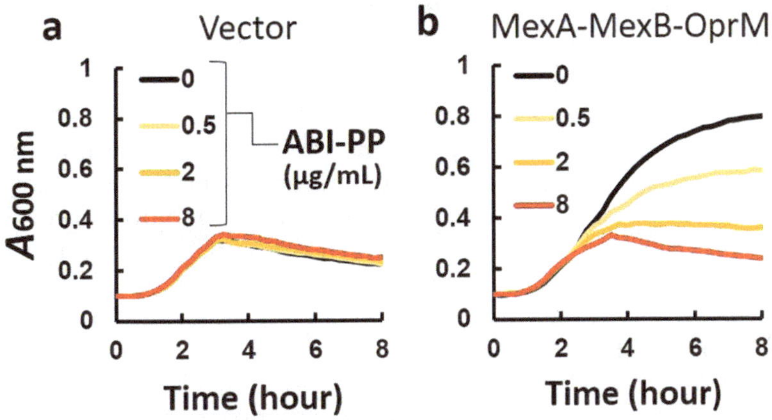

Fig. 2 Bacterial growth curves measured using both aztreonam (antibiotic) and ABI-PP (inhibitor). (**a**) The growth of *E. coli* MG1655Δ*acrB*Δ*tolC*/pMMB67HE in the presence of 0.125 μg/mL of aztreonam and various concentrations of ABI-PP. (**b**) The growth of *E. coli* MG1655Δ*acrB*Δ*tolC*/pMMB67HE-*mexA-mexB-oprM* strain in the presence of 0.125 μg/mL of aztreonam and various concentrations of ABI-PP. These tests were performed in triplicate. As the inhibitor concentration increases, the efflux pump is inhibited, and the antibiotic becomes more effective, leading to growth inhibition in the *E. coli* MG1655Δ*acrB*Δ*tolC*/pMMB67HE-*mexA-mexB-oprM* cells. ABI-PP concentrations and colors: 0, 0.5, 2, and 8 μg/mL as black, yellow, orange, and red, respectively

5. If OD_{600} is measured with another device using a test tube or cuvette instead of a plate reader, the OD_{600} values obtained may differ from those of the plate reader. In such cases, it is recommended to make a corresponding table of OD_{600} values.
6. The concentration of the antibiotic used in combination with the EPI should be at such a concentration that the growth of the MG1655Δ*acrB*Δ*tolC*/pMMB67HE-*mexA-mexB-oprM* strain (the strain expressing the efflux pump targeted by the inhibitor) is slightly suppressed by the antibiotic alone. Under these conditions, the effect of potential inhibitors can be readily observed.
7. By adding the inhibitor separately to the wells, compared to a method where antibiotics and inhibitors are added simultaneously, it becomes easier to visually confirm whether the inhibitor has been added to the well. This also helps in preventing errors such as forgetting to add the inhibitor or accidentally adding it twice.
8. Mixing at this point is often unnecessary due to sufficient mixing during the subsequent shaking in the plate reader.
9. If the 96-well plate is not a Nunc Edge plate and does not have grooves for sterile water at the corners, do not use the

outermost row of wells for growth curve measurement. Instead, use these wells to hold 200 μL of sterile water each to minimize evaporation of the culture medium during the measurement.

10. If a smooth growth curve is not obtained, adjust the presence or method of shaking (linear reciprocal or orbital motion) and the measurement conditions according to the characteristics of the strain, antibiotic, and inhibitor used.

Acknowledgments

This study was supported by the Research Program for CORE lab and CORE^2-A lab of "Dynamic Alliance for Open Innovation Bridging Human, Environment and Materials" in "Network Joint Research Center for Materials and Devices" from the Ministry of Education, Culture, Sports, Science and Technology of Japan (MEXT); Grant-in-Aid for Scientific Research (B) (KAKENHI JP23H02631), Scientific Research (C) (KAKENHI JP25K10343), Challenging Research (Exploratory) (KAKENHI JP22K19831), and Early-Career Scientists (KAKENHI JP21K16318 and JP23K14346) from the Japan Society for the Promotion of Science (JSPS); Japan Agency for Medical Research and Development (AMED); the Uehara Memorial Foundation; Yakult Bio-Science Foundation; the Sumitomo Foundation; Takeda Science Foundation; Kowa Life Science Foundation; Japan Antibiotics Research Association; Japanese Society of Chemotherapy; the Nippon Foundation-Osaka University Project for Infectious Disease Prevention.

Conflicts of Interest The authors have no conflicts of interest to declare that are relevant to the content of this chapter.

References

1. Nishino K, Yamaguchi A (2001) Analysis of a complete library of putative drug transporter genes in *Escherichia coli*. J Bacteriol 183:5803–5812
2. Nishino K, Latifi T, Groisman EA (2006) Virulence and drug resistance roles of multidrug efflux systems of *Salmonella enterica* serovar Typhimurium. Mol Microbiol 59:126–141
3. Blair JM, Webber MA, Baylay AJ, Ogbolu DO, Piddock LJ (2015) Molecular mechanisms of antibiotic resistance. Nat Rev Microbiol 13: 42–51
4. Blair JM, Richmond GE, Piddock LJ (2014) Multidrug efflux pumps in Gram-negative bacteria and their role in antibiotic resistance. Future Microbiol 9:1165–1177
5. Nikaido H (2009) Multidrug resistance in bacteria. Annu Rev Biochem 78:119–146
6. Poole K (2004) Efflux-mediated multiresistance in Gram-negative bacteria. Clin Microbiol Infect 10:12–26
7. Poole K (2005) Efflux-mediated antimicrobial resistance. J Antimicrob Chemother 56:20–51

8. Yamasaki S, Fujioka T, Hayashi K, Yamasaki S, Hayashi-Nishino M, Nishino K (2016) Phenotype microarray analysis of the drug efflux systems in *Salmonella enterica* serovar Typhimurium. J Infect Chemother 22:780–784
9. Yamasaki S, Nagasawa S, Fukushima A, Hayashi-Nishino M, Nishino K (2013) Cooperation of the multidrug efflux pump and lipopolysaccharides in the intrinsic antibiotic resistance of *Salmonella enterica* serovar Typhimurium. J Antimicrob Chemother 68:1066–1070
10. Yamasaki S, Wang LY, Hirata T, Hayashi-Nishino M, Nishino K (2015) Multidrug efflux pumps contribute to *Escherichia coli* biofilm maintenance. Int J Antimicrob Agents 45:439–441
11. Gillis RJ, White KG, Choi KH, Wagner VE, Schweizer HP, Iglewski BH (2005) Molecular basis of azithromycin-resistant *Pseudomonas aeruginosa* biofilms. Antimicrob Agents Chemother 49:3858–3867
12. Baugh S, Ekanayaka AS, Piddock LJ, Webber MA (2012) Loss of or inhibition of all multidrug resistance efflux pumps of *Salmonella enteri*ca serovar Typhimurium results in impaired ability to form a biofilm. J Antimicrob Chemother 67:2409–2417
13. Piddock LJ (2006) Multidrug-resistance efflux pumps – not just for resistance. Nat Rev Microbiol 4(8):629–636
14. Lomovskaya O, Bostian KA (2006) Practical applications and feasibility of efflux pump inhibitors in the clinic–a vision for applied use. Biochem Pharmacol 71:910–918
15. Pages JM, Amaral L (2009) Mechanisms of drug efflux and strategies to combat them: challenging the efflux pump of Gram-negative bacteria. Biochim Biophys Acta 1794:826–833
16. Yamasaki S, Koga N, Zwama M, Sakurai K, Nakashima R, Yamaguchi A, Nishino K (2022) Spatial characteristics of the efflux pump MexB determine inhibitor binding. Antimicrob Agents Chemother 66:e0067222
17. Yoshida K, Nakayama K, Ohtsuka M, Kuru N, Yokomizo Y, Sakamoto A, Takemura M, Hoshino K, Kanda H, Nitanai H, Namba K, Yoshida K, Imamura Y, Zhang JZ, Lee VJ, Watkins WJ (2007) MexAB-OprM specific efflux pump inhibitors in *Pseudomonas aeruginosa*. Part 7: highly soluble and in vivo active quaternary ammonium analogue D13-9001, a potential preclinical candidate. Bioorg Med Chem 15:7087–7097

Chapter 15

Cloning and Characterizing Multidrug Efflux Pumps

Martijn Zwama and Kunihiko Nishino

Abstract

This protocol describes how to clone and investigate multidrug efflux pumps from plasmids to be expressed in *E. coli* and *P. aeruginosa*. We want to provide a quick overview for those starting with RND-type efflux pump research or facing cloning issues by sharing an updated method we have recently been using to clone multiple functional pumps and complex components quickly and efficiently.

Key words Multidrug exporter, RND superfamily, Resistance–nodulation–cell division, Multidrug resistance, Antibiotic resistance, Molecular cloning, Antibiotics, Drug efflux, Functional analysis

1 Introduction

RND-type multidrug efflux pumps play an important role in multidrug-resistant infections, as they can recognize and expel a wide range of structurally different antibiotics [1, 2]. To study RND-type efflux pumps, including the other members of the tripartite complex (periplasmic adaptor protein, or PAP, and outer membrane protein, or OMP) that span the inner and outer membrane of Gram-negative bacteria, genes from these pumps can be cloned into plasmids and expressed to check their functionality. Once actively expressed from plasmids, it is easy to introduce and study mutations to better understand the function and mechanisms of the pumps [3–5]. Well-studied pumps include AcrB [1, 5–14] (*Escherichia coli* and *Salmonella*), MexY and MexB [10, 15–25] (*Pseudomonas aeruginosa*), and AdeB [26–30] (*Acinetobacter baumannii*). Here, we will give an overview of how (unstudied) multidrug efflux pumps can be cloned into plasmids and characterized by providing a detailed method based on procedures we have been performing in the lab and optimized by trial-and-error to, hopefully, help others starting or who are somehow having difficulty getting the right clones or functional pumps.

Kunihiko Nishino and Mikio Tanabe (eds.), *Bacterial Multidrug Exporters: Methods and Protocols*, Methods in Molecular Biology, vol. 3044, https://doi.org/10.1007/978-1-0716-5316-6_15,
© The Author(s), under exclusive license to Springer Science+Business Media, LLC, part of Springer Nature 2026

2 Materials

2.1 Construction of Plasmids and Transformation

1. Genomic (template) DNA, either from growing colonies on agar plates (colony PCR) or from using isolated genomic DNA. Reference strain examples are MG1655 or W3104 (*E. coli*), ATCC19606 or BAA-1794 (*A. baumannii*), PAO1 (*P. aeruginosa*), and ATCC14028s (*Salmonella typhimurium*).
2. Vector DNA. Select a proper plasmid to insert your gene(s), ensuring that the resistance marker is not a substrate of the pump you plan to investigate. We have used multiple plasmids to express AcrB and other (RND-type) drug efflux pumps in our lab (including pUCP20 [10], pQE70 [31, 32], pUC118/119 [31, 33–35], pTH18cr/19cr [35], pTrc99A [33, 35, 36], and pHSG398/399 [35, 37]), but often use pBAD33 for His-tagged AcrB [3, 9–12, 38] (and *H. influenzae* AcrRAB [38]), of which the expression can be controlled by arabinose (10 mM final concentration). Other plasmids for the expression of entire tripartite complexes have also been used, such as shuttle vector pMMBH67HE or pMMBH67EH [10, 15, 22], which can express efflux pump systems in both *E. coli* (no IPTG needed) and *P. aeruginosa*.
3. DNA polymerase. A high-fidelity (HF) DNA polymerase is recommended. Use the polymerase (e.g., PrimeSTAR GXL, TaKaRa) with reaction buffer and dNTP mixture, or use a ready-to-use master mix. We have recently been using KOD One 2× Master Mix (TOYOBO) and Q5 Hot Start High-Fidelity 2× Master Mix (NEB).
4. Cloning kit. We have recently been using In-Fusion to eliminate the need for restriction enzyme digestion, ligation, and phosphorylation: In-Fusion Snap Assembly Master Mix (TaKaRa) and Cloning Enhancer (TaKaRa). With the slightly altered method described later, we obtained a 100% success rate (no empty vectors), also when inserting multiple inserts.
5. Software. You can design primers manually or use automated software such as SnapGene (used in this chapter and described in the Methods section) to design and optimize constructs and primers.
6. Primers (oligonucleotides). For sequencing and PCR in general: ideally 18–24 bp, 40–60% GC content, GC-clamp at the end if possible, and a melting temperature (T_m) of 50–60 °C. For cloning with In-Fusion (TaKaRa): a 15 bp overlap between the insert(s) and template is needed and a melting temperature of the annealing part of the primers of 60 °C.
7. Nuclease-free water (NFW).

8. QIAquick PCR purification kit, or spin columns (e.g., Econo-Spin IIa), in combination with Buffer PE (alternatively 10 mM Tris–HCl, pH 7.5, 80% ethanol) and PB (QIAGEN).
9. Agarose (e.g., Nacalai Agarose for ≧1 kbp fragment).
10. Midori Green Direct nucleic acid stain (NIPPON Genetics).
11. Quick-Load (Purple) 1 kb Plus DNA Ladder (NEB).
12. Gel Loading Dye, Purple (6X) (NEB).
13. Chemically competent cells of *E. coli* JM109 (TaKaRa).
14. A water bath (42 °C).
15. Sterile 80% glycerol.
16. Electroporation cuvettes (Gene Pulser/MicroPulser 0.1 cm gap) (Bio-Rad) and an electroporation system such as the Gene Pulser Xcell (Bio-Rad).
17. LB broth: 1% tryptone, 0.5% yeast extract, 1% sodium chloride; LB agar: LB broth with 1.5% agar; SOC medium, 2% tryptone, 0.5% yeast extract, 0.05% sodium chloride, 0.24% magnesium sulfate (anhydrous), 0.00186% potassium chloride, 0.4% glucose. (Alternatively, use premade powders, e.g., from BD for LB, LB agar, and SOB).
18. Incubator (30–37 °C) with shaker.
19. Glass culture tubes.
20. QIAprep Spin Miniprep (QIAGEN) and spin columns.
21. NanoDrop (Thermo Scientific) or NanoPhotometer (Implen).
22. PCR thermal cycler.
23. QIAEX II Gel Extraction Kit (QIAGEN).

2.2 Expression of Efflux Pumps

1. Electrocompetent *E. coli* MG1655, C43(DE3), KAM3, deficient in *acrB*, *acrAB*, *acrBtolC*, or *acrABtolC* (depending on your research), or *P. aeruginosa* PAO1 deficient in one or more efflux pump genes (e.g., *mexXY*, *mexAB*, *oprM*, etc.).
2. LB broth: as described in Subheading 2.1, **item 17**; 2xYT broth: 1% yeast extract, 1.6% tryptone, 0.5% sodium chloride.
3. Sterile 10% glycerol (ice-cold).
4. Inducers: IPTG (e.g., typically 1 mM) or arabinose (e.g., typically 10 mM). Different IPTG and arabinose concentrations can be tested, and expression and activity can be checked to determine the optimal concentration.
5. Ultracentrifuge and rotor. We use a TLA 100.3 rotor, which fits centrifuge tubes of about 5–7 mL volume.
6. SDS–PAGE gel, percentage depending on the efflux pump size. We generally use 10% SDS, such as Mini-PROTEAN TGX Precast Gel 10% (Bio-Rad) or Extra PAGE One Precast Gel 10% (Nacalai), and SDS Running Buffer Solution (10X) (Nacalai).

7. Protein standards: Precision Plus (1) Protein All Blue Prestained Protein Standards (Bio-Rad) and (2) Unstained Protein Standards, Strep-tagged recombinant (Bio-Rad).
8. PVDF membranes such as Trans-Blot Turbo Mini 0.2 μm (Bio-Rad) and a transfer system such as the Trans-Blot Turbo (Bio-Rad) (*see* **Note 1**).
9. Antibodies: 1°, Anti-His-tag mAb (MBL D291–3); 2°, Mouse IgG HRP-linked whole Ab (Merck) and Precision Protein StrepTactin-HRP Conjugate (Bio-Rad).
10. Chemi-Lumi One Super (Nacalai).
11. Baffled 250 mL glass shaking flasks.
12. Buffers: Lysis Buffer: 50 mM Tris–HCl, pH 7.5, 1 mM $MgCl_2$, 0.5 mM EDTA-NaOH, pH 7.5, 10% glycerol; Membrane Buffer: as Lysis Buffer, but without glycerol and $MgCl_2$.
13. Protein Assay BCA Kit (Nacalai).

2.3 Evaluation of the Function of Efflux Pumps

1. Your newly created strains with the plasmid of interest, including negative controls (cells harboring an empty vector).
2. 96-well plates, sterile, flat-bottom with lid (Nunc).
3. Petri dishes (diameter 87.3 mm), usually 12 plates per antibiotic.
4. Cell stamp device "MIC Measuring Device Microplanter" (Sakuma, Japan).
5. LB and LB agar as described in Subheading 2.1, **item 17**.
6. Antibiotics (*see* **Note 2**).
7. Incubation: 30–37 °C incubator for liquid and plate (solid) MIC, and a plate reader (TECAN Infinite 200 PRO series) for liquid MIC or growth ability assays.

3 Methods

This section describes how to clone an efflux pump gene cluster into a vector. We will use the example of a recently prepared plasmid, where the efflux pump complex AdeABC genes from *A. baumannii* strain BAA-1794 were cloned into the pMMB67HE vector. We use this AdeABC construct as an example to show that non-native RBS designs function here. Additionally, we had to adjust the protocol to lower growth temperatures to achieve active expression. Furthermore, this construct can be used in both *E. coli* and *P. aeruginosa*. Here, we used our existing pMMB67HEmexXYoprM [15] as a template to amplify pMMB67HE, as this amplification product can be easily separated by size on an agarose gel from the template, and because we used the artificial ribosome binding site (RBS) region before the start codon (GAAGGA-TACTCTatg) used upstream of *mexX* for *adeA*, so that we only had to clone the genes themselves without the need to determine the upstream region before *adeA*. The intrinsic RBS of *adeB* and *adeC* (with a natural linker between them) was confirmed to be present using SnapGene. However, we typically include upstream regions of about 10–30 bp (interestingly, we needed to include the whole *acrR* gene upstream of acrA when cloning *H. influenzae* AcrAB to get active AcrB expression [38]).

3.1 Primer Design

1. First, obtain the template DNA sequences for both the vector and the insert.
   (a) Plasmid DNA sequences can be found on several websites, including AddGene (https://www.addgene.org/vector-database/) or the SnapGene database (https://www.snapgene.com/plasmids), or based on your own sequencing data.
   (b) Genomic DNA data can be obtained from NCBI (http://www.ncbi.nlm.nih.gov/genome), or if you ordered a specific (clinical isolate) from ATCC, you can download it from the ATCC website (https://genomes.atcc.org).
2. Next, primers must be designed to amplify the vector and insert (s) and to sequence the final construct before usage. Load the template DNA sequences into the SnapGene software (version 7.1.1 at the time of writing). If your insert genes come from a genomic DNA template instead of a plasmid, locate the genes in the genome by doing a (homology) search (Fig. 1). In this example case, we use the sequence of *adeA* from *A. baumannii* from the NCBI website (https://www.ncbi.nlm.nih.gov/nuccore/) searching "Acinetobacter baumannii adeA" (from the NCBI Gene page, you can download the FASTA file) and perform a homology search in multidrug-resistant *A. baumannii* ATCC BAA-1794 by clicking "find similar DNA sequences" (Fig. 1a). Annotated *.gbk* files can also be used to import the *.fasta* file into SnapGene. When you click the "Show translations" button, you can readily select the gene you wish to insert if there are annotations.
3. Once the gene (cluster) has been located, check the presence of an RBS and the upstream region you wish to include. Otherwise, as in this example case, select only the genes of interest (Fig. 1b). Copy this region.
4. From the vector window, click "Actions → In-Fusion Cloning → Insert fragment . . ." and select the region you want to be replaced in the Vector tab (Fig. 2a).
5. In the Fragment tab, select the source of the fragment to be your genomic DNA, and perform a search by pasting the copied genes you selected (and select the orientation you want your genes to be cloned into the vector).

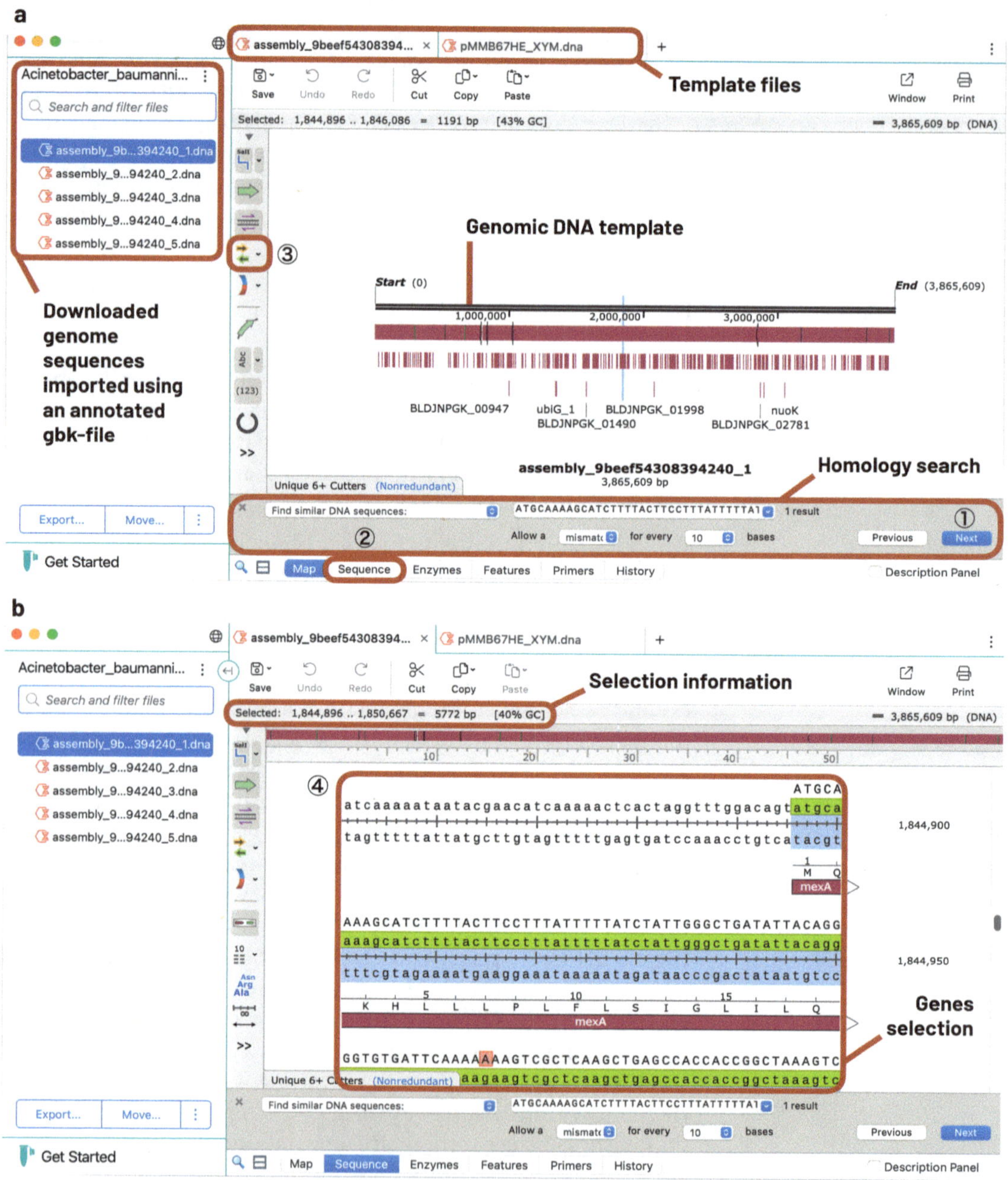

Fig. 1 SnapGene screenshots showing homology search (**a**) and gene selection (**b**). Two template DNA files are loaded in: the insert (genomic DNA "assembly_[. . .]") and the vector (plasmid DNA "pMMB67HE_XYM"). ① Search the gene of interest by homology search, ② View the nucleotide sequence, ③ When there are no annotations, press the "Show translations" button, and ④ Select the gene(s) including a potential upstream region you wish to clone into the vector

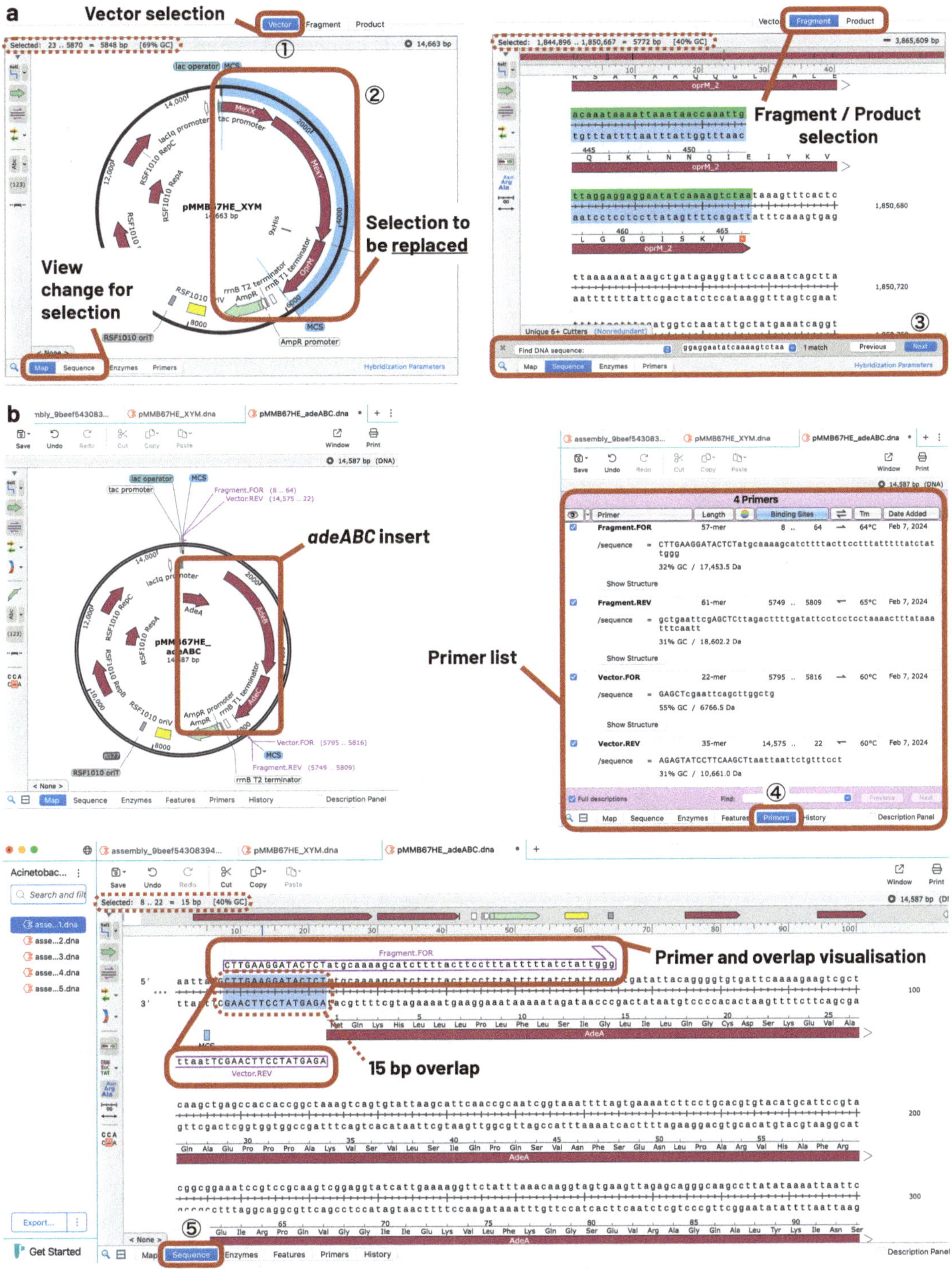

Fig. 2 SnapGene screenshots showing In-Fusion cloning (**a**) and primer design (**b**). Perform the In-Fusion cloning function. ① Start from the Vector tab, ② Select the region to be replaced or the location of insertion, ③ From the Fragment tab, search for your genes, select them, choose the insert orientation, and go to the Product tab and press "Choose Overlapping PCR Primers. . .," ④ ⑤ check and verify the primers created. Note that the annotated genes can have names different from the actual gene name (as in this case, *adeA* is called *mexA*, *adeB* is called *acrB*, and *adeC* is called *oprM_2*)

6. Click the Product tab and click "Choose Overlapping PCR Primers...," generally set to 60 °C and 15 bp overlap (this length can be increased if needed), and press Clone (*see* **Note 3**).
7. Now, the final product can be seen in "Map," the primers can be checked and copied from "Primers," and the sequence and overlaps can be checked from "Sequence" (Fig. 1b). You can perform a "find similar" search of the primers in the template DNA files to check potential nonspecific annealing sites.
8. Now you have your final construct, you can design primers for sequencing. You can use online tools, such as GenScript (https://www.genscript.com/tools/dna-sequencing-primer-design). However, we usually design primers by hand. Look at your sequence in SnapGene and create short primers as described in Subheading 2.1, **item 6**. If your sequencing read is about 1000 bp, you can aim for 100–200 bp overlaps between reads. You can go to "View → Set Origin..." if it makes it easier. Examples of primer design can be seen in Fig. 3.
9. You can select the sequence to check the T_m, length, and GC content (Fig. 3a).
10. Once you select your primer region, you can add the primer by clicking "Primers → Add primer..." (Fig. 3b).

3.2 Amplification

Amplify your vector and insert. For plasmid as template DNA, use a small amount, such as 100 pg–1 ng. Higher amounts can cause lower yields (no bands, nonspecific bands, or smears on the agarose gel) and unwanted template DNA as background after cloning and transformation. For genomic DNA as template DNA, we usually use colonies grown on an agar plate (this can be freshly made the day before or used from a stored plate in the fridge for a couple of weeks).

1. Prepare forward (FWD) and reverse (REV) primers for the vector and insert(s) at 5 μM. For plasmid template DNA, prepare a 100 pg–1 ng/μL solution.
2. Firstly, to a PCR tube, add 21 μL nuclease-free water (NFW) for plasmid templates or 22 μL NFW for colony PCR.
3. For plasmid DNA as a template, add 1 μL of DNA.
4. Add 1.5 μL FWD and REV primers to the tube.
5. For colony PCR, lightly touch a colony on an LB agar plate with a sterile toothpick. Dip the tip of the toothpick a few times in the tube. Add a small amount: if you are able to see the cells as a slightly cloudy mixture, you have added too much.
6. Lastly, add 25 μL 2× KOD One polymerase master mix and mix well.

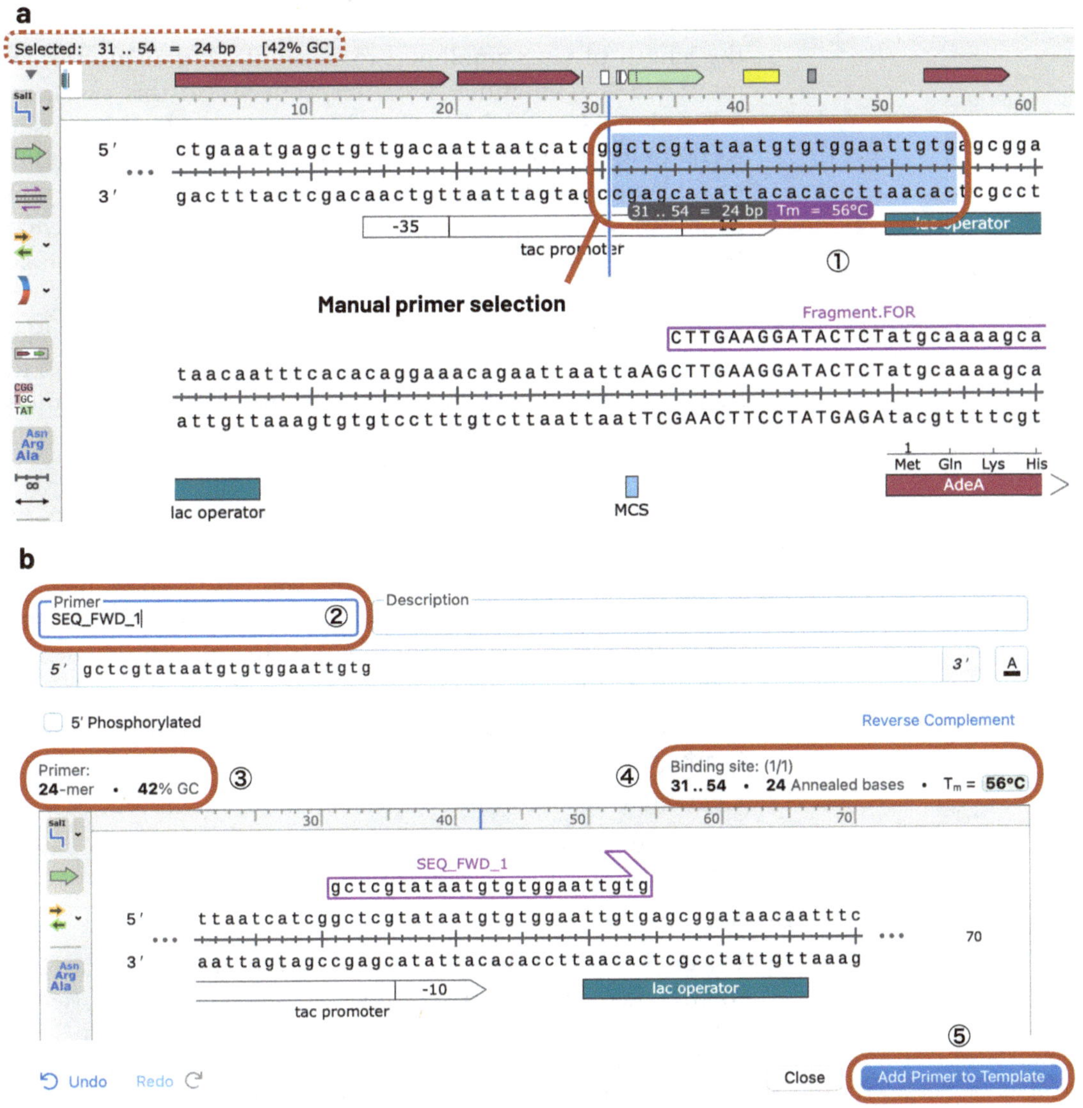

Fig. 3 SnapGene screenshots showing manual sequencing primer design. (a) ① Select the sequence to check your primer length, GC content and Tm, ② Set a primer name, (b) ③ Check the primer length and GC content, ④ Check the Tm and the number of binding sites, ⑤ Add primer to the template

7. Determine the PCR protocol. The speed depends on your polymerase (e.g., 20–30 s/kb for NEB's Q5, while 5–10 s/kb for ToYoBo's KOD One for amplimers up to 10 kb or larger than 10 kb). The annealing temperature depends on your polymerase setup: different concentrations of ions cause significant differences in the T_m of the primers and the template DNA. Use NEB's online tool (https://tmcalculator.neb.com/), and for KOD One, use the lowest T_m—5 °C (e.g., in this example case, the vector primers have a T_m of 60 °C, so use

55 °C, while the insert primers have a T_m of 64 and 65 °C, so use 59 °C). The polymerization (extension) time depends on the length of your amplimer; for example, if your vector and insert are 8.8 kb and 5.8 kb, respectively (in this case), use 8.8 × 5 = 44 s and 5.8 × 5 = 29 s, respectively (when using KOD One). If you have one PCR machine, you can round up to 60 s and use an annealing temperature of the lowest temperatures (55 °C) for both PCR reactions. Make sure to check the amplification on an agarose gel and alter the protocol accordingly.

8. Set up the PCR reactions (KOD One as an example). 25 cycles: 98 °C for 10 s, annealing temperatures for 5 s, and 68 °C for however long, depending on the length of your amplimers. The PCR amplification flowchart and overview can be seen in SnapGene by clicking the History button (Fig. 4).
9. While the PCR is running, prepare a 0.5 (−1.0)% agarose gel (*see* **Note 4**).
10. Purify the PCR reactions. After the run has finished, perform PCR purification by adding 250 μL of Buffer PB, transferring the mixture to a spin column, and centrifuging with a benchtop centrifuge (max speed).
11. Wash with 750 μL Buffer PE or homemade buffer (*see* Subheading 2.1, **item 8**) and centrifuge again.
12. Centrifuge again to make sure all ethanol is removed.
13. Elute the amplified DNA by adding 20–25 μL NFW (lower volume gives a higher concentration).
14. Measure the DNA concentration.
15. Check the PCR products. Aim to add around 200 ng of DNA to the agarose gel. For example, if the vector PCR yielded 193 ng/μL and the insert PCR 48 ng/μL, use 1 μL and 4 μL, respectively. Add 1 μL of Midori Green Direct, 1 μL 6× dye, and NFW to a final volume of 6 μL. If you have a lower DNA concentration, you can change the volumes (including the final volume) accordingly. We generally add 6 μL ladder with 2 μL Midori Green Direct.
16. Perform the electrophoresis: 100 V for 30 min or 135 V for 25 min.
17. After the run, carefully verify the length of the PCR products. The correct product (amplimer) sizes can be found in the History button in SnapGene (Fig. 4).
18. If you see one clear band of the expected length, you can continue with In-Fusion cloning. Because we will use Cloning Enhancer (TaKaRa) with the purified PCR products, we do not need DpnI restriction enzyme digestion or gel extraction (based on our experience) (*see* **Note 5**).

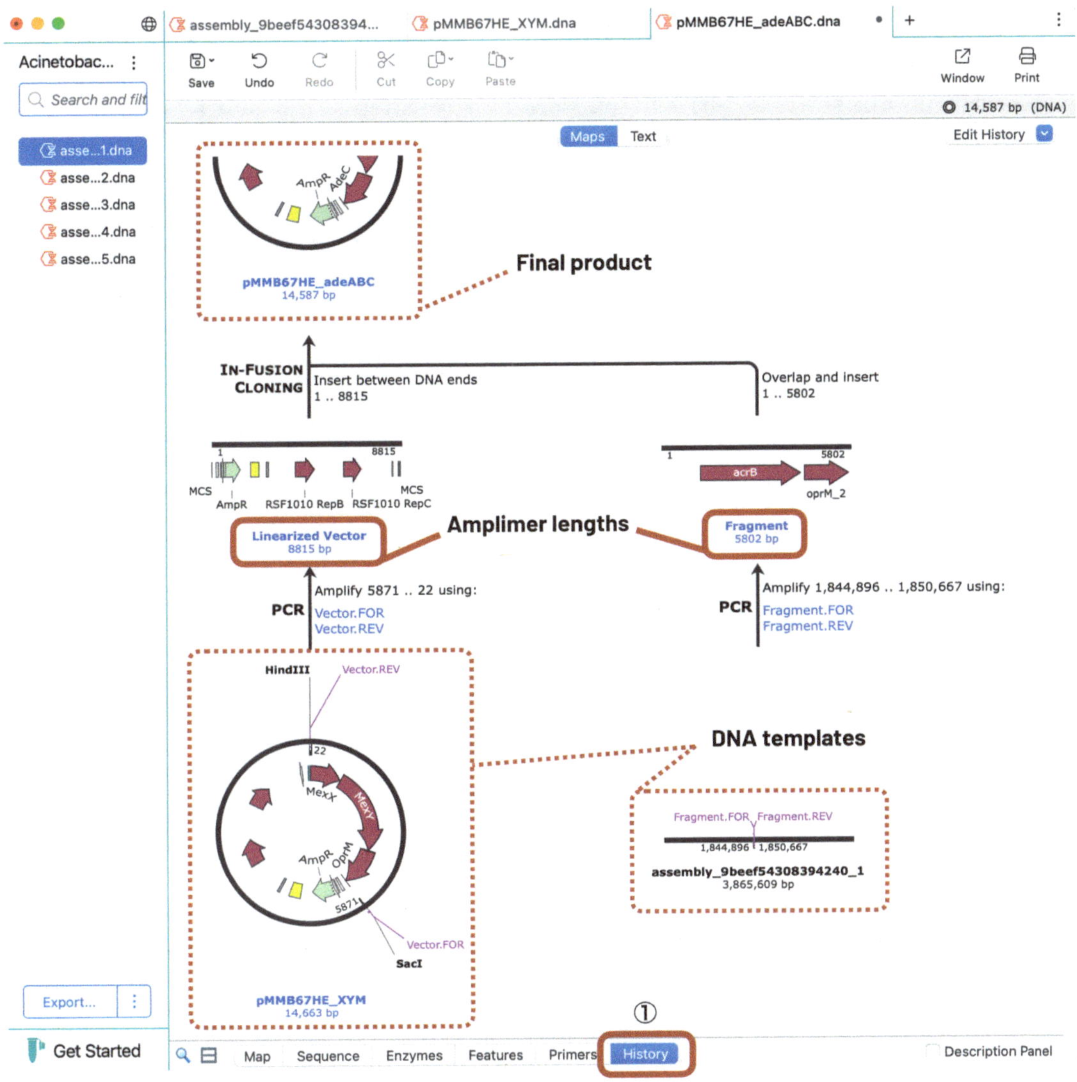

Fig. 4 SnapGene screenshots showing PCR and In-Fusion flowchart. ① Click on the History tab to visualize the workflow of the amplification and the In-Fusion reaction

3.3 Cloning, Transformation, and Sequencing

Now that you know your DNA concentrations (e.g., vector PCR 193 ng/μL and insert PCR 48 ng/μL) and the DNA lengths (e.g., vector 8.8 kb and insert 5.8 kb), start preparing the cloning (Fig. 5). As a trial, we have cloned two inserts into a vector with and without a cloning enhancer. Both with 1-step (the enhancer and In-Fusion reactions at the same time) and 2-step methods (In-Fusion separately after cloning, enhancing was performed and deactivated at 80 °C), and all with In-Fusion "HD" and newer In-Fusion "Snap" (TaKaRa). We found that In-Fusion "Snap"

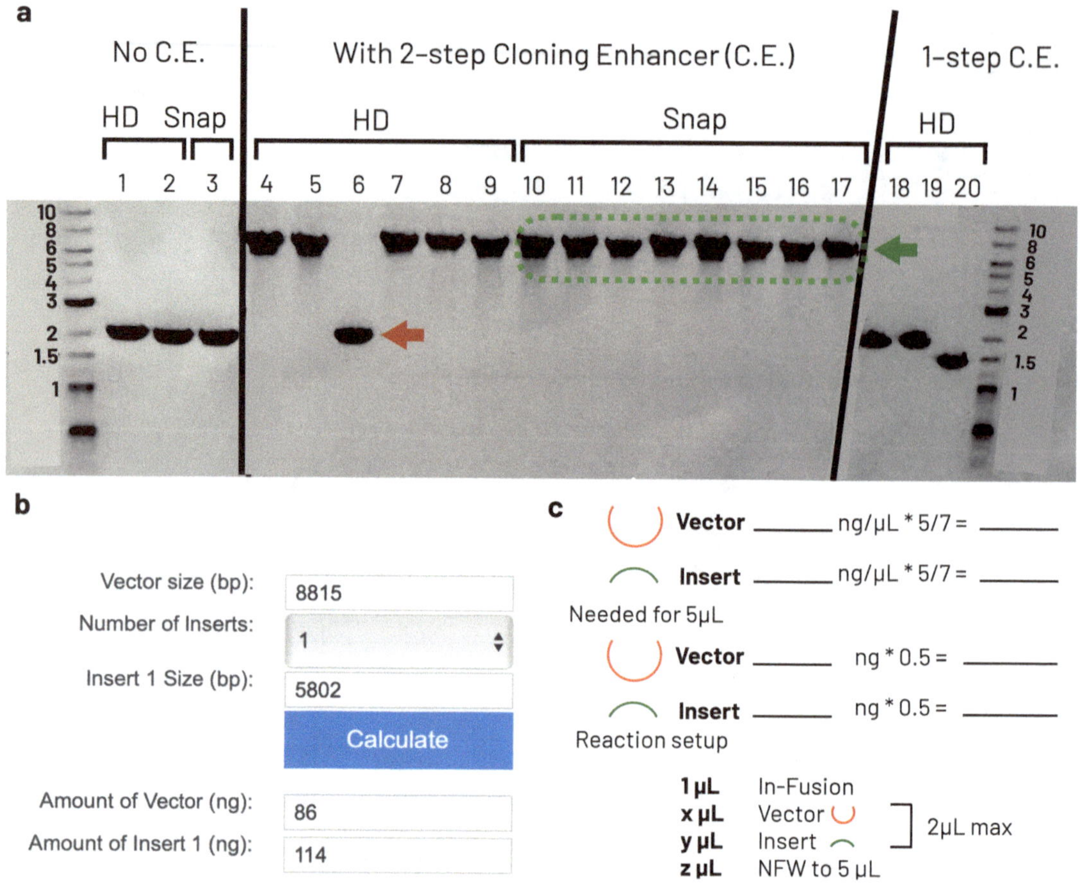

Fig. 5 In-Fusion cloning example and overview. (**a**) Example of cloning two inserts into an empty vector with In-Fusion. Different In-Fusion methods have been used (HD and Snap), as well as different protocols (without cloning enhancer (C.E.) and with C.E. (1-step and 2-step methods)). The 1-step method combined C.E. and In-Fusion in the same PCR tube, and the thermocycler was set to 37 °C (15 min) and then 50 °C (15 min). The empty vector is 1.9 kb (red arrow), and the vector + inserts are 6.6 kb (green arrow). (**b**) Snapshot of TaKaRa's In-Fusion molar ratio calculation website. (**c**) Template to fill out and calculate the amount of DNA needed for a 5 μL In-Fusion reaction (scale up if needed). (Refer to the official manual for more information)

using the 2-step protocol with cloning enhancer using purified DNA yields 100% success (Fig. 5a). In later cloning with other RND-type efflux pumps, we found the same results.

1. Head to the In-Fusion website to calculate the molar ratio needed for In-Fusion cloning (https://www.takarabio.com/learning-centers/cloning/primer-design-and-other-tools/in-fusion-molar-ratio-calculator). In this example, we need 86 and 114 ng of DNA for the vector and the insert, respectively (Fig. 5b). This depends on the lengths of the PCR products.
2. To prepare for the transformation step, preheat a water bath at 42 °C, SOC medium (500–1000 μL per transformation) at

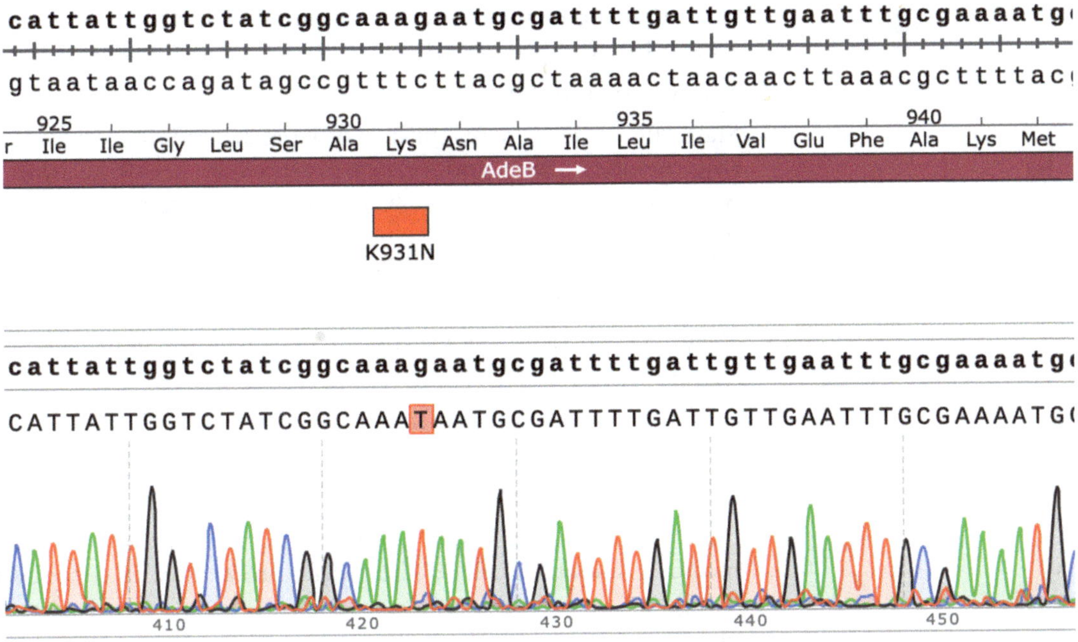

Fig. 6 Sequencing results of the single JM109 colony after transformation with the In-Fusion mix of pMMB67HEadeABC grown at 37 °C. The K931 residue in AdeB is one of the essential residues in the proton relay pathway of the RND-type transporter

30/37 °C, and LB agar plates supplemented with appropriate antibiotics at 30/37 °C (*see* **Note 6** and Fig. 6).

3. Now, before performing the cloning step, in a PCR tube, add 5 μL of the purified PCR and add 2 μL Cloning Enhancer (TaKaRa) (*see* **Note** 7).
4. Set the tube to the thermocycler: 37 °C for 20 min, 80 °C for 15 min, and cool down (4 °C infinitely). You can use 15 min of incubation instead of 20 when using less than 100 ng DNA.
5. Calculate the final adjusted concentration after adding the enhancer (diluted from 5 to 7 μL). In this example, 193 ng/mL*5/7 = 138 ng/μL and 48 ng/mL*5/7 = 34 ng/μL (Fig. 5c).
6. The 86 and 114 ng needed for In-Fusion are based on a 10 μL final volume. You can also lower the final volume to 5 or 7.5 μL if needed. For example, here, we can use 86*0.5 = 43 ng (~0.3 μL) vector and 114*0.5 = 57 ng (~1.7 μL) insert if we choose a final volume of 5 μL instead of 10 μL.
7. Prepare the In-Fusion reaction. In a PCR tube, add the desired amount of DNA as explained above.

8. For a 5 μL final volume, add 1 μL In-Fusion after adding NFW (such that the final volume is 5 μL) and the DNA. In this case: 2 μL NFW, 0.3 and 1.7 μL DNA, and 1 μL In-Fusion.
9. Mix well by pipetting.
10. Start the In-Fusion reaction. To a pre-heated thermocycler, add the PCR tube and run it for 15 min at 50 °C (after which cool it down to 4 °C). The cloned mixture is now ready to be transformed (we use JM109 chemically competent cells for initial transformation and to store the plasmid-containing strains).
11. Thaw 50(−100) μL of JM109 chemically competent cells on ice (about 5–10 min).
12. Add 2.5 μL of the cooled In-Fusion mixture (5 μL for 100 μL cells).
13. Mix by carefully flicking the tube with your finger.
14. Place on ice for 30 min. (Follow the instructions of your specific competent cells if they differ from what is described here.)
15. Place the cells in the 42 °C water bath for 45 s.
16. Place the cells on ice for 2 min.
17. Add 500 μL (or 1 mL for 100 μL cells) of pre-warmed SOC medium and shake at 30/37 °C for an hour.
18. Spin down the cells (5000× *g* for 5 min) and remove the supernatant to allow resuspension in about 100 μL remaining SOC.
19. Mix the cells and plate on pre-warmed LB agar plates, including the appropriate antibiotic (in this case, 100 μg/mL ampicillin).
20. Place the plate upside-down in the incubator and let the cells grow overnight (*see* **Note 8**).
21. The following day, pick a few colonies to make a glycerol stock and sequence the plasmid products (*see* **Note 9**).
22. Depending on the length of your insert (and thus the number of primers needed to sequence the product), make enough overnight culture to ensure you have enough plasmid DNA (3–20 mL LB medium).
23. When you incubate the plates at 37 °C, store the agar plates in the fridge until you continue with the pre-culture in the afternoon. When incubated at 30 °C, you can let the colonies grow longer until you make the pre-culture in the afternoon.
24. The following day, mix 800 μL of 80% sterile glycerol with 200 μL overnight culture and store at −80 °C.

25. Isolate the plasmid by QIAprep Spin Miniprep and elute the plasmid DNA with 30–50 μL NFW.
26. Determine the plasmid concentration.
27. Prepare the sequencing samples. Do this according to your sequencing protocol. In our case, we prepare 14 μL samples with 1.3 μL of 5 μM primer and 300–600 ng DNA.
28. When your DNA results come in, analyze them. Do this by opening your SnapGene plasmids file and clicking on "Tools → Align to Reference DNA Sequence → Align Imported Sequences. . . ." You can use text files, but we prefer importing chromatogram.*ab1* files to check the quality of the peaks and annotated sequences, especially at the beginning and the end of your sequence data (including dye blobs, low signals, and other distortions). An example of successful sequencing is shown in Fig. 7 for the pMMB67HE_adeABC example used for this chapter.

While your plasmid is being sequenced, you can prepare electrocompetent cells. You can now transform the sequenced plasmid into the host in which you wish to study the efflux pumps.

29. Make an overnight culture of the host you want to use.
30. The next day, add the overnight culture to fresh LB medium (e.g., 50 μL to 5 mL).
31. Pre-cool a centrifuge to spin down the cells (4 °C).
32. Determine the amount of LB needed: generally, make a 100× larger volume than the final volume. For example, if you want 200 μL cells for 2 × 100 μL or 4 × 50 μL cells, make a 20 mL culture.
33. Grow the cells until they reach OD_{600} of 0.25–0.35.
34. Place the cells on ice for 10 min.
35. Spin down the cells (7000× *g* for 10 min).
36. Discard the supernatant and add ice-cold sterile 10% glycerol.
37. Mix well by vortexing and repeat the centrifuge and wash step. For these wash steps, wash at least 4–5 times with glycerol. For this 20 mL example, add glycerol in steps with decreasing volume, such as 20 mL, 15 mL, 10 mL, 5 mL, and 2.5 mL.
38. Now, add the final volume (such as 200 μL) and mix the cells carefully.
39. Aliquot the needed amounts into separate tubes.
40. In liquid nitrogen, flash-freeze the cells.
41. Store the cells at −80 °C until usage. Alternatively, use the cells on the same day to transform your sequenced product.
42. Pre-cool electroporation cuvettes on ice.

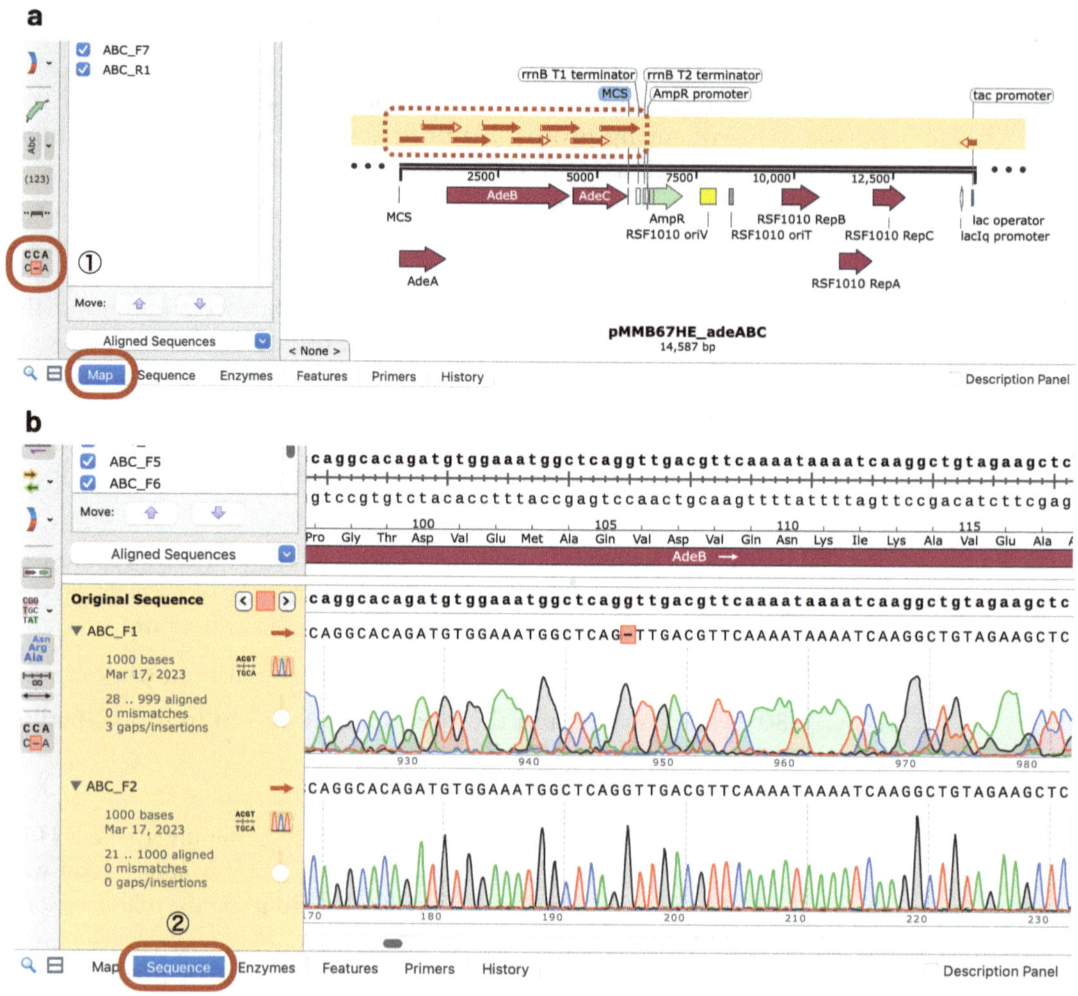

Fig. 7 Snapshot of sequencing results shown in SnapGene of a JM109 colony after transformation with the InFusion mix of pMMB67HEadeABC grown at 30 °C. Having sequences overlap the end of the previous sequence is needed to ensure proper sequence analysis, as seen by this "F1" and F2" sequence example. (a) You can toggle on and off the sequence information by clicking ① View alignments. (b) You can see the overlay and aligned/matched regions in the Map view, and you can analyse the data in detail in the ② Sequence tap

43. For transformation, add 1–100 ng of sequenced DNA (this can depend on your plasmid and host) and mix by flicking the tube with your fingers without warming the cells.
44. Pipette the cells (50–100 μL) into the cold cuvette and remove bubbles by tapping the cuvette on a hard surface.
45. Place the cuvette in the pulser and execute the pulse (1.8 kV, 2500 μF).

46. Remove the cuvette and immediately add 1 mL of pre-warmed SOC medium (LB medium will also work) and move the suspension to a 1.5 mL tube.
47. Shake for an hour (*see* **Note 6**).
48. Spin down the cells described in Subheading 3.3, **step 18**, and plate the cells on LB agar plates with the appropriate antibiotic (*see* **Note 10**).
49. Over the next 2 days, pick three colonies and make an overnight culture to create a glycerol stock. However, this time, 3 mL of LB is enough, as you do not need plasmid DNA for sequencing unless you desire to do so.

3.4 Expression and Functional Analysis

You have now created the strains you wish to use to analyze the efflux pumps. The next steps are functional analysis and expression level checking. It is not always necessary to check the expression levels, as cells with functional pumps should show a significantly lower susceptibility to antibiotics, from which you can conclude the pumps are expressed (compared to strains harboring only an empty vector). However, in many cases, you would want to check the expression levels, especially if you compare several pumps and if you want to make sure that the phenotype is attributed to the pump itself. We will skip the details of western blotting (as we use a specific assay using the iBind system; *see* **Note 1**) but will provide general information.

Liquid MIC is a great way to see the growth of bacteria by measuring OD_{600} in the presence of antibiotics (or inhibitors) and to see growth in detail. Additionally (not described in this chapter), we often perform solid MIC to get much MIC data at once. For solid MIC, we add different concentrations of antibiotics to LB agar plates, stamp 100× diluted overnight cultured cells on them, and check viability the next day after incubation. Here, we will briefly explain how to create a liquid MIC assay in a 96-well plate.

1. Make a pre-culture (3 mL LB with antibiotics) and shake overnight at 30/37 °C. Include a negative (e.g., Δ*acrAB*Δ*tolC*/pMMB67HE) and positive (e.g., Δ*acrAB*Δ*tolC*/pMMB67HEmexXYoprM) control (*see* **Note 6**).
2. Dilute the overnight culture (50 μL in 3 mL LB without antibiotics). Use more LB if needed for the final volume. Add IPTG or arabinose if needed.
3. Incubate while shaking at the appropriate temperature. Let the cells grow until OD_{600} of 0.5 (*see* **Note 6**).
4. While the cells are growing, prepare the antibiotic(s) stock(s).
5. To avoid affecting the LB concentration too much, add a relatively small volume of antibiotics. Furthermore, calculate a 2× concentration of the antibiotic since you make a 2× dilution

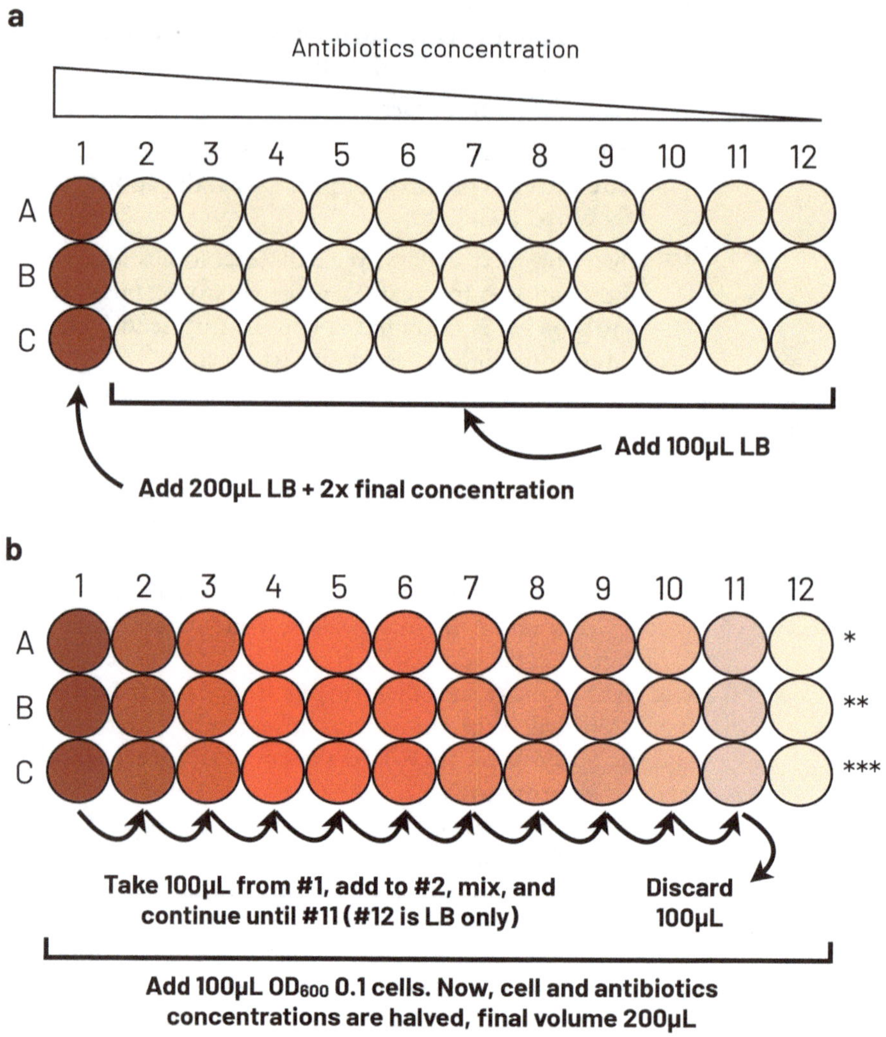

Fig. 8 A–C columns of a 96-well plate used for liquid MIC. You can use more lanes and plates for more strains and antibiotics to fit the desired analysis you wish to perform. (a) Make sure to make a 2x concentrated plate (including the concentration of inducers and potential inhibitors), as 100µL of cells (also in a 2x concentration) will be added to all wells. (b) 100µL serial dilutions will create a concentration gradient. Strains in this example: * Negative control, ** Positive control, *** Pump of interest

when adding the cells. The same goes for the inducer: the LB you add to the wells should have a 2× concentration of IPTG or arabinose. You can do the same when adding an inhibitor. Example: as shown in Fig. 8. If your final highest concentration is 512 µg/mL (Fig. 1, wells A1, B1, and C1 for three different strains), prepare a 200 µL 1024 µg/mL solution in LB. Do this by adding 8 µL of antibiotic stock (25.6 mg/mL) and 194 µL LB medium to the three wells. Add 100 µL of LB to the other wells (Fig. 1a). Next, pipette 100 µL from wells #1 to #2, mix, and repeat to pass forward the consecutively 2× diluted antibiotic mixture until you reach well #11. Simply discard 100 μL from wells #11 and keep #12 antibiotic-free as a control. Use a multichannel pipette to do this for multiple rows at once.

6. Now that you have your 2× concentrated liquid MIC plate prepared, prepare the cells. When they reach OD_{600} 0.5, make a dilution in a reservoir for multichannel pipettes such that the OD_{600} becomes 0.1. For example, 1 mL of cells and 4 mL of LB. Make more if you have more plates or lanes to fill.
7. With a multichannel pipette (preferably 12 tips), add 100 μL of cells from the reservoir to the respective row (Fig. 8b) and mix. Now, the cells have a final OD_{600} of 0.05, and the inducers and antibiotics have their desired final concentration.
8. You can incubate the cells overnight at 30 °C or 37 °C in an incubator and check the final OD_{600} the next day, or alternatively, measure the OD_{600} at intervals (e.g., 15 or 30 min) to visualize the growth of the cells (*see* **Note 6**).

Additionally, we will provide a very brief description of the preparation of Western blotting; however, as explained before, we will not go into the details of the specific system we use (*see* **Note 1**). Here, we describe how we grow the cells and harvest the membranes to be analyzed with Western blotting.

9. Make a pre-culture (3 mL LB with antibiotics) and shake overnight at 30/37 °C. Including a control for which the expression and detection are verified is recommended.
10. The next day, in 5 mL 2xYT broth, add 100 μL overnight culture and grow the cells at 30 °C or 37 °C until they reach OD_{600} of 0.7.
11. Add the 5 mL culture to 100 mL LB in a baffled 250 mL shaking flask and shake for 2.5 h at 30 °C or 37 °C. Make sure to add an inducer if needed.
12. After 2.5 h, harvest the cells (we use 2 × 50 mL tubes, spun at 4 °C at 8000× *g* for 20 min).
13. Resuspend the pellet in 7 mL Lysis Buffer and sonicate on ice.
14. Spin down (4 °C at 8000× *g* for 20 min).
15. Add to supernatant to an ultracentrifuge tube and completely fill (and balance) by adding Lysis Buffer.
16. Spin (we use 70,000 rpm in a small TLA 100.3 rotor (~200,000× *g*) for 30 min at 4 °C).
17. Repeat to wash.
18. Add 100–150 μL Membrane Buffer and dissolve well (the volume is very small it helps to let it sit on ice for a while).
19. Determine the protein concentration (we use a BCA kit, using 20× diluted samples).

20. Add 2 μg (or up to 10 μg if needed) samples on an SDS–PAGE gel. Make sure to add a marker that can be determined by Western blotting. We generally use Precision Plus standards from Bio-Rad, which can be detected with anti-strep-tag antibodies. We usually run the gels at 200 V for 35 min (c.v.) or 30 mA for 50 min (c.c.).
21. When the run has finished, perform transblotting. We use a PVDF membrane (10 V, 2.5 A, 20 min).
22. Now, block your membrane and perform Western blotting based on your systems and protocols, including antibiotic stocks (*see* **Note 1**).
23. Visualize the tagged protein with a chemiluminescence assay.

4 Notes

1. We have been using the iBind Western Device (Invitrogen), with iBind Cards and iBind Flex Solution Kit (Invitrogen). This system allows you to set up all wash and antibody buffers and let them run for 2.5 h without needing to change buffers manually.
2. Wide range of antibiotics from different classes, e.g., erythromycin/azithromycin/clarithromycin (macrolides), cloxacillin/meropenem/aztreonam (β-lactams/monobactams), ethidium bromide/acriflavine/crystal violet/benzalkonium (planar aromatic cations), novobiocin, SDS, deoxycholic acid, kanamycin (aminoglycoside), minocycline, rifampicin, and nalidixic acid/norfloxacin/ciprofloxacin/moxifloxacin (quinolones).
3. This chapter does not describe the addition of the His-tag. However, a His-tag (e.g., CATCATCACCATCACCAT (His_6) or CACCATCACCATCACCATCACCATCAC (His_9)) can be added later if needed by PCR. If the vector template DNA has a His-tag, this can simply be used to be included in the final product (e.g., if you already have an RND pump gene with a His-tag in a plasmid, design the primers in SnapGene by selecting the RND-gene region to be replaced except for the His-tag, and the gene to be inserted not to have the stop-codon). Be careful that you include a few base pairs upstream of the His-tag to prevent multiple binding sites to the template. If you wish to keep other features from an original plasmid (e.g., promotor regions, signal peptides, ribosome binding sites, affinity tags, and more) with multiple genes, they can also be replaced one by one using In-Fusion cloning. In this case, sequence the genes after transformation in JM109, after which the final construct can be transformed into the host to be used for the functional analysis.

4. The PCR can take under an hour to a couple of hours, depending on the length of your amplimers and the polymerase you use. When the PCR reaction is done, and you wish to continue the next day, you can store the PCR overnight at 4–10 °C. However, with this protocol using KOD One, you can perform amplification, purification, cloning, and transformation on the same day.
5. However, if you see multiple bands, cutting the band from the gel and performing gel extraction is recommended. In that case, elute the DNA from the beads with a low volume (such as 20 μL) to increase the concentration. For gene amplification, it is also recommended that a plasmid with a different resistance cassette be used to decrease the chance of incorrect cloning or transformation.
6. 37 °C (instead of 30 °C) is usually used. However, transformation proved difficult for the described pMMB67HE plasmids harboring the *adeABC* (BAA–1794) gene cluster. We could not get colonies except for one, which interestingly had an inactive mutation (K932N) in the *adeB* gene (Fig. 6). Changing the incubation temperature to 30 °C may have allowed for better folding in *E. coli* cells and resulted in hundreds of successful colonies.
7. The official protocol recommends unpurified PCR, but we found spin column-purified DNA works significantly better. Perhaps because our polymerase buffer is incompatible with the cloning enhancer.
8. As explained, we used 30 °C in this specific example; however, you can simply use 37 °C, which works in most cases. We wanted to mention this example, as it helped us express this specific protein from this specific plasmid.
9. We typically pick three colonies: a small, medium, and a big colony.
10. When we use Δ*tolC* cells (including Δ*acrAB*Δ*tolC*), we generally use a lower concentration of antibiotics, as the cells are very susceptible. For example, 50 μg/mL instead of 100 μg/mL ampicillin often yields more colonies.

Acknowledgments

This work was supported by Grants-in-Aid for Scientific Research (Early-Career Scientists, B and C) (Kakenhi 23K14346, 23H02631 and 25K10343) from the Japan Society for the Promotion of Science (JSPS), The Uehara Memorial Foundation, Yakult Bio-Science Foundation, The Nippon Foundation-Osaka

University Project for Infectious Disease Prevention, Japan Agency for Medical Research and Development (AMED) and the Sumitomo Foundation.

Conflicts of Interest The authors have no conflicts of interest to declare that are relevant to the content of this chapter.

References

1. Zwama M, Yamaguchi A (2018) Molecular mechanisms of AcrB-mediated multidrug export. Res Microbiol 169:372–383
2. Yamasaki S, Zwama M, Yoneda T, Hayashi-Nishino M, Nishino K (2023) Drug resistance and physiological roles of RND multidrug efflux pumps in *Salmonella enterica*, *Escherichia coli* and *Pseudomonas aeruginosa*. Microbiology 169:001322
3. Zwama M, Nishino K (2022) Proximal binding pocket Arg717 substitutions in *Escherichia coli* AcrB cause clinically relevant divergencies in resistance profiles. Antimicrob Agents Ch 66: e0239221
4. Zwama M, Nishino K (2021) Ever-adapting RND efflux pumps in gram-negative multidrug-resistant pathogens: a race against time. Antibiotics 10:774
5. Blair JMA, Bavro VN, Ricci V, Modi N, Cacciotto P, Kleinekathöfer U, Ruggerone P, Vargiu AV, Baylay AJ, Smith HE, Brandon Y, Galloway D, Piddock LJV (2015) AcrB drug-binding pocket substitution confers clinically relevant resistance and altered substrate specificity. Proc Natl Acad Sci 112:3511–3516
6. Murakami S, Nakashima R, Yamashita E, Yamaguchi A (2002) Crystal structure of bacterial multidrug efflux transporter AcrB. Nature 419:587–593
7. Seeger MA, Schiefner A, Eicher T, Verrey F, Diederichs K, Pos KM (2006) Structural asymmetry of AcrB trimer suggests a peristaltic pump mechanism. Science 313:1295–1298
8. Murakami S, Nakashima R, Yamashita E, Matsumoto T, Yamaguchi A (2006) Crystal structures of a multidrug transporter reveal a functionally rotating mechanism. Nature 443: 173–179
9. Nakashima R, Sakurai K, Yamasaki S, Nishino K, Yamaguchi A (2011) Structures of the multidrug exporter AcrB reveal a proximal multisite drug-binding pocket. Nature 480: 565–569
10. Nakashima R, Sakurai K, Yamasaki S, Hayashi K, Nagata C, Hoshino K, Onodera Y, Nishino K, Yamaguchi A (2013) Structural basis for the inhibition of bacterial multidrug exporters. Nature 500:102–106
11. Zwama M, Yamasaki S, Nakashima R, Sakurai K, Nishino K, Yamaguchi A (2018) Multiple entry pathways within the efflux transporter AcrB contribute to multidrug recognition. Nat Commun 9:124
12. Zwama M, Hayashi K, Sakurai K, Nakashima R, Kitagawa K, Nishino K, Yamaguchi A (2017) Hoisting-loop in bacterial multidrug exporter AcrB is a highly flexible hinge that enables the large motion of the subdomains. Front Microbiol 8:2095
13. Sennhauser G, Amstutz P, Briand C, Storchenegger O, Grütter MG (2007) Drug export pathway of multidrug exporter AcrB revealed by DARPin inhibitors. PLoS Biol 5:e7
14. Eicher T, Seeger MA, Anselmi C, Zhou W, Brandstätter L, Verrey F, Diederichs K, Faraldo-Gómez JD, Pos KM (2014) Coupling of remote alternating-access transport mechanisms for protons and substrates in the multidrug efflux pump AcrB. elife 3:213
15. Mokhonov VV, Mokhonova EI, Akama H, Nakae T (2004) Role of the membrane fusion protein in the assembly of resistance-nodulation-cell division multidrug efflux pump in *Pseudomonas aeruginosa*. Biochem Biophys Res Commun 322:483–489
16. Islam S, Jalal S, Wretlind B (2004) Expression of the MexXY efflux pump in amikacin-resistant isolates of *Pseudomonas aeruginosa*. Clin Microbiol Infec 10:877–883
17. Lau CHF, Hughes D, Poole K (2014) MexY-promoted aminoglycoside resistance in *Pseudomonas aeruginosa*: involvement of a putative proximal binding pocket in aminoglycoside recognition. MBio 5:e01068
18. Yoneyama H, Maseda H, Kamiguchi H, Nakae T (2000) Function of the membrane fusion protein MexA of the MexAB-OprM efflux pump in *Pseudomonas aeruginosa* without an anchoring membrane. J Biol Chem 275: 4628–4634

19. Sennhauser G, Bukowska MA, Briand C, Grütter MG (2009) Crystal structure of the multidrug exporter MexB from *Pseudomonas aeruginosa*. J Mol Biol 389:134–145
20. Ramaswamy VK, Vargiu AV, Malloci G, Dreier J, Ruggerone P (2018) Molecular determinants of the promiscuity of MexB and MexY multidrug transporters of *Pseudomonas aeruginosa*. Front Microbiol 9:1144
21. Sakurai K, Yamasaki S, Nakao K, Nishino K, Yamaguchi A, Nakashima R (2019) Crystal structures of multidrug efflux pump MexB bound with high-molecular-mass compounds. Sci Rep 9:4359
22. Yamasaki S, Koga N, Zwama M, Sakurai K, Nakashima R, Yamaguchi A, Nishino K (2022) Spatial characteristics of the efflux pump MexB determine inhibitor binding. Antimicrob Agents Ch 66:e0067222
23. Masuda N, Sakagawa E, Ohya S, Gotoh N, Tsujimoto H, Nishino T (2000) Substrate specificities of MexAB-OprM, MexCD-OprJ, and MexXY-OprM efflux pumps in *Pseudomonas aeruginosa*. Antimicrob Agents Ch 44:3322–3327
24. Aires JR, Kohler T (1999) Involvement of an active efflux system in the natural resistance of *Pseudomonas aeruginosa* to aminoglycosides. Antimicrob Agents Chemother 43:2624–2628
25. Poole K, Krebes K, McNally C, Neshat S (1993) Multiple antibiotic resistance in *Pseudomonas aeruginosa*: evidence for involvement of an efflux operon. J Bacteriol 175:7363–7372
26. Su CC, Morgan CE, Kambakam S, Rajavel M, Scott H, Huang W, Emerson CC, Taylor DJ, Stewart PL, Bonomo RA, Yu EW (2019) Cryo-electron microscopy structure of an *Acinetobacter baumannii* multidrug efflux pump. MBio 10:e01295–e01219
27. Morgan CE, Glaza P, Leus IV, Trinh A, Su CC, Cui M, Zgurskaya HI, Yu EW (2021) Cryoelectron microscopy structures of AdeB illuminate mechanisms of simultaneous binding and exporting of substrates. MBio 12:e03690–e03620
28. Ornik-Cha A, Wilhelm J, Kobylka J, Sjuts H, Vargiu AV, Malloci G, Reitz J, Seybert A, Frangakis AS, Pos KM (2021) Structural and functional analysis of the promiscuous AcrB and AdeB efflux pumps suggests different drug binding mechanisms. Nat Commun 12:6919
29. Magnet S, Courvalin P, Lambert T (2001) Resistance-nodulation-cell division-type efflux pump involved in aminoglycoside resistance in *Acinetobacter baumannii* strain BM4454. Antimicrob Agents Ch 45:3375–3380
30. Sugawara E, Nikaido H (2014) Properties of AdeABC and AdeIJK efflux systems of *Acinetobacter baumannii* compared with those of the AcrAB-TolC system of *Escherichia coli*. Antimicrob Agents Ch 58:7250–7257
31. Kobayashi N, Nishino K, Yamaguchi A (2001) Novel macrolide-specific ABC-type efflux transporter in *Escherichia coli*. J Bacteriol 183:5639–5644
32. Nishino K, Yamaguchi A (2001) Overexpression of the response regulator EvgA of the two-component signal transduction system modulates multidrug resistance conferred by multidrug resistance transporters. J Bacteriol 183:1455–1458
33. Nishino K, Yamaguchi A (2001) Analysis of a complete library of putative drug transporter genes in *Escherichia coli*. J Bacteriol 183:5803–5812
34. Horiyama T, Yamaguchi A, Nishino K (2010) TolC dependency of multidrug efflux systems in *Salmonella enterica* serovar Typhimurium. J Antimicrob Chemoth 65:1372–1376
35. Kobayashi N, Tamura N, van Veen HW, Yamaguchi A, Murakami S (2014) β-Lactam selectivity of multidrug transporters AcrB and AcrD resides in the proximal binding pocket. J Biol Chem 289:10680–10690
36. Horiyama T, Nishino K (2014) AcrB, AcrD, and MdtABC multidrug efflux systems are involved in enterobactin export in *Escherichia coli*. PLoS One 9:e108642
37. Nishino K, Yamada J, Hirakawa H, Hirata T, Yamaguchi A (2003) Roles of TolC-dependent multidrug transporters of *Escherichia coli* in resistance to β-lactams. Antimicrob Agents Ch 47:3030–3033
38. Zwama M, Yamaguchi A, Nishino K (2019) Phylogenetic and functional characterisation of the *Haemophilus influenzae* multidrug efflux pump AcrB. Commun Biol 2:340

Chapter 16

Silkworm Infection Model for Evaluating Bacterial Virulence Genes

Chikara Kaito, Kinuka Hongu, Ayano Kuriu, Kazuya Ishikawa, and Kazuyuki Furuta

Abstract

Experiments using animal infection models are necessary to analyze the pathogenicity of bacterial genes in animals. Infection models using silkworms make it easy to conduct infection experiments using many animals and are effective for identifying and evaluating the functions of genes involved in the pathogenicity of various bacterial species. In this chapter, we introduce the experimental methods using the silkworm infection model.

Key words Silkworm, Animal infection model, Virulence genes, Pathogenic bacteria, Survival rate curve, *Staphylococcus aureus*

1 Introduction

Bacterial infections are one of the leading causes of death worldwide. In particular, the increase in drug-resistant bacteria shows no sign of slowing, making it urgent to understand the mechanisms of infection and develop new drugs to treat them [1]. To evaluate the pathogenicity of bacteria that infect humans and evaluate drugs to treat infectious diseases, it is necessary to conduct infection experiments using animal infection models. In many studies, mammals, such as mice, have been used as infection models; however, mammalian models are not suitable for infection experiments using a large number of individuals owing to ethical and cost considerations. To circumvent this problem, methods using non-mammalian infection models have been developed [2]. For example, zebrafish [3, 4], nematodes [5], fruit flies [6], silkworms [7], honeyworms (*Galleria mellonella* larvae) [8], and crickets [9] have been used.

Silkworms are the larvae of the lepidopteran insect *Bombyx mori*, which has been domesticated through thousands of years of breeding. They do not bite human hands or escape from their

Kunihiko Nishino and Mikio Tanabe (eds.), *Bacterial Multidrug Exporters: Methods and Protocols*, Methods in Molecular Biology, vol. 3044, https://doi.org/10.1007/978-1-0716-5316-6_16,
© The Author(s), under exclusive license to Springer Science+Business Media, LLC, part of Springer Nature 2026

cages, and their body size is just right, allowing samples to be injected with a medical needle without machines, making it easy to conduct infection experiments by injection [10]. Infection experiments using silkworms have identified virulence genes in several bacterial species, including *Staphylococcus aureus*, group A *Streptococcus*, *Bacillus subtilis*, and *Escherichia coli* [11–14]. In *S. aureus*, cell wall proteins involved in cell adhesion, genes that regulate the expression of various virulence factors, and genes involved in resistance to radical-dependent killing by immune cells contribute to the virulence in silkworms [11, 15–18]. Additionally, in *E. coli*, cell surface factors that act on resistance to antibiotics and antimicrobial peptides are involved in the virulence in silkworms [13, 19–21]. Furthermore, the silkworm infection model has been used to discover new antibiotics and evaluate the toxicity of chemical compounds in animals [22–24]. In this chapter, we describe the evaluation of the pathogenicity of the bacterial strains using a silkworm infection model.

2 Materials

2.1 Rearing of Silkworms

1. Third-instar sleeping stage silkworms (Ehimesansyu, Ehime, Japan).
2. Safety cabinet.
3. Incubator (e.g., MIR-554, PHC).
4. Disposable packaging box with a hole (OP-136R, Denkapolymer, Tokyo, Japan).
5. Disposable packaging box without a hole (2774464, CHUO KAGAKU Co., Saitama, Japan).
6. Artificial diets (Silkmate2S, NOSAN Co., Kanagawa, Japan).
7. Kimwipe paper (62011, Nippon Paper Creasia Co., Tokyo, Japan).
8. Latex gloves.

2.2 Bacterial Culture and Preparation of Injection Samples

1. 50 mL disposable tube (e.g., Thermo Fisher).
2. Bacterial colonies formed on an agar plate.
3. Liquid medium (e.g., Lysogeny broth and tryptic soy broth).
4. Agar medium (e.g., Lysogeny broth agar and tryptic soy broth agar).
5. Shaking incubator (e.g., BR-3000LF, TAITEC, Saitama, Japan).
6. Absorption spectrophotometer (e.g., DU-730, Beckman).
7. 1.5 mL microtube.

8. Centrifuge machine for 50 mL tube (e.g., CR-21 N, Eppendorf Himac Technologies, Ibaraki, Japan).
9. 0.9% NaCl.

2.3 Injection into Silkworms

1. Tuberculin syringe (SS-01 T, Terumo, Japan).
2. 27-gauge needle (NN-2719S, Terumo).
3. Latex gloves.
4. Fifth-instar silkworm larva.
5. Disposable packaging box without a hole (2774464, CHUO KAGAKU Co., Saitama, Japan).
6. Kimwipe paper (62011, Nippon Paper Creasia Co., Tokyo, Japan).
7. Rubber band.
8. Incubator (e.g., MIR-554, PHC).

2.4 Analysis of Survival Curves

1. Prism software (GraphPad Software, Boston, MA).
2. PC installed with R.

3 Methods

3.1 Rearing of Silkworms

1. Perform all procedures by wearing sanitized gloves to prevent silkworm contamination with the pathogen in a safety cabinet or clean bench.
2. Purchase the third-instar sleeping stage silkworms (*see* **Note 1**) (Fig. 1a).
3. Incubate them in an incubator at 27 °C to wait for them to be hatched into the fourth-instar silkworms.
4. After the silkworms hatch to the fourth instar, divide them into 30 individuals in a perforated disposable packaging box with a hole lined with a Kimwipe paper.
5. Feed fourth-instar silkworms with artificial diets, and raise them at 27 °C until they reach the fourth-instar sleeping stage in 4–5 days (Fig. 1b–d).
6. Once the silkworms reach the fourth-instar sleeping stage (Fig. 1e), divide them into 30 silkworms in a disposable packaging box without a hole and keep them at 27 °C without feeding for 1 day.
7. Once the silkworms molt to the fifth instar (Fig. 1f), feed them with artificial diets and keep them at 27 °C for 1 day to use in the infection experiment.

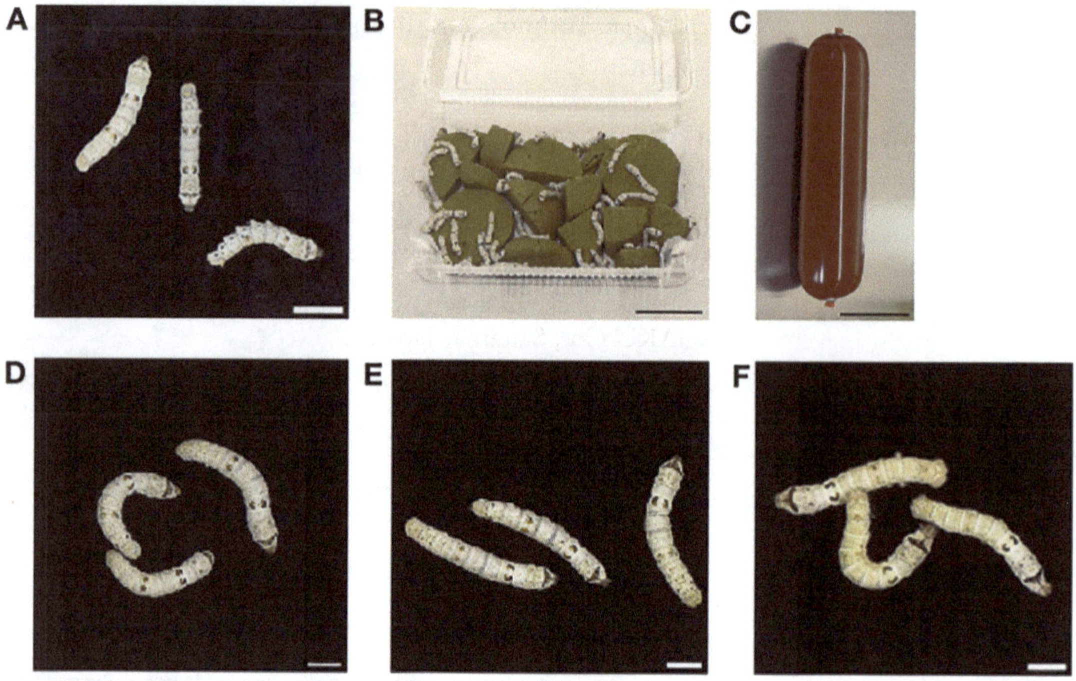

Fig. 1 Silkworms in the different instars. Third-instar sleep-stage silkworms (**a**), fourth-instar silkworms fed artificial diets (**b**), artificial diets (Silkmate2S) (**c**), fourth-instar silkworms (**d**), fourth-instar sleep-stage silkworms (**e**), and fifth-instar silkworms (**f**) are presented. The scale bars indicate 1 cm (**a**, **d**, **e**, and **f**), 5 cm (**b**), and 10 cm (**c**)

3.2 Bacterial Culture and Preparation of Injection Samples

1. Inoculate a single colony formed on the agar medium into 5 mL of liquid medium in a 50-mL disposable tube.
2. Culture it overnight at 37 °C with shaking.
3. Use tryptic soy broth medium for *S. aureus*, and use lysogeny broth medium for *E. coli*.
4. Dilute the cultured bacterial liquid to an appropriate ratio, and measure the optical density at 600 nm (OD_{600}) using an absorption spectrophotometer.
5. Dilute the culture liquid by 10^{-6} with saline, plate 50 μL on an agar medium, and culture it overnight to determine the colony-forming unit (CFU).
6. Repeat the procedures #4 and #5 several times before the day of the infection experiment to determine the relationship between the OD_{600} of the culture liquid and the CFU (Fig. 2).
7. On the day before the infection experiment, culture the bacterial strains in the same manner as described above.
8. On the day of the infection experiment, measure the OD_{600} of the overnight bacterial culture solution and calculate the CFU in the solution from the standard curve obtained in #6 (Fig. 2).

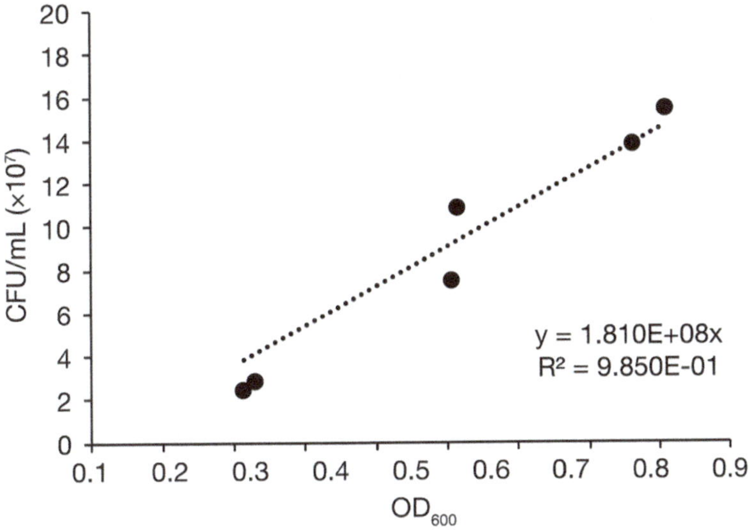

Fig. 2 Relationship between OD_{600} and CFU of bacterial overnight culture. Overnight cultures of *S. aureus* NCTC8325-4 were diluted, and the OD_{600} and CFU were determined. A linear approximation was performed using data from two biological replicates

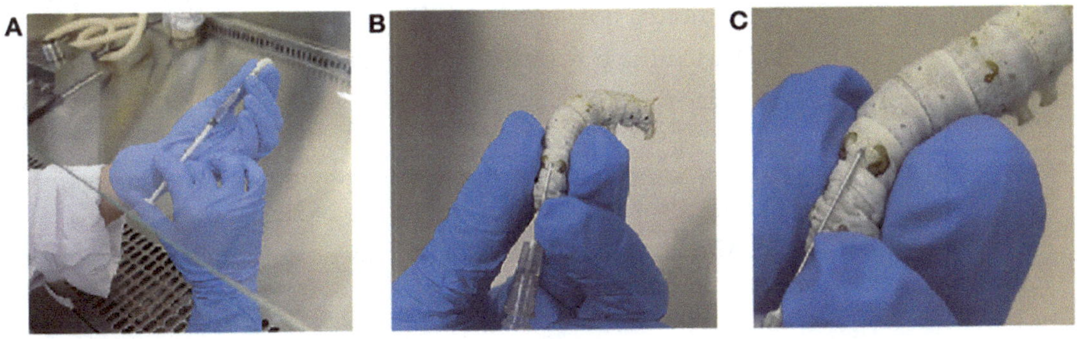

Fig. 3 Silkworm injection procedure. An overall view of the injection (**a**), finger position to hold a silkworm (**b**), and injection position of a silkworm (**c**) are presented

9. Centrifuge the overnight culture solution at 10,400× *g* for 10 min and suspend the precipitated bacteria in an appropriate amount of 0.9% NaCl for use in the infection experiment (*see* **Note 2**).

3.3 Injection into Silkworms

1. Perform the injection in a safety cabinet.
2. Suck the prepared bacterial solution sample into a tuberculin syringe connected to a 27-gauge needle.
3. Expel air from the syringe by flicking it with a finger.
4. Support the fifth-instar larva with three fingers (Fig. 3).

Fig. 4 Dead silkworms caused by bacterial infection. Silkworms were injected with saline (0.9% NaCl) (**a**) or *S. aureus* NCTC8325-4 bacterial solution (5×10^6 CFU) (**b**) and incubated for 60 h at 27 °C. The scale bars indicate 2 cm

5. Inject the bacterial solution into the back of the fifth-instar larva (Fig. 3). Hold the silkworm to prevent needle movement during injection.
6. When evaluating the efficacy of drugs, inject drug samples into the silkworms (*see* **Note 3**).
7. Place the injected silkworm in a new disposable packaging box without a hole lined with Kimwipe paper.
8. Fasten the box with a rubber band.
9. Place the box in an incubator set at 27 °C (*see* **Note 4**). Do not provide food.
10. Confirm silkworm survival by poking with a pipette tip to check for movement (Fig. 4) (*see* **Note 5**).
11. Autoclave and dispose of all silkworms in their boxes after the experiment.

3.4 Analysis of Survival Curves

3.4.1 Time–Survival Rate Curve

1. Measure the survival rate every 24 or 12 h after the bacterial infection and plot the survival rate against the time (*see* **Note 6**).
2. Test significant differences between the survival curves using the log-rank test in GraphPad Prism software (Fig. 5a).

3.4.2 Bacterial Dose–Survival Rate Curve

1. Inject silkworms with two or threefold serial dilutions of bacterial solution.
2. Measure the survival rate at 24 or 48 h after injection. The time required to calculate the survival rate varies depending on the nature of the pathogen, at which the correspondence between the number of injected bacteria and the survival rate is clearly observed.

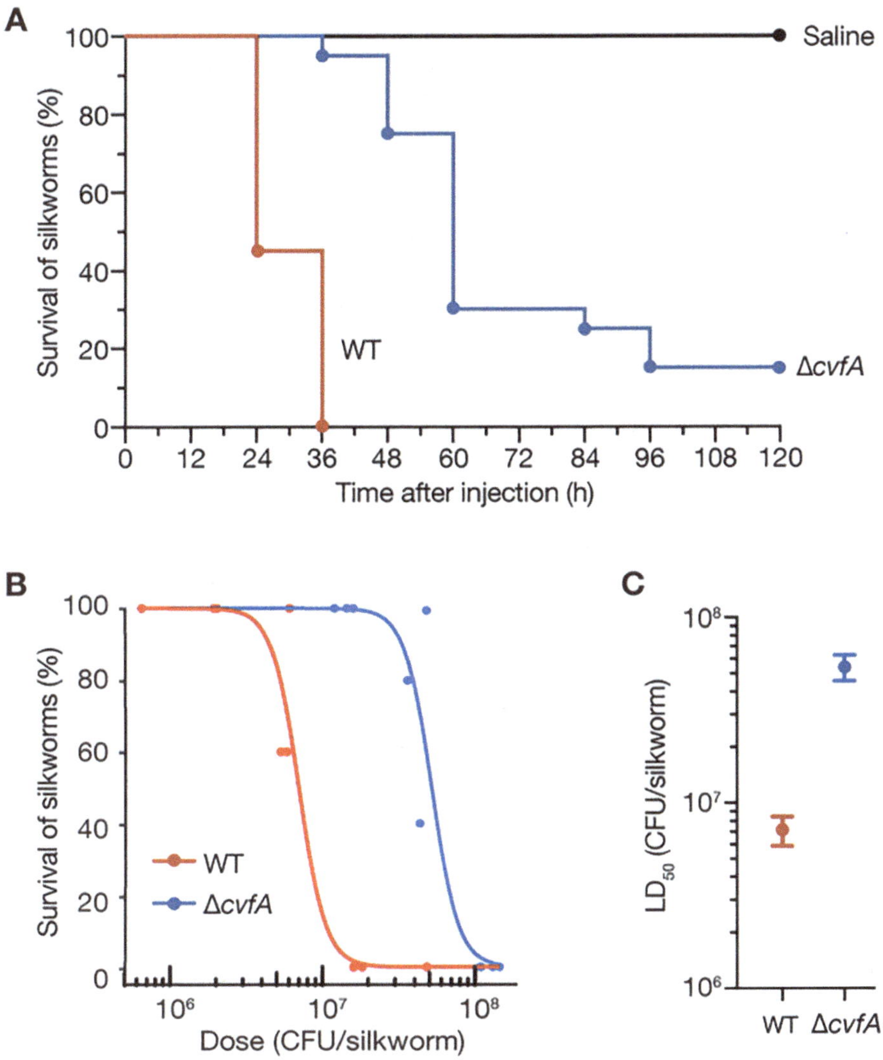

Fig. 5 Two types of survival curves. (**a**) Time–survival rate curve of *S. aureus* against silkworms is presented. *S. aureus cvfA* gene regulates the expression of various virulence factors and contributes to virulence against silkworms [11, 28, 29]. Silkworms were injected with 5×10^6 CFU of *S. aureus* NCTC8325-4 (WT) or the *cvfA*-deleted mutant (△*cvfA*), or saline, and the survival was monitored every 12 h after the injection. Data are pooled from two independent experiments ($n = 20$). A log-rank test was performed between WT and △*cvfA* ($p < 0.0001$). (**b**) Bacterial dose–survival rate curve of *S. aureus* against silkworms is presented. Silkworms ($n = 5$) were injected with serially diluted solutions of *S. aureus* NCTC8325-4 (WT) or the *cvfA*-deleted mutant (△*cvfA*), and the survival was monitored at 24 h after the injection. Data are pooled from three independent experiments. A likelihood ratio test was performed between WT and *cvfA* ($p < 0.0001$). (**c**) LD_{50} with standard error was determined from (**b**)

3. Perform logistic regression analysis on the relationship between the number of bacteria injected and the survival rate to calculate the LD_{50} and its standard error by using R with the glm function from the MASS library [25] (Fig. 5b, c).
4. Test significant differences between the survival curves using the likelihood ratio test with the anova function in R (Fig. 5b).

4 Notes

1. Silkworms can be purchased from various vendors. Different silkworm strains have different susceptibility to bacteria. We use a strain of KINSYU female × SHOWA male. Silkworms can be purchased at various stages; however, they are raised in the laboratory from the third-instar stage to acclimatize them to the breeding environment.
2. If the culture supernatant is non-toxic, the overnight bacterial culture solution can be directly used in the infection experiment without centrifugation. The hemolytic toxin in the culture supernatant of *S. aureus* exhibits killing activity against silkworms. A tenfold diluted solution of the culture supernatant shows almost no killing activity, whereas an undiluted solution shows killing activity [26].
3. In addition to drugs, to evaluate therapeutic agents [27], silkworms can be injected with isopropyl β-D-1-thiogalactopyranoside to induce gene expression or antibiotics to prevent plasmid loss from bacterial cells [28].
4. To enhance bacterial pathogenicity for weakly pathogenic strains, place the injected silkworms at 37 °C. For example, *E. coli* exhibits weak killing activity at 27 °C but higher pathogenicity at 37 °C [21]. Saline-injected silkworms typically survive at 37 °C for approximately 5 days.
5. As the bacterial infection progresses, the silkworm's body turns black, and its movements become sluggish. When the silkworm dies, it turns black further because of the melanin reaction (Fig. 4).
6. Survival curves are divided into two types: time–survival rate and number of injected bacteria–survival rate curves. The number of injected bacteria–survival rate curves is suitable for the quantitative evaluation of the pathogenicity of the injected pathogen because it allows the calculation of LD_{50}, but it is time-consuming because it requires a large number of samples.

Conflicts of Interest

The authors have no conflicts of interest to declare that are relevant to the content of this chapter.

References

1. Collaborators GBDAR (2024) Global burden of bacterial antimicrobial resistance 1990–2021: a systematic analysis with forecasts to 2050. Lancet 404(10459):1199–1226
2. Kaito C, Murakami K, Imai L, Furuta K (2020) Animal infection models using non-mammals. Microbiol Immunol 64(9):585–592
3. Neely MN, Pfeifer JD, Caparon M (2002) Streptococcus-zebrafish model of bacterial pathogenesis. Infect Immun 70(7):3904–3914
4. Davis JM, Clay H, Lewis JL, Ghori N, Herbomel P, Ramakrishnan L (2002) Real-time visualization of mycobacterium-macrophage interactions leading to initiation of granuloma formation in zebrafish embryos. Immunity 17(6):693–702
5. Tan MW, Mahajan-Miklos S, Ausubel FM (1999) Killing of *Caenorhabditis elegans* by *Pseudomonas aeruginosa* used to model mammalian bacterial pathogenesis. Proc Natl Acad Sci USA 96(2):715–720
6. Needham AJ, Kibart M, Crossley H, Ingham PW, Foster SJ (2004) *Drosophila melanogaster* as a model host for *Staphylococcus aureus* infection. Microbiology (Reading) 150(Pt 7): 2347–2355
7. Kaito C, Akimitsu N, Watanabe H, Sekimizu K (2002) Silkworm larvae as an animal model of bacterial infection pathogenic to humans. Microb Pathog 32(4):183–190
8. Mylonakis E, Moreno R, El Khoury JB, Idnurm A, Heitman J, Calderwood SB et al (2005) *Galleria mellonella* as a model system to study *Cryptococcus neoformans* pathogenesis. Infect Immun 73(7):3842–3850
9. Kochi Y, Miyashita A, Tsuchiya K, Mitsuyama M, Sekimizu K, Kaito C (2016) A human pathogenic bacterial infection model using the two-spotted cricket, *Gryllus bimaculatus*. FEMS Microbiol Lett 363(15)
10. Kaito C (2016) Understanding of bacterial virulence using the silkworm infection model. Drug Discov Ther 10(1):30–33
11. Kaito C, Kurokawa K, Matsumoto Y, Terao Y, Kawabata S, Hamada S et al (2005) Silkworm pathogenic bacteria infection model for identification of novel virulence genes. Mol Microbiol 56(4):934–944
12. Matsumoto Y, Kaito C, Morishita D, Kurokawa K, Sekimizu K (2007) Regulation of exoprotein gene expression by the *Staphylococcus aureus cvfB* gene. Infect Immun 75(4): 1964–1972
13. Miyashita A, Iyoda S, Ishii K, Hamamoto H, Sekimizu K, Kaito C (2012) Lipopolysaccharide O-antigen of enterohemorrhagic *Escherichia coli* O157:H7 is required for killing both insects and mammals. FEMS Microbiol Lett 333(1):59–68
14. Ishikawa K, Shirakawa R, Takano D, Kosaki T, Furuta K, Kaito C (2022) Knockout of *ykcB*, a putative glycosyltransferase, leads to reduced susceptibility to vancomycin in *Bacillus subtilis*. J Bacteriol 204(12):e0038722
15. Miyazaki S, Matsumoto Y, Sekimizu K, Kaito C (2012) Evaluation of *Staphylococcus aureus* virulence factors using a silkworm model. FEMS Microbiol Lett 326(2):116–124.
16. Imae K, Saito Y, Kizaki H, Ryuno H, Mao H, Miyashita A et al (2016) Novel nucleoside diphosphatase contributes to *Staphylococcus aureus* virulence. J Biol Chem 291(36): 18608–18619
17. Kyuma T, Kimura S, Hanada Y, Suzuki T, Sekimizu K, Kaito C (2015) Ribosomal RNA methyltransferases contribute to *Staphylococcus aureus* virulence. FEBS J 282(13):2570–2584
18. Kyuma T, Kizaki H, Ryuno H, Sekimizu K, Kaito C (2015) 16S rRNA methyltransferase KsgA contributes to oxidative stress resistance and virulence in *Staphylococcus aureus*. Biochimie 119:166–174
19. Nasu H, Shirakawa R, Furuta K, Kaito C (2022) Knockout of *mlaA* increases *Escherichia coli* virulence in a silkworm infection model. PLoS One 17(7):e0270166
20. Murakami K, Nasu H, Fujiwara T, Takatsu N, Yoshida N, Furuta K et al (2021) The absence of Osmoregulated periplasmic Glucan confers antimicrobial resistance and increases virulence in *Escherichia coli*. J Bacteriol 203(12): e0051520
21. Kaito C, Yoshikai H, Wakamatsu A, Miyashita A, Matsumoto Y, Fujiyuki T et al (2020) Non-pathogenic *Escherichia coli* acquires virulence by mutating a growth-

essential LPS transporter. PLoS Pathog 16(4): e1008469

22. Hamamoto H, Urai M, Ishii K, Yasukawa J, Paudel A, Murai M et al (2015) Lysocin E is a new antibiotic that targets menaquinone in the bacterial membrane. Nat Chem Biol 11(2): 127–133
23. Hamamoto H, Tonoike A, Narushima K, Horie R, Sekimizu K (2009) Silkworm as a model animal to evaluate drug candidate toxicity and metabolism. Comp Biochem Physiol C Toxicol Pharmacol 149(3):334–339
24. Usui K, Nishida S, Sugita T, Ueki T, Matsumoto Y, Okumura H et al (2016) Acute oral toxicity test of chemical compounds in silkworms. Drug Discov Ther 10(1):57–61
25. Venables WN, Ripley BD, Venables WN (2002) Modern applied statistics with S, 4th edn. Springer, New York. xi, 495 pp
26. Hossain MS, Hamamoto H, Matsumoto Y, Razanajatovo IM, Larranaga J, Kaito C et al (2006) Use of silkworm larvae to study pathogenic bacterial toxins. J Biochem 140(3): 439–444
27. Hamamoto H, Kurokawa K, Kaito C, Kamura K, Manitra Razanajatovo I, Kusuhara H et al (2004) Quantitative evaluation of the therapeutic effects of antibiotics using silkworms infected with human pathogenic microorganisms. Antimicrob Agents Chemother 48(3):774–779
28. Nagata M, Kaito C, Sekimizu K (2008) Phosphodiesterase activity of CvfA is required for virulence in *Staphylococcus aureus*. J Biol Chem 283(4):2176–2184
29. Numata S, Nagata M, Mao H, Sekimizu K, Kaito C (2014) CvfA protein and polynucleotide phosphorylase act in an opposing manner to regulate *Staphylococcus aureus* virulence. J Biol Chem 289(12):8420–8431

INDEX

Kunihiko Nishino and Mikio Tanabe (eds.), *Bacterial Multidrug Exporters: Methods and Protocols*, Methods in Molecular Biology, vol. 3044, https://doi.org/10.1007/978-1-0716-5316-6,

© The Editor(s) (if applicable) and The Author(s), under exclusive license to Springer Science+Business Media, LLC, part of Springer Nature 2026